AF769790

Introductory Particle Physics

Introductory Particle Physics

Frederick Gates

NY RESEARCH PRESS
New York

Published by NY Research Press
118-35 Queens Blvd., Suite 400,
Forest Hills, NY 11375, USA
www.nyresearchpress.com

Introductory Particle Physics
Frederick Gates

© 2019 NY Research Press

International Standard Book Number: 978-1-63238-706-6 (Hardback)

This book contains information obtained from authentic and highly regarded sources. All chapters are published with permission under the Creative Commons Attribution Share Alike License or equivalent. A wide variety of references are listed. Permissions and sources are indicated; for detailed attributions, please refer to the permissions page. Reasonable efforts have been made to publish reliable data and information, but the authors, editors and publisher cannot assume any responsibility for the validity of all materials or the consequences of their use.

The publisher's policy is to use permanent paper from mills that operate a sustainable forestry policy. Furthermore, the publisher ensures that the text paper and cover boards used have met acceptable environmental accreditation standards.

Trademark Notice: Registered trademark of products or corporate names are used only for explanation and identification without intent to infringe.

Cataloging-in-Publication Data

Introductory particle physics / Frederick Gates.
p. cm.
Includes bibliographical references and index.
ISBN 978-1-63238-706-6
1. Particles (Nuclear physics). 2. Nuclear physics. I. Gates, Frederick.
QC793.2 .I58 2019
539.7--dc23

Printed in China.

Contents

Preface .... **VII**

Chapter 1 **A Brief Introduction to Particle Physics** **1**
- i. Particle 1
- ii. Particle Physics 3
- iii. Computational Particle Physics 9

Chapter 2 **Subatomic Particles and its Types** **16**
- i. Subatomic Particle 16
- ii. Electron 20
- iii. Proton 46
- iv. Neutron 69

Chapter 3 **Quarks and Hadrons** **121**
- i. Quark 121
- ii. Hadron 144
- iii. Color Charge 168

Chapter 4 **Lepton: An Elementary Particle** **177**
- i. Lepton 177
- ii. Neutrino 185
- iii. Lepton Number 191

Chapter 5 **Particle Physics: Models and Theories** **193**
- i. Standard Model 193
- ii. Gauge Boson 200
- iii. Higgs Boson 203
- iv. Supersymmetry 227
- v. Grand Unified Theory 244
- vi. String Theory 250

Permissions

Index

Preface

Particle physics studies the fundamental constituents of matter and radiation and their interactions. Fundamental particles exhibit wave-particle duality and are represented using quantum state vectors in Hilbert space. All elementary particles and their interactions are understood using a quantum field theory, which is called the Standard Model. This model accounts for 61 particles, among which are 24 fermions, 8 gluons, the photon and W^-, W^+ and Z bosons. The Standard Model further describes the fundamental interactions of these particles, such as electromagnetic, strong and weak interactions. Particle physics has enriched human lives with the production of medical isotopes for use in external beam radiotherapy and PET imaging, development of touchscreen technology and the World Wide Web. This book is a valuable compilation of topics, ranging from the basic to the most complex theories and principles in the field of particle physics. The topics included herein are of utmost significance and bound to provide incredible insights to readers. This textbook is appropriate for particle physicists, high-energy physicists, phenomenologists, field theorists, students and other experts associated with this field.

Given below is the chapter wise description of the book:

Chapter 1- As a discipline of physics, particle physics studies the nature of the fundamental particles, which form matter and radiation. This is an introductory chapter, which will introduce briefly all the significant aspects of particle physics as well as key concepts of computational particle physics such as lattice field theory, automatic calculation of particle interaction or decay, and event generator.

Chapter 2- Subatomic particles are the particles smaller than atoms. They are of two types namely, elementary particles, and composite particles. This chapter discusses the sub-atomic particles namely, the electron, the proton and the neutron. It also discusses the concepts of electron rest mass, classical electron radius, electron excitation, proton decay, emission and magnetic moment, proton-to-electron mass ratio, neutron detection and neutron emission, etc. The topics elaborated in this chapter will provide a better perspective about subatomic particles.

Chapter 3- Quarks are the fundamental constituents of matter. These have intrinsic properties of electric charge, mass, spin and color charge. They combine to form hadrons, which are composite particles. These hadrons are either baryons, such as protons and neutrons, or mesons, such as pions. This chapter closely examines such key concepts of quarks and hadrons to provide an extensive understanding of the subject.

Chapter 4- Leptons are elementary particles, with half-integer spin. Its two main classes are charged leptons and neutral leptons. The topics elaborated in this chapter will help in providing a better perspective about the different kinds of neutrinos, such as electron neutrino, muon neutrino, and tau neutrino. It also sheds light on the concept of lepton number.

Chapter 5- Fundamental particles and their interactions are understood from a quantum field theoretic model called the Standard Model. This chapter includes a detailed explanation of the Standard Model and various other theories associated to particle physics like the Grand Unified Theory and String theory. It also closely examines some fundamental particles as given by the Standard Model such as gauge bosons and the Higgs boson, in order to provide an extensive understanding of the subject.

Indeed, my job was extremely crucial and challenging as I had to ensure that every chapter is informative and structured in a student-friendly manner. I am thankful for the support provided by my family and colleagues during the completion of this book.

Frederick Gates

A Brief Introduction to Particle Physics

As a discipline of physics, particle physics studies the nature of the fundamental particles, which form matter and radiation. This is an introductory chapter, which will introduce briefly all the significant aspects of particle physics as well as key concepts of computational particle physics such as lattice field theory, automatic calculation of particle interaction or decay, and event generator.

Particle

A particle refers to a quantity of matter that is used by scientists to construct theories about their field of study. There is no particular size restriction on defining a particle. Astronomers can define particles to be stars in the night sky, while physicists can define particles to be electrons. It mostly depends on the scientific field and theory under development.

Conceptual Properties

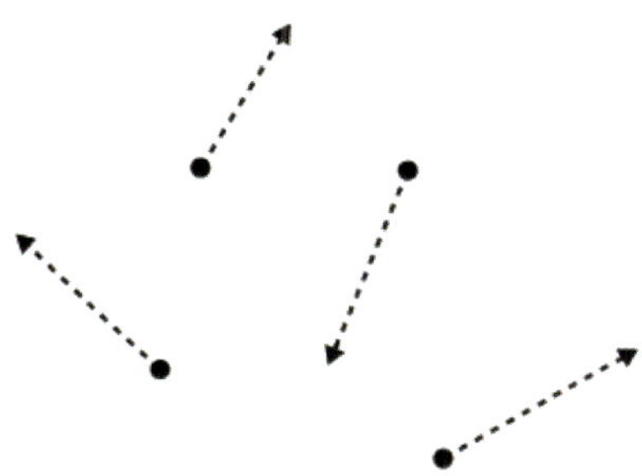

Particles are often represented as dots. This figure could represent the movement of atoms in a gas, people in crowds or stars in the night sky.

The concept of particles is particularly useful when modelling nature, as the full treatment of many phenomena can be complex and also involve difficult computation. It can be used to make simplifying assumptions concerning the processes involved. Francis Sears and Mark Zemansky, in *University Physics*, give the example of calculating the landing location and speed of a baseball thrown in the air. They gradually strip the baseball of most of its properties, by first idealizing it as a rigid smooth sphere, then by neglecting rotation, buoyancy and friction, ultimately reducing the problem to the ballistics of a classical point particle. The treatment of large numbers of particles is the realm of statistical physics.

Size

Galaxies are so large that stars can be considered particles relative to them

The term "particle" is usually applied differently to three classes of sizes. The term *macroscopic particle*, usually refers to particles much larger than atoms and molecules. These are usually abstracted as point-like particles, even though they have volumes, shapes, structures, etc. Examples of macroscopic particles would include powder, dust, sand, pieces of debris during a car accident, or even objects as big as the stars of a galaxy.

Another type, *microscopic particles* usually refers to particles of sizes ranging from atoms to molecules, such as carbon dioxide, nanoparticles, and colloidal particles. These particles are studied in chemistry, as well as atomic and molecular physics. The smallest of particles are the *subatomic particles*, which refer to particles smaller than atoms. These would include particles such as the constituents of atoms – protons, neutrons, and electrons – as well as other types of particles which can only be produced in particle accelerators or cosmic rays. These particles are studied in particle physics.

Because of their extremely small size, the study of microscopic and subatomic particles fall in the realm of quantum mechanics. They will exhibit phenomena demonstrated in the particle in a box model, including wave–particle duality, and whether particles can be considered distinct or identical is an important question in many situations.

Composition

Particles can also be classified according to composition. *Composite particles* refer to particles that have composition – that is particles which are made of other particles. For example, a carbon-14 atom is made of six protons, eight neutrons, and six electrons. By contrast, *elementary particles* (also called *fundamental particles*) refer to particles that are not made of other particles. According to our current understanding of the world, only a very small number of these exist, such as leptons, quarks, and gluons. However it is possible that some of these might turn up to be composite particles after all, and merely appear to be elementary for the moment. While composite particles can very often be considered *point-like*, elementary particles are truly *punctual*.

A proton is composed of three quarks.

Stability

Both elementary (such as muons) and composite particles (such as uranium nuclei), are known to undergo particle decay. Those that do not are called stable particles, such as the electron or a helium-4 nucleus. The lifetime of stable particles can be either infinite or large enough to hinder attempts to observe such decays. In the latter case, those particles are called "observationally stable". In general, a particle decays from a high-energy state to a lower-energy state by emitting some form of radiation, such as the emission of photons.

Types of Particle

Throughout the whole of the known universe there are only 2 types of particle. Particles that make up matter, and particle that carry force. They are the only 2 types found so far. Now you may be thinking that, yes there may only be two categories but I bet they're filled with hundreds of different subgroups and types. Thankfully this isn't the case, particles follow specific rules and once you known them everything gets a lot easier.

The two groups are called Fermions and Bosons.

Particle Physics

Particle physics is a branch of physics that studies the elementary constituents of matter and radiation, and the interactions between them.

It is also called "high energy physics", because many elementary particles do not occur under normal circumstances in nature, but can be created and detected during energetic collisions of other particles, as is done in particle accelerators.

Modern particle physics research is focused on subatomic particles, which have less structure than atoms.

These include atomic constituents such as electrons, protons, and neutrons (protons and neutrons are actually composite particles, made up of quarks), particles produced by radiative and scattering processes, such as photons, neutrinos, and muons, as well as a wide range of exotic particles.

Strictly speaking, the term particle is a misnomer because the dynamics of particle physics are governed by quantum mechanics.

As such, they exhibit wave-particle duality, displaying particle-like behavior under certain experimental conditions and wave-like behavior in others (more technically they are described by state vectors in a Hilbert space).

All the particles and their interactions observed to date can be described by a quantum field theory called the Standard Model.

The Standard Model has 40 species of elementary particles (24 fermions, 12 vector bosons, and 4 scalars), which can combine to form composite particles, accounting for the hundreds of other species of particles discovered since the 1960s.

Subatomic Particles

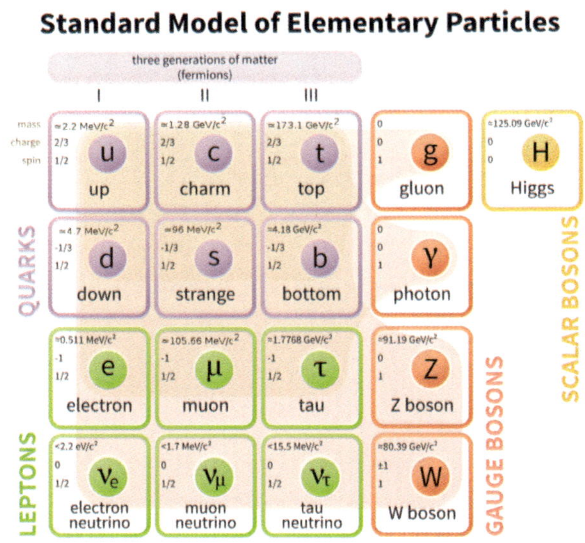

The particle content of the Standard Model of Physics

Modern particle physics research is focused on subatomic particles, including atomic constituents such as electrons, protons, and neutrons (protons and neutrons are composite particles called baryons, made of quarks), produced by radioactive and scattering processes, such as photons, neutrinos, and muons, as well as a wide range of exotic particles. Dynamics of particles is also governed by quantum mechanics; they exhibit wave–particle duality, displaying particle-like behaviour under certain experimental conditions and wave-like behaviour in others. In more technical terms, they are described by quantum state vectors in a Hilbert space, which is also treated in quantum field theory. Following the convention of particle physicists, the term *elementary particles* is applied to those particles that are, according to current understanding, presumed to be indivisible and not composed of other particles.

Elementary Particles					
	Types	**Generations**	**Antiparticle**	**Colours**	**Total**
Quarks	2	3	Pair	3	36
Leptons			Pair	None	12
Gluons	1	1	Own	8	8
Photon			Own	None	1
Z Boson			Own		1
W Boson			Pair		2
Higgs			Own		1
Total number of (known) elementary particles:					**61**

All particles and their interactions observed to date can be described almost entirely by a quantum field theory called the Standard Model. The Standard Model, as currently formulated, has 61 elementary particles. Those elementary particles can combine to form composite particles, accounting for the hundreds of other species of particles that have been discovered since the 1960s. The Standard Model has been found to agree with almost all the experimental tests conducted to date. However, most particle physicists believe that it is an incomplete description of nature and that a more fundamental theory awaits discovery. In recent years, measurements of neutrino mass have provided the first experimental deviations from the Standard Model.

Standard Model

The current state of the classification of all elementary particles is explained by the Standard Model. It describes the strong, weak, and electromagnetic fundamental interactions, using mediating gauge bosons. The species of gauge bosons are eight gluons, W^-, W^+, and Z Bosons, and the photon. The Standard Model also contains 24 fundamental fermions(12 particles and their associated anti-particles), which are the constituents of all matter.Finally, the Standard Model also predicted the existence of a type of boson known as the Higgs boson. Early in the morning on 4 July 2012, physicists with the Large Hadron Collider at CERN announced they had found a new particle that behaves similarly to what is expected from the Higgs boson.

What are Fundamental Particles?

Fundamental, or elementary, particles are the smallest constituents of something and aren't made up of even smaller particles. Particle physics addresses both matter and force in terms of particles, so fundamental particles come in two types as well. Bosons are force-carriers: if a particle is acted upon by a force, then it can also be understood as the particle interacting with the particulate carriers of that force. For example: if the weak force is acting on a particle, then the particle is interacting with W± and Z bosons, which are carriers of the weak force. Fermions are matter-makers: they come together under the influence of others like them as well as bosons to make up matter (or antimatter).

However, a deeper classification of bosons and fermions alludes to their collective properties when they accumulate in large numbers. All particles have a spin quantum number associated with their angular momentum, expressed as integer fractions (0, 1/2, 1, 3/2, ...). All fermions have half-integer spins (1/2, 3/2, ...) and their behaviour is thus defined by the rules of Fermi-Dirac statistics. All bosons have integer spins (0, 1, 2, ...) and their behaviour is thus defined by the rules of Bose-Einstein statistics.

One commonly cited difference between the statistics is that bosons don't obey Pauli's exclusion principle while fermions do. The principle states that, given all fermions can be accorded four quantum numbers that define the particles' states, no two particles can have the same four numbers at the same time. And because bosons can possess such identical quantum numbers, they can form Bose-Einstein condensates.

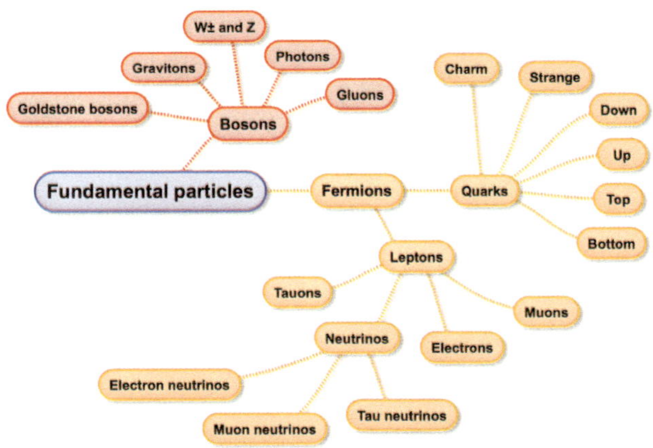

What are the Fundamental Forces?

There are four forces that occur naturally. In the descending order of their strength, they are the strong nuclear force (holds nuclei together), the electromagnetic force (acts between charged particles), the weak nuclear force (responsible for radioactivity) and gravitation (acts between things having mass). They are respectively mediated by gluons, photons, W±/Z bosons and the hypothetical gravitons. At higher energies, the electromagnetic and weak nuclear forces unify into one common electroweak force. Some physicists think that at even higher energies, beyond those we can access with the current state of experimental physics, all four forces will unify to form one 'grand unified' force. The theories that aspire to account for such a unification are called grand unified theories.

Centrifugal Mechanism of Acceleration

Centrifugal acceleration of astroparticles to relativistic energies might take place in rotating astrophysical objects. It is strongly believed that active galactic nuclei and pulsars have rotating magnetospheres, therefore, they potentially can drive charged particles to high and ultra-high energies. It is a proposed explanation for ultra-high-energy cosmic rays (UHECRs) and extreme-energy cosmic rays (EECRs) exceeding the Greisen Zatsepin–Kuzmin limit.

Acceleration to High Energies

It is well known that the magnetospheres of AGNs and pulsars are characterized by strong magnetic fields that force charged particles to follow the field lines. If the magnetic field is rotating (which is the case for such astrophysical objects), the particles will inevitably undergo centrifugal acceleration. The pioneering work by Machabeli & Rogava was a thought experiment in which a bead moves inside a straight rotating pipe. Dynamics of the particle were analyzed both analytically and numerically and it was shown that if the rigid rotation is maintained for a sufficiently long time energy of the bead will asymptotically increase. In particular, Rieger & Mannheim, building on the theory of Machabeli & Rogava, showed that the Lorentz factor of the bead behaves as

$$\gamma = \frac{\gamma_0}{1-\Omega^2 r^2 / c^2}$$

where γ_0 is the initial Lorentz factor, Ω is the angular velocity of rotation, r is the radial coordinate of the particle, and c is the speed of light. From this behavior it is evident that radial motion will exhibit a nontrivial character. In due course of motion the particle will reach the light cylinder surface (a hypothetical area where the linear velocity of rotation exactly equals the speed of light), leading to the increase of the poloidalcomponent of velocity. On the other hand, the total velocity cannot exceed the speed of light, therefore, the radial component must decrease. This means that the centrifugal force changes its sign.

The Lorentz factor of the particle tends to infinity if the rigid rotation is maintained. This means that in reality the energy has to be limited by certain processes. Generally speaking, there are two major mechanisms: The inverse Compton scattering (ICS) and the so-called breakdown of the bead on the wire (BBW) mechanism. For jet-like structures in an AGN it has been shown that, for a wide range of inclination angles of field lines with respect to the rotation axis, ICS is the dominant mechanism efficiently limiting the maximum attainable Lorentz factors of electrons $\gamma_{ICS}^{max} \sim 10^8$. On the other hand, it was shown that the BBW becomes dominant for relatively low luminosity AGN $L < 8\times 10^{40} erg/s$, leading to $\gamma_{BBW}^{max} \sim 10^7$.

The centrifugal effects are more efficient in millisecond pulsars as the rotation rate is quite high. Osmanov & Rieger considered the centrifugal acceleration of charged particles in the light cylinder area of the Crab-like pulsars. It has been shown that electrons might achieve the Lorentz factors $\gamma_{KN}^{max} \sim 10^7$ via inverse Compton Klein–Nishina up-scattering.

Acceleration to Very High and Ultra-high Energies

Although the direct centrifugal acceleration has limitations, as analysis shows the effects of rotation still might play an important role in the processes of acceleration of charged particles. Generally speaking, it is believed that the centrifugal relativistic effects may induce plasma waves, which under certain conditions might be unstable efficiently pumping energy from the background flow. On the second stage energy of wave-modes can be transformed into energy of plasma particles, leading to consequent acceleration.

In rotating magnetospheres the centrifugal force acts differently in different locations, leading to generation of Langmuir waves, or plasma oscillations via the parametric instability. One can show that this mechanism efficiently works in the magnetospheres of AGN and pulsars.

Considering Crab-like pulsars it has been shown that by means of the Landau damping the centrifugally induced electrostatic waves efficiently lose energy transferring it to electrons. It is found that energy gain by electrons is given by

$$\in \approx \frac{n_p F_{reac} \delta r}{n_{GJ}},$$

where $\delta r \sim c/\Gamma, \Gamma$ is the increment of the instability, $F_{reac} \approx 2mc\Omega\xi(r)^{-3}$, $\xi(r) = \left(1 - \Omega^2 r^2 / c^2\right)^{1/2}$,n_p is the plasma number density, m is the electron›s mass and n_{GJ} is the Goldreich-Julian density. One can show that for typical parameters of the Crab-like pulsars, the particles might gain energies of the order of $100s$ of $TeVs$ or even $PeVs$. In case of millisecond newly born pulsars, the electrons might be accelerated to even higher energies of $10^{18} eV$.

By examining the magnetospheres of AGNs, the acceleration of protons takes place through the Langmuir collapse. As it is shown this mechanism is strong enough to guarantee efficient acceleration of particles to ultra-high energies via the Langmuir damping

$$\epsilon_p\left(eV\right) \approx 6.4 \times 10^{17} \times M_8^{-5/2} \times L_{42}^{5/2}$$

where $L_{42} \equiv L / 10^{42} erg / s$ is the normalized luminosity of AGN, $M_8 \equiv M / (10^8 M_{\odot})$ is its normalized mass and $M_{\odot}$ is the Solar mass. As it is evident, for a convenient set of parameters one can achieve enormous energies of the order of $10^{21} eV$, so AGNs become cosmic Zevatrons.

Computational Particle Physics

Computational particle physics refers to the methods and computing tools developed in and used by particle physics research. Like computational chemistry or computational biology, it is, for particle physics both a specific branch and an interdisciplinary field relying on computer science, theoretical and experimental particle physics and mathematics. The main fields of computational particle physics are: lattice field theory(numerical computations), automatic calculation of particle interaction or decay (computer algebra) and event generators (stochastic methods).

Lattice Field Theory

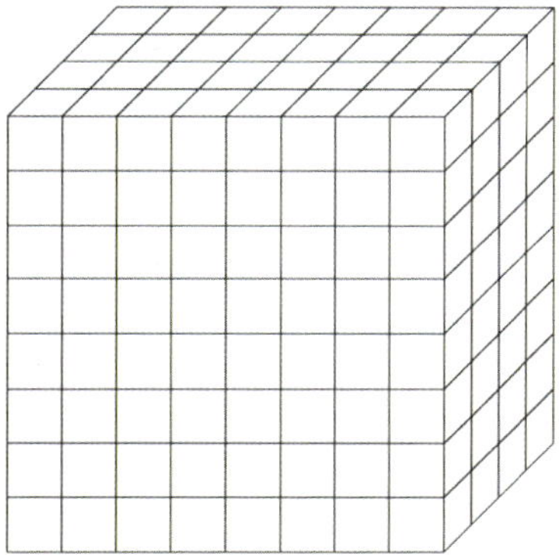

Figure

Lattice field theory is an area of theoretical physics, specifically quantum field theory, which deals with field theories defined on a spatial or space-time lattice.

The theoretical description of the fundamental constituents of matter and the interactions between them is based on quantum field theory. The basic ingredients of field theory are fields. They are functions ϕ which associate to each point x of space-time a quantity $\phi(x)$.

In the case of classical field theories, $\phi(x)$ usually is an element of a finite dimensional real or complex manifold, which in many cases is a linear space. Prominent examples are:

- The real scalar field $\phi(x) \in \mathrm{R}$,
- The complex scalar field $\phi(x) \in \mathrm{C}$,
- The n-vector field $\phi(x) \in \mathrm{R}^n$,
- The photon field $A_\mu(x) \in \mathrm{R}$, where μ=0,1,2,3 is a Lorentz index,
- The Yang-Mills field $A_\mu(x) = \sum_b A^b{}_\mu(x) T_b$, whose components are elements of the Lie algebra of a compact Lie group with generators T_b.

In contrast, in the operator formulation of quantum field theory the fields are operators acting in a Hilbert space. (More precisely, quantum fields $\phi(x)$ are operator valued distributions, which means that integrals $\int f(x)\phi(x)dx$ with suitable test functions $f(x)$ are operators.)

The physical content of a field theory depends essentially on the Lagrangian $L(\phi(x), \partial^{n}\phi(x))$, which is a function of $\phi(x)$ and its derivatives. The Lagrangian determines the field equations, which comprise the interactions. If the strength of an interaction is given by a small parameter g, it is possible to calculate physical quantities approximately to a satisfactory accuracy by means of perturbation theory, which amounts to a power series expansion in g. This is, for example, the case in quantum electrodynamics (QED), where the interaction is proportional to the fine structure constant $\alpha \approx 1/137$, and many interesting observables can be obtained as power series in α. There are, however, important cases, where it turned out that perturbation theory is inadequate for the calculation of physical quantities. The most prominent example is the low-energy regime of Quantum Chromodynamics (QCD), the theory of the strong interactions of elementary particles.

Not only Quantum Chromodynamics, but also other components of the Standard Model of elementary particle physics and moreover theories of physics beyond the Standard Model supply us with non-perturbative problems. An important step to answer such questions has been made by K. Wilson in 1974 (Wilson, 1974). He introduced a formulation of Quantum Chromodynamics on a space-time lattice, which allows the application of various non-perturbative techniques. This discretization will be explained in detail below. It leads to mathematically well-defined problems, which are (at least in principle) solvable. It should also be pointed out that the introduction of a space-time lattice can be taken as a starting point for a mathematically clean approach to quantum field theory, so-called constructive quantum field theory.

In modern quantum field theory, the introduction of a space-time lattice is part of an approach different from the operator formalism. This is lattice field theory. Its main ingredients are

- Functional integrals,
- Euclidean field theory and
- The space-time discretization of fields.

Lattice field theory has turned out to be very successful for the non-perturbative calculation of physical quantities.

Automatic Calculation of Particle Interaction or Decay

The automatic calculation of particle interaction or decay is part of the computational particle physics branch. It refers to computing tools that help calculating the complex particle interactions as studied in high-energy physics, astroparticle physics and cosmology. The goal of the automation is to handle the full sequence of calculations in an automatic (programmed) way: from the Lagrangian expression describing the physics model up to the cross-sections values and to the event generator software.

Particle accelerator or colliders produce collisions (interactions) of particle (like the electronor the proton). The colliding particles form the *Initial State*. In the collision, particles can be annihilated or/and exchanged producing possibly different sets of particles, the *Final States*. The Initial and Final States of the interaction relate through the so-called scattering matrix (S-matrix).

For example, at LEP, $e^+ + e^- \rightarrow e^+ + e^-$, $or\, e^+ + e^- \rightarrow \mu^+ + \mu^-$ are processes where the *initial state* is an electron and a positron colliding to produce an electron and a positron or two muons of opposite charge: the *final states*. In these simple cases, no automatic packages are needed and cross-section analytical expression can be easily derived at least for the lowest approximation: the Born approximation also called the leading order or the tree level (as Feynman diagrams have only trunk and branches, no loops).

But particle physics is now requiring much more complex calculations like at LHC $pp \rightarrow n_{jets}$ where p are protons and n_{jets} is the number of jets of particles initiated by proton constituents (quarks and gluons). The number of subprocesses describing a given process is so large that automatic tools have been developed to mitigate the burden of hand calculations.

Interactions at higher energies open a large spectrum of possible final states and consequently increase the number of processes to compute.

High precision experiments impose the calculation of higher order calculation, namely the inclusion of subprocesses where more than one virtual particle can be created and annihilated during the interaction lapse creating so-called *loops* which induce much more involved calculations.

Finally new theoretical models like the supersymmetry model (MSSM in its minimal version) predict a flurry of new processes.

The automatic packages, once seen as mere teaching support, have become, this last 10 years an essential component of the data simulation and analysis suite for all experiments. They help constructing event generators and are sometimes viewed as *generators of event generators* or *Meta-generators*.

A particle physics model is essentially described by its Lagrangian. To simulate the production of events through event generators, 3 steps have to be taken. The Automatic Calculation project is to create the tools to make those steps as automatic (or programmed) as possible:

I Feynman rules, coupling and mass generation

- LanHEP is an example of Feynman rules generation.
- Some model needs an additional step to compute, based on some parameters, the mass and coupling of new predicted particles.

II Matrix element code generation: Various methods are used to automatically produce the matrix element expression in a computer language (Fortran, C/C++). They use values (i.e. for the masses) or expressions (i.e. for the couplings) produced by step I or model specific libraries constructed *by hands* (usually heavily relying on Computer algebra languages). When this expression is integrated (usually numerically) over the internal degrees of freedom it will provide the total and differential cross-sections for a given set of initial parameters like the *initial state* particle energies and polarization.

III Event generator code generation: This code must them be interfaced to other packages to fully provide the actual *final state*. The various effects or phenomenon that need to be implemeted are:

- Initial state radiation and beamstrahlung for e^+e^- initial states.
- Parton distribution functions describing the actual content in terms of gluons and quarks of the p or p-bar initial state particles
- Parton showering describing the way final state quarks or gluons due to the QCD confinement generate additional quark/gluon pairs generating a so-called shower of partons before transforming into hadrons.
- Hadronization describing how the final quark pairs/triplets form the visible and detectable hadrons.
- Underlying event takes care of the way the rest, in term of constituent, of the initial protons also contribute to any given event.

The interplay or *matching* of the precise matrix element calculation and the approximations resulting from the simulation of the *parton shower* gives rise to further complications, either within a given level of precision like at leading order (LO) for the production of n jets or between two levels of precision when tempting to connect matrix element computed at next-to-leading (NLO) (1-loop) or next-to-next-leading order (NNLO) (2-loops) with LO partons shower package.

But the only correct way is to match packages at the same level theoretical accuracy like the NLO matrix element calculation with NLO parton shower packages. This is currently in development.

Event Generator

Event generators are software libraries that generate simulated high-energy particle physicsevents. They randomly generate events as those produced in particle accelerators, collider experiments or the early universe. Events come in different types called *processes* as discussed in the Automatic calculation of particle interaction or decay article.

Despite the simple structure of the tree-level perturbative quantum field theory description of the collision and decay processes in an event, the observed high-energy process usually contains significant amount of modifications, like photon and gluon *bremsstrahlung* or loop diagram corrections, that usually are too complex to be easily evaluated in real calculations directly on the diagrammatic level. Furthermore, the non-perturbative nature of QCD bound states makes it necessary to include information that is well beyond the reach of perturbative quantum field theory, and also beyond present ability of computation in lattice QCD. And in collisional systems more complex than a few leptons and hadrons (e.g. heavy-ion collisions), the collective behavior of the system would involve a phenomenological description that also cannot be easily obtained from the fundamental field theory by a simple calculus.

Use in Simulations

As said above, the experimental calibration involves processes that usually are too complicated to be easily evaluated in calculations directly, so any realistic test of the underlying physical process in a particle accelerator experiment, therefore, requires an adequate inclusion of these complex behaviors surrounding the actual process. Based on the fact that in most processes a factorization of the full process into individual problems is possible (which means a negligible effect from interference), these individual processes are calculated separately, and the probabilistic branching between them are performed using Monte Carlo methods.

The final-state particles generated by event generators can be fed into the detector simulation, allowing a precise prediction and verification for the entire system of experimental setup. However, as the detector simulation is usually a complex and computationally expensive task, simple event analysis techniques are also performed directly on event generator results.

Some automatic software packages exist, that help in constructing event generators and are sometimes viewed as *generators of event generators* or *meta-generators*.

Partly due to historic reasons, most event generators are written in FORTRAN 77, with a few C++ generators slowly emerging in recent years. The Particle Data Group maintains a standard for designating Standard Model particles and resonances with integer codes in event generators (also known as the "PDG code").

Processes

A typical hadronic event generator simulates the following subprocesses:

- Initial-state composition and substructure
- Initial-state showers
- The hard process
- Resonance decay
- Final-state showers
- Accompanying semi-hard processes
- Hadronization and further decay

A typical heavy-ion event generator usually can be less strict in simulating the rare and rather negligible processes found in a hadronic generator, but would need to simulate the following subprocesses, in addition to those in a hadronic generator:

- Nuclear initial-state
- High multiplicity, soft processes
- In-medium energy loss
- Collective behavior of the medium *(not handled properly by any generators so far)*

References

- R. Eisberg; R. Resnick (1985). "Photons—Particlelike Properties of Radiation". Quantum Physics of Atoms, Molecules, Solids, Nuclei, and Particles (2nd ed.). John Wiley & Sons. pp. 26–54. ISBN 978-0-471-87373-0
- G. Coppola; F. La Barbera; M. Capaccioli (2009). "Sérsic galaxy with Sérsic halo models of early-type galaxies: A tool for N-body simulations". Publications of the Astronomical Society of the Pacific. 121 (879): 437. arXiv:0903.4758. Bibcode:2009PASP..121..437C. doi:10.1086/599288
- J. Dubinski (2003). "Galaxy Dynamics and Cosmology on Mckenzie". Canadian Institute for Theoretical Astrophysics. Retrieved 2011-02-24
- Nakamura, K (1 July 2010). "Review of Particle Physics". Journal of Physics G: Nuclear and Particle Physics. 37 (7A): 075021. Bibcode:2010JPhG...37g5021N. doi:10.1088/0954-3899/37/7A/075021
- F. Reif (1965). "Statistical Description of Systems of Particles". Fundamentals of Statistical and Thermal Physics. McGraw-Hill. pp. 47ff. ISBN 978-0-07-051800-1
- Graps (20 March 2000). "N-Body / Particle Simulation Methods". Archived from the original on 5 April 2001. Retrieved 2011-02-24

- G. Coppola; F. La Barbera; M. Capaccioli (2009). "Sérsic galaxy with Sérsic halo models of early-type galaxies: A tool for N-body simulations". Publications of the Astronomical Society of the Pacific. 121 (879): 437. arXiv:0903.4758. Bibcode:2009PASP..121..437C. doi:10.1086/599288
- Boos, E.E; et al. (1994). "Automatic calculation in high-energy physics by Grace/Chanel and CompHEP". International Journal of Modern Physics C. 5 (4): 615. Bibcode:1994IJMPC...5..615B. doi:10.1142/S0129183194000787
- Braibant, S.; Giacomelli, G.; Spurio, M. (2009). Particles and Fundamental Interactions: An Introduction to Particle Physics. Springer. pp. 313–314. ISBN 978-94-007-2463-1

Subatomic Particles and its Types

Subatomic particles are the particles smaller than atoms. They are of two types namely, elementary particles, and composite particles. This chapter discusses the sub-atomic particles namely, the electron, the proton and the neutron. It also discusses the concepts of electron rest mass, classical electron radius, electron excitation, proton decay, emission and magnetic moment, proton-to-electron mass ratio, neutron detection and neutron emission, etc. The topics elaborated in this chapter will provide a better perspective about subatomic particles.

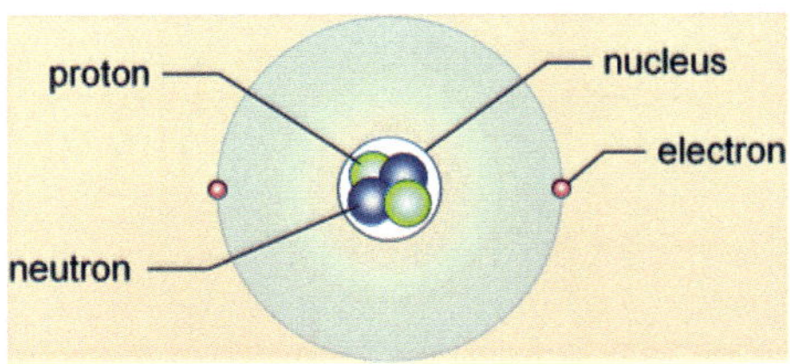

Subatomic particle, also called elementary particle, any of various self-contained units of matter or energy that are the fundamental constituents of all matter. Subatomic particles include electrons, the negatively charged, almost massless particles that nevertheless account for most of the size of the atom, and they include the heavier building blocks of the small but very dense nucleus of the atom, the positively charged protons and the electrically neutral neutrons. But these basic atomic components are by no means the only known subatomic particles. Protons and neutrons, for instance, are themselves made up of elementary particles called quarks, and the electron is only one member of a class of elementary particles that also includes the muon and the neutrino. More-unusual subatomic particles—such as the positron, the antimatter counterpart of the electron—have been detected and characterized in cosmic ray interactions in Earth's atmosphere. The field of subatomic particles has expanded dramatically with the construction of powerful particle accelerators to study high-energy collisions of electrons, protons, and other particles with matter. As particles collide at high energy, the collision energy becomes available for the creation of subatomic particles such as mesons and hyperons. Finally, completing the revolution that began in the early 20th century with theories of the equivalence of matter and energy, the study of subatomic particles has been transformed by the discovery that the actions of forces are due to the exchange of "force" particles such as photons and gluons. More than 200 subatomic particles have been detected—most of them highly unstable, existing for less than a millionth of a second—as a result of collisions produced in cosmic ray reactions or particle accelerator experiments. Theoretical and experimental research in particle physics, the study of subatomic particles and their properties, has given scientists a clearer understanding of the nature of matter and energy and of the origin of the universe.

Large Hadron Collider
The Large Hadron Collider (LHC), the world's most powerful particle accelerator. At the LHC, located underground in Switzerland, physicists study subatomic particles.

The current understanding of the state of particle physics is integrated within a conceptual framework known as the Standard Model. The Standard Model provides a classification scheme for all the known subatomic particles based on theoretical descriptions of the basic forces of matter.

Classification

By Statistics

Any subatomic particle, like any particle in the three-dimensional space that obeys the laws of quantum mechanics, can be either a boson (with integer spin) or a fermion (with odd half-integer spin).

By Composition

The elementary particles of the Standard Model include:

- Six "flavors" of quarks: up, down, bottom, top, strange, and charm;
- Six types of leptons: electron, electron neutrino, muon, muon neutrino, tau, tau neutrino;
- Twelve gauge bosons (force carriers): the photon of electromagnetism, the three W and Z bosons of the weak force, and the eight gluons of the strong force;
- The Higgs boson.

Various extensions of the Standard Model predict the existence of an elementary graviton particle and many other elementary particles.

Composite subatomic particles (such as protons or atomic nuclei) are bound states of two or more elementary particles. For example, a proton is made of two up quarks and one down quark, while the atomic nucleus of helium-4 is composed of two protons and two neutrons. The neutron is made of two down quarks and one up quark. Composite particles include all hadrons: these include baryons (such as protons and neutrons) and mesons (such as pions and kaons).

By Mass

In special relativity, the energy of a particle at rest equals its mass times the speed of light squared, $E = mc^2$. That is, mass can be expressed in terms of energy and vice versa. If a particle has a frame of reference in which it lies at rest, then it has a positive rest mass and is referred to as *massive*.

All composite particles are massive. Baryons (meaning "heavy") tend to have greater mass than mesons (meaning "intermediate"), which in turn tend to be heavier than leptons (meaning "lightweight"), but the heaviest lepton (the tau particle) is heavier than the two lightest flavours of baryons (nucleons). It is also certain that any particle with an electric charge is massive.

All massless particles (particles whose invariant mass is zero) are elementary. These include the photon and gluon, although the latter cannot be isolated.

The Divisible Atom

The physical study of subatomic particles became possible only during the 20th century, with the development of increasingly sophisticated apparatuses to probe matter at scales of 10^{-15} metre and less (that is, at distances comparable to the diameter of the proton or neutron). Yet the basic philosophy of the subject now known as particle physics dates to at least 500 bce, when the Greek philosopher Leucippus and his pupil Democritus put forward the notion that matter consists of invisibly small, indivisible particles, which they called atoms. For more than 2,000 years the idea of atoms lay largely neglected, while the opposing view that matter consists of four elements—earth, fire, air, and water—held sway. But by the beginning of the 19th century the atomic theory of matter had returned to favour, strengthened in particular by the work of John Dalton, an English chemist whose studies suggested that each chemical element consists of its own unique kind of atom. As such, Dalton's atoms are still the atoms of modern physics. By the close of the century, however, the first indications began to emerge that atoms are not indivisible, as Leucippus and Democritus had imagined, but that they instead contain smaller particles.

In 1896 the French physicist Henri Becquerel discovered radioactivity, and in the following year J.J. Thomson, a professor of physics at the University of Cambridge in England, demonstrated the existence of tiny particles much smaller in mass than hydrogen, the lightest atom. Thomson had discovered the first subatomic particle, the electron. Six years later Ernest Rutherford and Frederick Soddy, working at McGill University in Montreal, found that radioactivity occurs when atoms of one type transmute into those of another kind. The idea of atoms as immutable, indivisible objects had become untenable.

The basic structure of the atom became apparent in 1911, when Rutherford showed that most of the mass of an atom lies concentrated at its centre, in a tiny nucleus. Rutherford postulated that the atom resembled a miniature solar system, with light, negatively charged electrons orbiting the dense, positively charged nucleus, just as the planets orbit the Sun. The Danish theorist Niels Bohr refined this model in 1913 by incorporating the new ideas of quantization that had been developed by the German physicist Max Planck at the turn of the century. Planck had theorized that electromagnetic radiation, such as light, occurs in discrete bundles, or "quanta," of energy now known as photons. Bohr postulated that electrons circled the nucleus in orbits of fixed size and energy and that an electron could jump from one orbit to another only by emitting or absorbing specific quanta of energy. By thus incorporating quantization into his theory of the atom, Bohr introduced one of the basic elements of modern particle physics and prompted wider acceptance of quantization to explain atomic and subatomic phenomena.

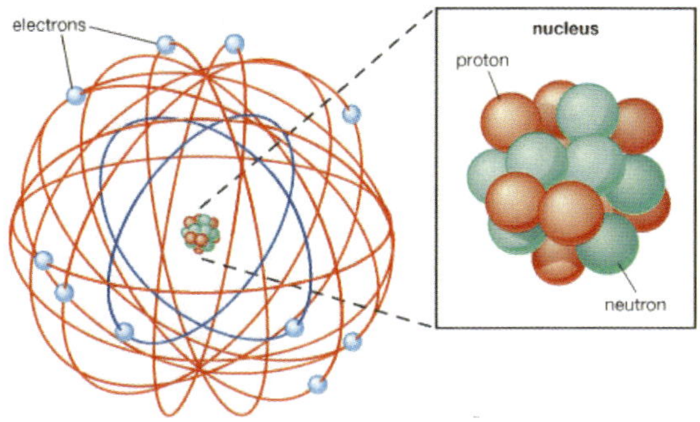

Rutherford atomic model

Diagram of the Rutherford atomic model. Physicist Ernest Rutherford envisioned the atom as a miniature solar system, with electrons orbiting around a massive nucleus, and as mostly empty space, with the nucleus occupying only a very small part of the atom. The neutron had not been discovered when Rutherford proposed his model, which had a nucleus consisting only of protons.

Size

Subatomic particles play two vital roles in the structure of matter. They are both the basic building blocks of the universe and the mortar that binds the blocks. Although the particles that fulfill these different roles are of two distinct types, they do share some common characteristics, foremost of which is size.

The small size of subatomic particles is perhaps most convincingly expressed not by stating their absolute units of measure but by comparing them with the complex particles of which they are a part. An atom, for instance, is typically 10^{-10} metre across, yet almost all of the size of the atom is unoccupied "empty" space available to the point-charge electrons surrounding the nucleus. The distance across an atomic nucleus of average size is roughly 10^{-14} metre—only $^{1}/_{10,000}$ the diameter of the atom. The nucleus, in turn, is made up of positively charged protons and electrically neutral neutrons, collectively referred to as nucleons, and a single nucleon has a diameter of about 10^{-15} metre—that is, about $^1/_{10}$ that of the nucleus and $^1/_{100,000}$ that of the atom. (The distance across the nucleon, 10^{-15} metre, is known as a fermi, in honour of the Italian-born physicist Enrico Fermi, who did much experimental and theoretical work on the nature of the nucleus and its contents.)

The sizes of atoms, nuclei, and nucleons are measured by firing a beam of electrons at an appropriate target. The higher the energy of the electrons, the farther they penetrate before being deflected by the electric charges within the atom. For example, a beam with an energy of a few hundred electron volts (eV) scatters from the electrons in a target atom. The way in which the beam is scattered (electron scattering) can then be studied to determine the general distribution of the atomic electrons.

At energies of a few hundred megaelectron volts (MeV; 10^6 eV), electrons in the beam are little affected by atomic electrons; instead, they penetrate the atom and are scattered by the positive nucleus. Therefore, if such a beam is fired at liquid hydrogen, whose atoms contain only single protons in their nuclei, the pattern of scattered electrons reveals the size of the proton. At energies greater than a gigaelectron volt (GeV; 10^9 eV), the electrons penetrate within the protons and neutrons, and their scattering patterns reveal an inner structure. Thus, protons and neutrons are no more indivisible than atoms are; indeed, they contain still smaller particles, which are called quarks.

Quarks are as small as or smaller than physicists can measure. In experiments at very high energies, equivalent to probing protons in a target with electrons accelerated to nearly 50,000 GeV, quarks appear to behave as points in space, with no measurable size; they must therefore be smaller than 10^{-18} metre, or less than $^1/_{1,000}$ the size of the individual nucleons they form. Similar experiments show that electrons too are smaller than it is possible to measure.

Electron

An electron is a negatively charged subatomic particle. It can be either free (not attached to any atom), or bound to the nucleus of an atom. Electrons in atoms exist in spherical shells of various radii, representing energy levels. The larger the spherical shell, the higher the energy contained in the electron.

In electrical conductors, current flow results from the movement of electrons from atom to atom individually, and from negative to positive electric poles in general. In semiconductor materials, current also occurs as a movement of electrons. But in some cases, it is more illustrative to envision the current as a movement of electron deficiencies from atom to atom. An electron-deficient atom in a semiconductor is called a hole. Holes "move" from positive to negative electric poles in general.

The charge on a single electron is considered as the unit electrical charge. It is assigned negative polarity. The charge on an electron is equal, but opposite, to the positive charge on a proton or hole. Electrical charge quantity is not usually measured in terms of the charge on a single electron, because this is an extremely small charge. Instead, the standard unit of electrical charge quantity is the coulomb, symbolized by C, representing about 6.24×10^{18} electrons. The electron charge, symbolized by e, is about 1.60×10^{-19} C. The mass of an electron at rest, symbolized m_e, is approximately 9.11×10^{-31} kilogram (kg). Electrons moving at an appreciable fraction of the speed of light, for example in a particle accelerator, have greater mass because of relativistic effects.

Characteristics

Classification

In the Standard Model of particle physics, electrons belong to the group of subatomic particles called leptons, which are believed to be fundamental or elementary particles. Electrons have the lowest mass of any charged lepton (or electrically charged particle of any type) and belong to the first-generation of fundamental particles. The second and third generation contain charged leptons, the muon and the tau, which are identical to the electron in charge, spin and interactions, but are more massive. Leptons differ from the other basic constituent of matter, the quarks, by their lack of strong interaction. All members of the lepton group are fermions, because they all have half-odd integer spin; the electron has spin $\frac{1}{2}$.

Fundamental Properties

The invariant mass of an electron is approximately 9.109×10^{-31} kilograms, or 5.489×10^{-4} atomic mass units. On the basis of Einstein's principle of mass–energy equivalence, this mass corresponds to a rest energy of 0.511 MeV. The ratio between the mass of a proton and that of an electron is about 1836. Astronomical measurements show that the proton-to-electron mass ratio has held the same value, as is predicted by the Standard Model, for at least half the age of the universe.

Electrons have an electric charge of -1.602×10^{-19} coulombs, which is used as a standard unit of charge for subatomic particles, and is also called the elementary charge. This elementary charge has a relative standard uncertainty of 2.2×10^{-8}. Within the limits of experimental accuracy, the electron charge is identical to the charge of a proton, but with the opposite sign. As the symbol *e* is used for the elementary charge, the electron is commonly symbolized by e^-, where the minus sign indicates the negative charge. The positron is symbolized by e^+ because it has the same properties as the electron but with a positive rather than negative charge.

The electron has an intrinsic angular momentum or spin of $\frac{1}{2}$. This property is usually stated by referring to the electron as a spin-$\frac{1}{2}$ particle. For such particles the spin magnitude is $\frac{\sqrt{3}}{2}\hbar$. while the result of the measurement of a projection of the spin on any axis can only be $\pm\frac{\hbar}{2}$. In addition to spin, the electron has an intrinsic magnetic moment along its spin axis. It is approximately equal to one Bohr magneton, which is a physical constant equal to $9.27400915(23)\times10^{-24}$ joules per tesla. The orientation of the spin with respect to the momentum of the electron defines the property of elementary particles known as helicity.

The electron has no known substructure and it is assumed to be a point particle with a point charge and no spatial extent. In classical physics, the angular momentum and magnetic moment of an object depend upon its physical dimensions. Hence, the concept of a dimensionless electron possessing these properties contrasts to experimental observations in Penning traps which point to finite non-zero radius of the electron. A possible explanation of this paradoxical situation is given below in the "Virtual particles" subsection by taking into consideration the Foldy-Wouthuysen transformation.

The issue of the radius of the electron is a challenging problem of the modern theoretical physics. The admission of the hypothesis of a finite radius of the electron is incompatible to the premises of the theory of relativity. On the other hand, a point-like electron (zero radius) generates serious mathematical difficulties due to the self-energy of the electron tending to infinity.

Observation of a single electron in a Penning trap suggests the upper limit of the particle's radius to be 10^{-22} meters. The upper bound of the electron radius of 10^{-18} meters can be derived using the uncertainty relation in energy.

There *is* also a physical constant called the "classical electron radius", with the much larger value of 2.8179×10^{-15} m, greater than the radius of the proton. However, the terminology comes from a simplistic calculation that ignores the effects of quantum mechanics; in reality, the so-called classical electron radius has little to do with the true fundamental structure of the electron.

There are elementary particles that spontaneously decay into less massive particles. An example is the muon, with a mean lifetime of 2.2×10^{-6} seconds, which decays into an electron, a muon neutrino and an electron antineutrino. The electron, on the other hand, is thought to be stable on theoretical grounds: the electron is the least massive particle with non-zero electric charge, so its decay would violate charge conservation. The experimental lower bound for the electron's mean lifetime is 6.6×10^{28} years, at a 90% confidence level.

Quantum Properties

As with all particles, electrons can act as waves. This is called the wave–particle duality and can be demonstrated using the double-slit experiment.

The wave-like nature of the electron allows it to pass through two parallel slits simultaneously, rather than just one slit as would be the case for a classical particle. In quantum mechanics, the wave-like property of one particle can be described mathematically as a complex-valued function, the wave function, commonly denoted by the Greek letter psi (ψ). When the absolute value of this function is squared, it gives the probability that a particle will be observed near a location—a probability density.

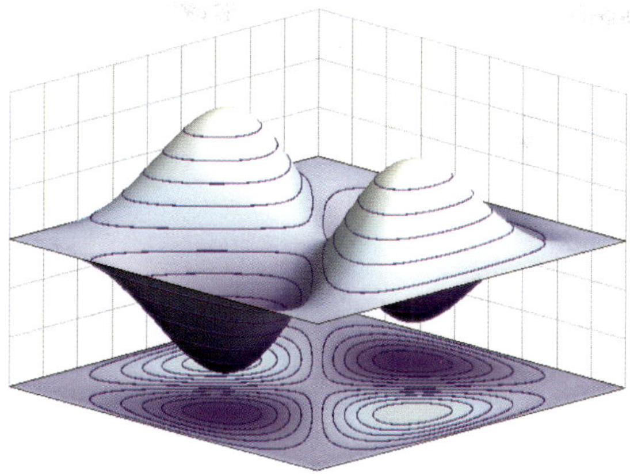

Example of an antisymmetric wave function for a quantum state of two identical fermions in a 1-dimensional box. If the particles swap position, the wave function inverts its sign.

Electrons are identical particles because they cannot be distinguished from each other by their intrinsic physical properties. In quantum mechanics, this means that a pair of interacting electrons must be able to swap positions without an observable change to the state of the system. The wave function of fermions, including electrons, is antisymmetric, meaning that it changes sign when two electrons are swapped; that is, $\psi(r_1, r_2) = -\psi(r_2, r_1)$, where the variables r_1 and r_2 correspond to the first and second electrons, respectively. Since the absolute value is not changed by a sign swap, this corresponds to equal probabilities. Bosons, such as the photon, have symmetric wave functions instead.

In the case of antisymmetry, solutions of the wave equation for interacting electrons result in a zero probability that each pair will occupy the same location or state. This is responsible for the Pauli exclusion principle, which precludes any two electrons from occupying the same quantum state. This principle explains many of the properties of electrons. For example, it causes groups of bound electrons to occupy different orbitals in an atom, rather than all overlapping each other in the same orbit.

Virtual Particles

In a simplified picture, every photon spends some time as a combination of a virtual electron plus its antiparticle, the virtual positron, which rapidly annihilate each other shortly thereafter. The combination of the energy variation needed to create these particles, and the time during which they exist, fall under the threshold of detectability expressed by the Heisenberg uncertainty relation, $\Delta E \cdot \Delta t \geq \hbar$. In effect, the energy needed to create these virtual particles, ΔE, can be "borrowed" from the vacuum for a period of time, Δt, so that their product is no more than the reduced Planck constant, $\hbar \approx 6.6\times10^{-16}$ eV·s. Thus, for a virtual electron, Δt is at most 1.3×10^{-21} s.

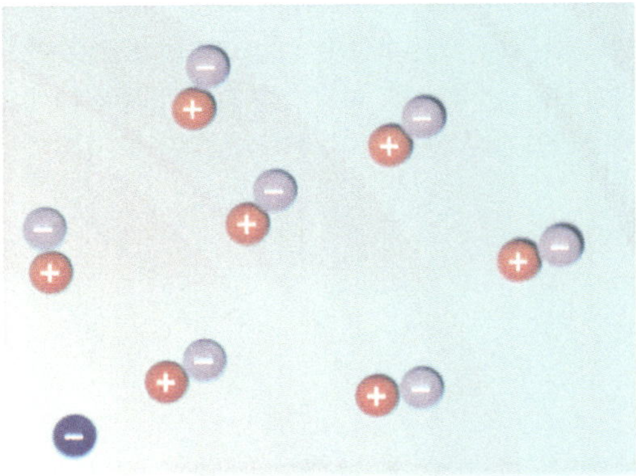

A schematic depiction of virtual electron–positron pairs appearing at random near an electron (at lower left)

While an electron–positron virtual pair is in existence, the coulomb force from the ambient electric field surrounding an electron causes a created positron to be attracted to the original electron, while a created electron experiences a repulsion. This causes what is called vacuum polarization. In effect, the vacuum behaves like a medium having a dielectric permittivity more than unity. Thus the effective charge of an electron is actually smaller than its true value, and the charge decreases with increasing distance from the electron. This polarization was confirmed experimentally in 1997 using the Japanese TRISTAN particle accelerator. Virtual particles cause a comparable shielding effect for the mass of the electron.

The interaction with virtual particles also explains the small (about 0.1%) deviation of the intrinsic magnetic moment of the electron from the Bohr magneton (the anomalous magnetic moment). The extraordinarily precise agreement of this predicted difference with the experimentally determined value is viewed as one of the great achievements of quantum electrodynamics.

The apparent paradox of a point particle electron having intrinsic angular momentum and magnetic moment can be explained by the formation of virtual photons in the electric field generated by the electron. These photons cause the electron to shift about in a jittery fashion (known as zitterbewegung), which results in a net circular motion with precession. This motion produces both the spin and the magnetic moment of the electron. In atoms, this creation of virtual photons explains the Lamb shift observed in spectral lines.

Interaction

An electron generates an electric field that exerts an attractive force on a particle with a positive charge, such as the proton, and a repulsive force on a particle with a negative charge. The strength of this force in nonrelativistic approximation is determined by

Coulomb's inverse square law. When an electron is in motion, it generates a magnetic field. The Ampère-Maxwell law relates the magnetic field to the mass motion of electrons (the current) with respect to an observer. This property of induction supplies the magnetic field that drives an electric motor. The electromagnetic field of an arbitrary moving charged particle is expressed by the Liénard–Wiechert potentials, which are valid even when the particle's speed is close to that of light (relativistic).

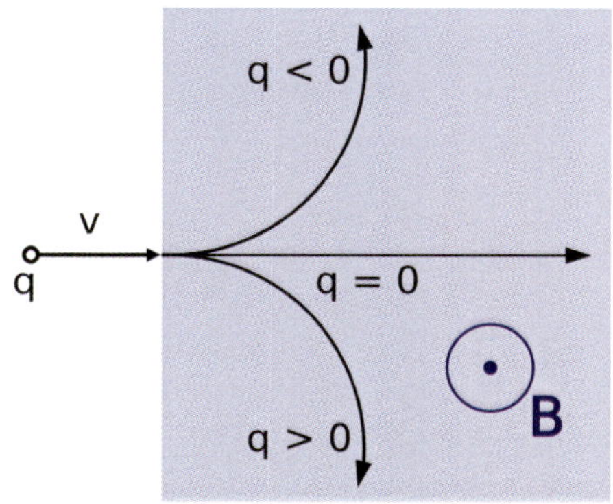

A particle with charge q (at left) is moving with velocity v through a magnetic field B that is oriented toward the viewer. For an electron, q is negative so it follows a curved trajectory toward the top.

When an electron is moving through a magnetic field, it is subject to the Lorentz force that acts perpendicularly to the plane defined by the magnetic field and the electron velocity. This centripetal force causes the electron to follow a helical trajectory through the field at a radius called the gyroradius. The acceleration from this curving motion induces the electron to radiate energy in the form of synchrotron radiation. The energy emission in turn causes a recoil of the electron, known as the Abraham–Lorentz–Dirac Force, which creates a friction that slows the electron. This force is caused by a back-reaction of the electron's own field upon itself.

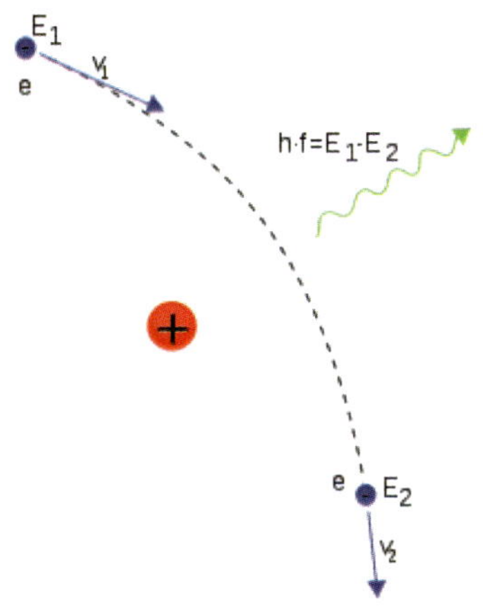

Here, Bremsstrahlung is produced by an electron e deflected by the electric field of an atomic nucleus. The energy change $E_2 - E_1$ determines the frequency f of the emitted photon.

Photons mediate electromagnetic interactions between particles in quantum electrodynamics. An isolated electron at a constant velocity cannot emit or absorb a real photon;

doing so would violate conservation of energy and momentum. Instead, virtual photons can transfer momentum between two charged particles. This exchange of virtual photons, for example, generates the Coulomb force. Energy emission can occur when a moving electron is deflected by a charged particle, such as a proton. The acceleration of the electron results in the emission of Bremsstrahlung radiation.

An inelastic collision between a photon (light) and a solitary (free) electron is called Compton scattering. This collision results in a transfer of momentum and energy between the particles, which modifies the wavelength of the photon by an amount called the Compton shift. The maximum magnitude of this wavelength shift is $h/m_e c$, which is known as the Compton wavelength. For an electron, it has a value of 2.43×10^{-12} m. When the wavelength of the light is long (for instance, the wavelength of the visible light is 0.4–0.7 μm) the wavelength shift becomes negligible. Such interaction between the light and free electrons is called Thomson scattering or linear Thomson scattering.

The relative strength of the electromagnetic interaction between two charged particles, such as an electron and a proton, is given by the fine-structure constant. This value is a dimensionless quantity formed by the ratio of two energies: the electrostatic energy of attraction (or repulsion) at a separation of one Compton wavelength, and the rest energy of the charge. It is given by $\alpha \approx 7.297353\times10^{-3}$, which is approximately equal to $\frac{1}{137}$.

When electrons and positrons collide, they annihilate each other, giving rise to two or more gamma ray photons. If the electron and positron have negligible momentum, a positronium atom can form before annihilation results in two or three gamma ray photons totalling 1.022 MeV. On the other hand, a high-energy photon can transform into an electron and a positron by a process called pair production, but only in the presence of a nearby charged particle, such as a nucleus.

In the theory of electroweak interaction, the left-handed component of electron's wavefunction forms a weak isospin doublet with the electron neutrino. This means that during weak interactions, electron neutrinos behave like electrons. Either member of this doublet can undergo a charged current interaction by emitting or absorbing a W and be converted into the other member. Charge is conserved during this reaction because the W boson also carries a charge, canceling out any net change during the transmutation. Charged current interactions are responsible for the phenomenon of beta decay in a radioactive atom. Both the electron and electron neutrino can undergo a neutral current interaction via a Z^o exchange, and this is responsible for neutrino-electron elastic scattering.

Atoms and Molecules

An electron can be *bound* to the nucleus of an atom by the attractive Coulomb force. A system of one or more electrons bound to a nucleus is called an atom. If the number of electrons is different from the nucleus' electrical charge, such an atom is called an ion.

The wave-like behavior of a bound electron is described by a function called an atomic orbital. Each orbital has its own set of quantum numbers such as energy, angular momentum and projection of angular momentum, and only a discrete set of these orbitals exist around the nucleus. According to the Pauli exclusion principle each orbital can be occupied by up to two electrons, which must differ in their spin quantum number.

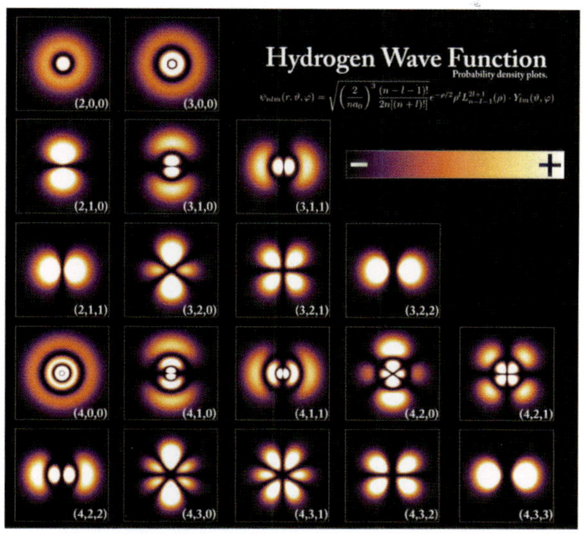

Probability densities for the first few hydrogen atom orbitals, seen in cross-section. The energy level of a bound electron determines the orbital it occupies, and the color reflects the probability of finding the electron at a given position.

Electrons can transfer between different orbitals by the emission or absorption of photons with an energy that matches the difference in potential. Other methods of orbital transfer include collisions with particles, such as electrons, and the Auger effect. To escape the atom, the energy of the electron must be increased above its binding energy to the atom. This occurs, for example, with the photoelectric effect, where an incident photon exceeding the atom's ionization energy is absorbed by the electron.

The orbital angular momentum of electrons is quantized. Because the electron is charged, it produces an orbital magnetic moment that is proportional to the angular momentum. The net magnetic moment of an atom is equal to the vector sum of orbital and spin magnetic moments of all electrons and the nucleus. The magnetic moment of the nucleus is negligible compared with that of the electrons. The magnetic moments of the electrons that occupy the same orbital (so called, paired electrons) cancel each other out.

The chemical bond between atoms occurs as a result of electromagnetic interactions, as described by the laws of quantum mechanics. The strongest bonds are formed by the sharing or transfer of electrons between atoms, allowing the formation of molecules. Within a molecule, electrons move under the influence of several nuclei, and occupy molecular orbitals; much as they can occupy atomic orbitals in isolated atoms. A fundamental factor in these molecular structures is the existence of electron pairs. These are electrons with opposed spins, allowing them to occupy the same molecular orbital without violating the Pauli exclusion principle (much like in atoms). Different molecular orbitals have different spatial distribution of the electron density. For instance, in bonded pairs (i.e. in the pairs that actually bind atoms together) electrons can be found with the maximal probability in a relatively small volume between the nuclei. By contrast, in non-bonded pairs electrons are distributed in a large volume around nuclei.

Conductivity

A lightning discharge consists primarily of a flow of electrons. The electric potential needed for lightning can be generated by a triboelectric effect.

If a body has more or fewer electrons than are required to balance the positive charge of the nuclei, then that object has a net electric charge. When there is an excess of electrons, the object is said to be negatively charged. When there are fewer electrons than the number of protons in nuclei, the object is said to be positively charged. When the number of electrons and the number of protons are equal, their charges cancel each other and the object is said to be electrically neutral. A macroscopic body can develop an electric charge through rubbing, by the triboelectric effect.

Independent electrons moving in vacuum are termed *free* electrons. Electrons in metals also behave as if they were free. In reality the particles that are commonly termed electrons in metals and other solids are quasi-electrons—quasiparticles, which have the same electrical charge, spin, and magnetic moment as real electrons but might have a different mass. When free electrons—both in vacuum and metals—move, they produce a net flow of charge called an electric current, which generates a magnetic field. Likewise a current can be created by a changing magnetic field. These interactions are described mathematically by Maxwell's equations.

At a given temperature, each material has an electrical conductivity that determines the value of electric current when an electric potential is applied. Examples of good conductors include metals such as copper and gold, whereas glass and Teflon are poor conductors. In any dielectric material, the electrons remain bound to their respective atoms and the material behaves as an insulator. Most semiconductors have a variable level of conductivity that lies between the extremes of conduction and insulation. On the other hand, metals have an electronic band structure containing partially filled electronic bands. The presence of such bands allows electrons in metals to behave as if they were free or delocalized electrons. These electrons are not associated with specific atoms, so when an electric field is applied, they are free to move like a gas (called Fermi gas) through the material much like free electrons.

Because of collisions between electrons and atoms, the drift velocity of electrons in a conductor is on the order of millimeters per second. However, the speed at which a change of current at one point in the material causes changes in currents in other parts of the material, the velocity of propagation, is typically about 75% of light speed. This occurs because electrical signals propagate as a wave, with the velocity dependent on the dielectric constant of the material.

Metals make relatively good conductors of heat, primarily because the delocalized electrons are free to transport thermal energy between atoms. However, unlike electrical conductivity, the thermal conductivity of a metal is nearly independent of temperature. This is expressed mathematically by the Wiedemann–Franz law, which states that the ratio of thermal conductivity to the electrical conductivity is proportional to the temperature. The thermal disorder in the metallic lattice increases the electrical resistivity of the material, producing a temperature dependence for electric current.

When cooled below a point called the critical temperature, materials can undergo a phase transition in which they lose all resistivity to electric current, in a process known as superconductivity. In BCS theory, this behavior is modeled by pairs of electrons entering a quantum state known as a Bose–Einstein condensate. These Cooper pairs have their motion coupled to nearby matter via lattice vibrations called phonons, thereby avoiding the collisions with atoms that normally create electrical resistance. (Cooper pairs have a radius of roughly 100 nm, so they can overlap each other.) However, the mechanism by which higher temperature superconductors operate remains uncertain.

Electrons inside conducting solids, which are quasi-particles themselves, when tightly confined at temperatures close to absolute zero, behave as though they had split into three other quasiparticles: spinons, orbitons and holons. The former carries spin and magnetic moment, the next carries its orbital location while the latter electrical charge.

Motion and Energy

According to Einstein's theory of special relativity, as an electron's speed approaches the speed of light, from an observer's point of view its relativistic mass increases, thereby making it more and more difficult to accelerate it from within the observer's frame of reference. The speed of an electron can approach, but never reach, the speed of light in a vacuum, c. However, when relativistic electrons—that is, electrons moving at a speed close to c—are injected into a dielectric medium such as water, where the local speed of light is significantly less than c, the electrons temporarily travel faster than light in the medium. As they interact with the medium, they generate a faint light called Cherenkov radiation.

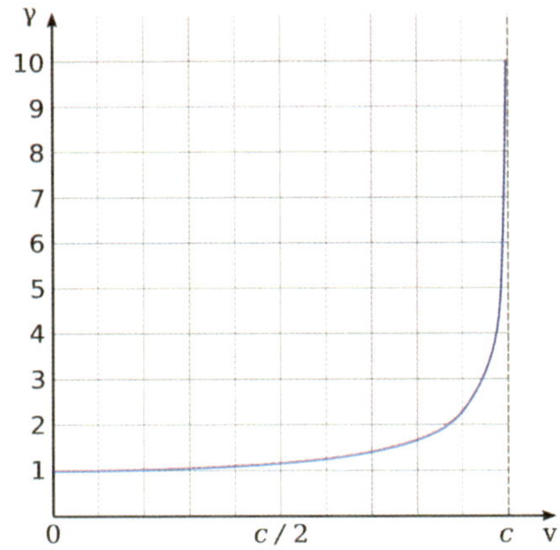

Lorentz factor as a function of velocity. It starts at value 1 and goes to infinity as v approaches c.

The effects of special relativity are based on a quantity known as the Lorentz factor, defined as $\gamma = 1/\sqrt{1 - v^2/c^2}$ where v is the speed of the particle. The kinetic energy K_e of an electron moving with velocity v is:

$$K_e = (\gamma - 1)m_e c^2,$$

where m_e is the mass of electron. For example, the Stanford linear accelerator can accelerate an electron to roughly 51 GeV. Since an electron behaves as a wave, at a given velocity it has a characteristic de Broglie wavelength. This is given by $\lambda_e = h/p$ where h is the Planck constant and p is the momentum. For the 51 GeV electron above, the wavelength is about 2.4×10^{-17} m, small enough to explore structures well below the size of an atomic nucleus.

Formation

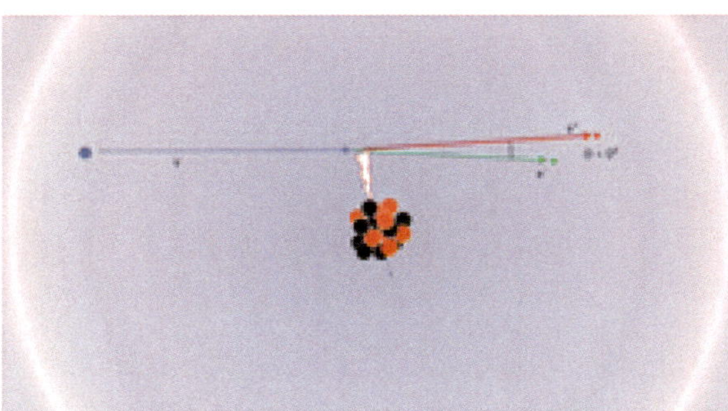

Pair production of an electron and positron, caused by the close approach of a photon with an atomic nucleus.

The lightning symbol represents an exchange of a virtual photon, thus an electric force acts. The angle between the particles is very small.

The Big Bang theory is the most widely accepted scientific theory to explain the early stages in the evolution of the Universe. For the first millisecond of the Big Bang, the temperatures were over 10 billion kelvins and photons had mean energies over a million electronvolts. These photons were sufficiently energetic that they could react with each other to form pairs of electrons and positrons. Likewise, positron-electron pairs annihilated each other and emitted energetic photons:

$$\gamma + \gamma \leftrightarrow e^{+} + e^{-}$$

An equilibrium between electrons, positrons and photons was maintained during this phase of the evolution of the Universe. After 15 seconds had passed, however, the temperature of the universe dropped below the threshold where electron-positron formation could occur. Most of the surviving electrons and positrons annihilated each other, releasing gamma radiation that briefly reheated the universe.

For reasons that remain uncertain, during the annihilation process there was an excess in the number of particles over antiparticles. Hence, about one electron for every billion electron-positron pairs survived. This excess matched the excess of protons over antiprotons, in a condition known as baryon asymmetry, resulting in a net charge of zero for the universe. The surviving protons and neutrons began to participate in reactions with each other—in the process known as nucleosynthesis, forming isotopes of hydrogen and helium, with trace amounts of lithium. This process peaked after about five minutes. Any leftover neutrons underwent negative beta decay with a half-life of about a thousand seconds, releasing a proton and electron in the process,

$$n \rightarrow p + e^{-} + \overline{v}_e$$

For about the next 300000–400000 years, the excess electrons remained too energetic to bind with atomic nuclei. What followed is a period known as recombination, when neutral atoms were formed and the expanding universe became transparent to radiation.

Roughly one million years after the big bang, the first generation of stars began to form. Within a star, stellar nucleosynthesis results in the production of positrons from the fusion of atomic nuclei. These antimatter particles immediately annihilate with electrons, releasing gamma rays. The net result is a steady reduction in the number of electrons, and a matching increase in the number of neutrons. However, the process of stellar evolution can result in the synthesis of radioactive isotopes. Selected isotopes can subsequently undergo negative beta decay, emitting an electron and antineutrino from the nucleus. An example is the cobalt-60 (^{60}Co) isotope, which decays to form nickel-60 (^{60}Ni).

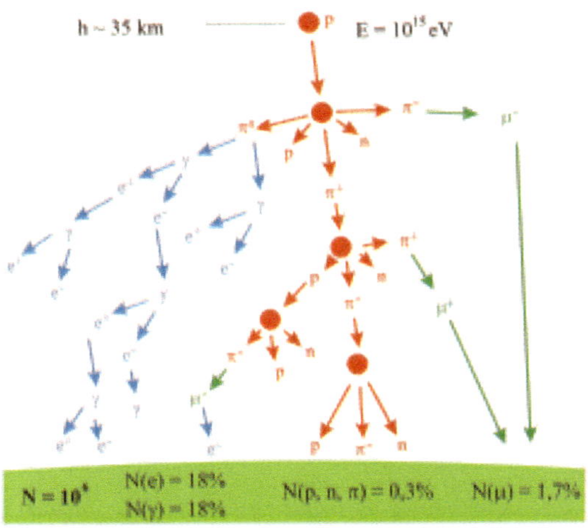

An extended air shower generated by an energetic cosmic ray striking the Earth's atmosphere

At the end of its lifetime, a star with more than about 20 solar masses can undergo gravitational collapse to form a black hole.According to classical physics, these massive stellar objects exert a gravitational attraction that is strong enough to prevent anything, even electromagnetic radiation, from escaping past the Schwarzschild radius. However, quantum mechanical effects are believed to potentially allow the emission of Hawking radiation at this distance. Electrons (and positrons) are thought to be created at the event horizon of these stellar remnants.

When a pair of virtual particles (such as an electron and positron) is created in the vicinity of the event horizon, random spatial positioning might result in one of them to appear on the exterior; this process is called quantum tunnelling. The gravitational potential of the black hole can then supply the energy that transforms this virtual particle into a real particle, allowing it to radiate away into space. In exchange, the other member of the pair is given negative energy, which results in a net loss of mass-energy by the black hole. The rate of Hawking radiation increases with decreasing mass, eventually causing the black hole to evaporate away until, finally, it explodes.

Cosmic rays are particles traveling through space with high energies. Energy events as high as 3.0×10^{20} eV have been recorded. When these particles collide with nucleons in the Earth's atmosphere, a shower of particles is generated, including pions. More than half of the cosmic radiation observed from the Earth›s surface consists of muons. The particle called a muon is a lepton produced in the upper atmosphere by the decay of a pion.

$$\pi^- \rightarrow \mu^- + \bar{\nu}_\mu$$

A muon, in turn, can decay to form an electron or positron.

$$\mu^- \rightarrow e^- + \bar{\nu}_e + \nu_\mu$$

Observation

Aurorae are mostly caused by energetic electrons precipitating into the atmosphere.

Remote observation of electrons requires detection of their radiated energy. For example, in high-energy environments such as the corona of a star, free electrons form a plasma that radiates energy due to Bremsstrahlung radiation. Electron gas can undergo plasma oscillation, which is waves caused by synchronized variations in electron density, and these produce energy emissions that can be detected by using radio telescopes.

The frequency of a photon is proportional to its energy. As a bound electron transitions between different energy levels of an atom, it absorbs or emits photons at characteristic frequencies. For instance, when atoms are irradiated by a source with a broad spectrum, distinct absorption lines appear in the spectrum of transmitted radiation. Each element or molecule displays a characteristic set of spectral lines, such as the hydrogen spectral series. Spectroscopic measurements of the strength and width of these lines allow the composition and physical properties of a substance to be determined.

In laboratory conditions, the interactions of individual electrons can be observed by means of particle detectors, which allow measurement of specific properties such as energy, spin and charge. The development of the Paul trap and Penning trap allows charged particles to be contained within a small region for long durations. This enables precise measurements of the particle properties. For example, in one instance a Penning trap was used to contain a single electron for a period of 10 months. The magnetic moment of the electron was measured to a precision of eleven digits, which, in 1980, was a greater accuracy than for any other physical constant.

The first video images of an electron's energy distribution were captured by a team at Lund University in Sweden, February 2008. The scientists used extremely short flashes of light, called attosecond pulses, which allowed an electron›s motion to be observed for the first time.

The distribution of the electrons in solid materials can be visualized by angle-resolved photoemission spectroscopy (ARPES). This technique employs the photoelectric effect to measure the reciprocal space—a mathematical representation of periodic structures that is used to infer the original structure. ARPES can be used to determine the direction, speed and scattering of electrons within the material.

Plasma Applications

Particle Beams

During a NASA wind tunnel test, a model of the Space Shuttle is targeted by a beam of electrons, simulating the effect of ionizing gases during re-entry.

Electron beams are used in welding. They allow energy densities up to 10^7 W·cm^{-2} across a narrow focus diameter of 0.1–1.3 mm and usually require no filler material. This welding technique must be performed in a vacuum to prevent the electrons from interacting with the gas before reaching their target, and it can be used to join conductive materials that would otherwise be considered unsuitable for welding.

Electron-beam lithography (EBL) is a method of etching semiconductors at resolutions smaller than a micrometer. This technique is limited by high costs, slow performance, the need to operate the beam in the vacuum and the tendency of the electrons to scatter in solids. The last problem limits the resolution to about 10 nm. For this reason, EBL is primarily used for the production of small numbers of specialized integrated circuits.

Electron beam processing is used to irradiate materials in order to change their physical properties or sterilize medical and food products. Electron beams fluidise or quasi-melt glasses without significant increase of temperature on intensive irradiation: e.g. intensive electron radiation causes a many orders of magnitude decrease of viscosity and stepwise decrease of its activation energy.

Linear particle accelerators generate electron beams for treatment of superficial tumors in radiation therapy. Electron therapy can treat such skin lesions as basal-cell carcinomas because an electron beam only penetrates to a limited depth before being absorbed, typically up to 5 cm for electron energies in the range 5–20 MeV. An electron beam can be used to supplement the treatment of areas that have been irradiated by X-rays.

Particle accelerators use electric fields to propel electrons and their antiparticles to high energies. These particles emit synchrotron radiation as they pass through magnetic fields. The dependency of the intensity of this radiation upon spin polarizes the electron beam—a process known as the Sokolov–Ternov effect. Polarized electron beams can be useful for various experiments. Synchrotron radiation can also cool the electron beams to reduce the momentum spread of the particles. Electron and positron beams are collided upon the particles' accelerating to the required energies; particle detectors observe the resulting energy emissions, which particle physics studies.

Imaging

Low-energy electron diffraction (LEED) is a method of bombarding a crystalline material with a collimated beam of electrons and then observing the resulting diffraction patterns to determine the structure of the material. The required energy of the electrons is typically in the range 20–200 eV. The reflection high-energy electron diffraction (RHEED) technique uses the reflection of a beam of electrons fired at various low angles to characterize the surface of crystalline materials. The beam energy is typically in the range 8–20 keV and the angle of incidence is 1–4°.

The electron microscope directs a focused beam of electrons at a specimen. Some electrons change their properties, such as movement direction, angle, and relative phase and energy as the beam interacts with the material. Microscopists can record these changes in the electron beam to produce atomically resolved images of the material. In blue light, conventional optical microscopes have a diffraction-limited resolution of about 200 nm. By comparison, electron microscopes are limited by the de Broglie wavelength of the electron. This wavelength, for example, is equal to 0.0037 nm for electrons accelerated across a 100,000-volt potential. The Transmission Electron Aberration-Corrected Microscope is capable of sub-0.05 nm resolution, which is more than enough to resolve individual atoms. This capability makes the electron microscope a useful laboratory instrument for high resolution imaging. However, electron microscopes are expensive instruments that are costly to maintain.

Two main types of electron microscopes exist: transmission and scanning. Transmission electron microscopes function like overhead projectors, with a beam of electrons passing through a slice of material then being projected by lenses on a photographic slide or a charge-coupled device. Scanning electron microscopes rasteri a finely focused electron beam, as in a TV set, across the studied sample to produce the image. Magnifications range from 100× to 1,000,000× or higher for both microscope types. The scanning tunneling microscope uses quantum tunneling of electrons from a sharp metal tip into the studied material and can produce atomically resolved images of its surface.

Other Applications

In the free-electron laser (FEL), a relativistic electron beam passes through a pair of undulators that contain arrays of dipole magnets whose fields point in alternating directions. The electrons emit synchrotron radiation that coherently interacts with the same electrons to strongly amplify the radiation field at the resonance frequency. FEL can emit a coherent high-brilliance electromagnetic radiation with a wide range of frequencies, from microwaves to soft X-rays. These devices are used in manufacturing, communication, and in medical applications, such as soft tissue surgery.

Electrons are important in cathode ray tubes, which have been extensively used as display devices in laboratory instruments, computer monitors and television sets. In a photomultiplier tube, every photon striking the photocathode initiates an avalanche of electrons that produces a detectable current pulse. Vacuum tubes use the flow of electrons to manipulate electrical signals, and they played a critical role in the development of electronics technology. However, they have been largely supplanted by solid-state devices such as the transistor.

Electron Rest Mass

The electron rest mass, symbolized m_e, is the mass of an electron as measured when its speed is zero relative to an observer. The electron has extremely small mass -- it's almost 2000 times less massive than the proton -- at approximately 9.10956×10^{-31} kilogram (kg) or 9.10956×10^{-28} gram (g). Every known electron at rest has the same mass as every other known electron at rest.

When an electron is accelerated to high speed, its mass increases because of relativistic effects. This increase is not significant until the speed of the particle, relative to the observer, becomes an appreciable fraction of the speed of light . If v is the speed of an electron in meters per second (m/s) and c is the speed of light (approximately 2.99792×10^{8} m/s), then the mass of an electron in motion, m_{e*}, is

$$m_{e*} = m_e / (1 - v^2/c^2)^{1/2} = 9.10956 \times 10^{-31} / (1 - v^2/c^2)^{1/2} \text{ kg.}$$

Determination

Since the electron mass determines a number of observed effects in atomic physics, there are potentially many ways to determine its mass from an experiment, if the values of other physical constants are already considered known.

Historically, the mass of the electron was determined directly from combining two measurements. The mass-to-charge ratio of the electron was first estimated by Arthur Schuster in 1890 by measuring the deflection of "cathode rays" due to a known magnetic field in a cathode ray tube. This was before it was even known what cathode rays actually were! It was seven years later that J. J. Thomson showed that cathode rays consist of streams of particles, to be called electrons, and made more precise measurements of their mass-to-charge ratio again using a cathode ray tube.

The second measurement was of the charge of the electron. This was determined with a precision of better than 1% by Robert A. Millikan in his famous oil drop experiment in 1909. Together with the mass-to-charge ratio, the electron mass was thereby determined with reasonable precision. The value of mass that was found for the electron was initially met with surprise by physicists, since it was so small (less than .1%) compared to the known mass of a hydrogen atom.

The electron rest mass can be calculated from the Rydberg constant R_∞ and the fine structure constant α obtained through spectroscopic measurements. Using the definition of the Rydberg constant:

$$R_\infty = \frac{m_e c \alpha^2}{2h},$$

thus

$$m_e = \frac{2R_\infty h}{c\alpha^2}$$

where c is the speed of light and h is the Planck constant. The relative uncertainty, 5×10^{-8} in the 2006 CODATA recommended value, is due entirely to the uncertainty in the value of the Planck constant.

The electron relative atomic mass can be measured directly in a Penning trap. It can also be inferred from the spectra of antiprotonic helium atoms (helium atoms where one of the electrons has been replaced by an antiproton) or from measurements of the electron g-factor in the hydrogenic ions $^{12}C^{5+}$ or $^{16}O^{7+}$.

The electron relative atomic mass is an adjusted parameter in the CODATA set of fundamental physical constants, while the electron rest mass in kilograms is calculated from the values of the Planck constant, the fine structure constant and the Rydberg constant.

Relationship to other Physical Constants

The electron mass is used to calculate the Avogadro constant N_A:

$$N_A = \frac{M_u A_r(e)}{m_e} = \frac{M_u A_r(e) c\alpha^2}{2R_\infty h}$$

Hence it is also related to the atomic mass constant m_u:

$$m_u = \frac{N_A}{M_u} = \frac{A_r(e)}{m_e} = \frac{A_r(e) c\alpha^2}{2R_\infty h}$$

where M_u is molar mass constant (defined in SI) and $A_r(e)$ is a directly measured quantity, the relative mass of electron.

Note that m_u is defined in terms of $A_r(e)$, and not the other way round, and so the name "electron mass in atomic mass units" for $A_r(e)$ involves a circular definition (at least in terms of practical measurements).

The electron relative atomic mass also enters into the calculation of all other relative atomic masses. By convention, relative atomic masses are quoted for neutral atoms, but the actual measurements are made on positive ions, either in a mass spectrometer or a Penning trap. Hence the mass of the electrons must be added back on to the measured values before tabulation. A correction must also be made for the mass equivalent of the binding energy E_b. Taking the simplest case of complete ionization of all electrons, for a nuclide X of atomic number Z,

$$A_r(X) = A_r(X^{Z+}) + ZA_r(e) - E_b / m_u c^2$$

As relative atomic masses are measured as ratios of masses, the corrections must be applied to both ions: fortunately, the uncertainties in the corrections are negligible, as illustrated below for hydrogen 1 and oxygen 16.

	^{1}H	^{16}O
relative atomic mass of the X^{Z+} ion	1.007 276 466 77(10)	15.990 528 174 45(18)
relative atomic mass of the Z electrons	0.000 548 579 909 43(23)	0.004 388 639 2754(18)
correction for the binding energy	−0.000 000 014 5985	−0.000 002 194 1559
relative atomic mass of the neutral atom	1.007 825 032 07(10)	15.994 914 619 57(18)

The principle can be shown by the determination of the electron relative atomic mass by Farnham *et al.* at the University of Washington (1995). It involves the measurement of the frequencies of the cyclotron radiation emitted by electrons and by $^{12}C^{6+}$ ions in a Penning trap. The ratio of the two frequencies is equal to six times the inverse ratio of the masses of the two particles (the heavier the particle, the lower the frequency of the cyclotron radiation; the higher the charge on the particle, the higher the frequency):

$$\frac{\nu_c(^{12}C^{6+})}{\nu_c(e)} = \frac{6A_r(e)}{A_r(^{12}C^{6+})} = 0.00027436518589(58)$$

As the relative atomic mass of $^{12}C^{6+}$ ions is very nearly 12, the ratio of frequencies can be used to calculate a first approximation to $A_r(e)$, $5.486\,303\,7178\times10^{-4}$. This approximate value is then used to calculate a first approximation to $A_r(^{12}C^{6+})$, knowing that $E_b(^{12}C)/m_u c^2$ (from the sum of the six ionization energies of carbon) is $1.105\,8674\times10^{-6}$: $A_r(^{12}C^{6+}) \approx 11.996\,708\,723\,6367$. This value is then used to calculate a new approximation to $A_r(e)$, and the process repeated until the values no longer vary (given the relative uncertainty of the measurement, 2.1×10^{-9}): this happens by the fourth cycle of iterations for these results, giving $A_r(e) = 5.485\,799\,111(12)\times10^{-4}$ for these data.

Electron Diffraction

Electron diffraction, interference effects owing to the wavelike nature of a beam of electrons when passing near matter. According to the proposal (1924) of the French physicist Louis de Broglie, electrons and other particles have wavelengths that are inversely proportional to their momentum. Consequently, high-speed electrons have short wavelengths, a range of which are comparable to the spacings between atomic layers in crystals. A beam of such high-speed electrons should undergo diffraction, a characteristic wave effect, when directed through thin sheets of material or when reflected from the faces of crystals. Electron diffraction, in fact, was observed (1927) by C.J. Davisson and L.H. Germer in New York and by G.P. Thomson in Aberdeen, Scot. The wavelike nature of electron beams was thereby experimentally established, thus supporting an underlying principle of quantum mechanics.

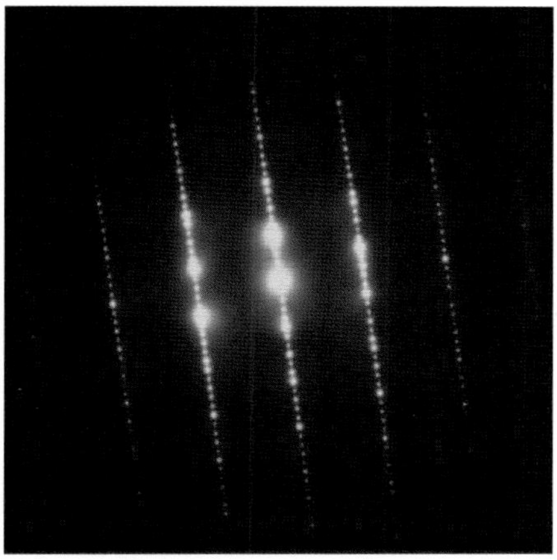

Typical electron diffraction pattern obtained in a transmission electron microscope with a parallel electron beam.

As an analytic method, electron diffraction is used to identify a substance chemically or to locate the position of atoms in a substance. This information can be read from the patterns that are formed when various portions of the diffracted electron beam cross each other and by interference make a regular arrangement of impact positions, some where many electrons reach and some where few or no electrons reach. Some advanced analytical techniques, such as LEEDX (low-energy electron diffraction), depend on these diffraction patterns to examine solids, liquids, and gases.

Theory

Louis de Broglie originated the idea that moving electrons may exhibit both particle and wave nature. He proposed that, analogous to photons, the wavelength of the electron is given by:

$$\lambda = \frac{h}{p} = \frac{h}{m v}$$

where h is Planck's constant and $p = mv$ is the nonrelativistic electron momentum.

This experiment is an attempt to see whether electrons really do act like waves or whether they always act like particles.

If electrons behave as particles (rigid spheres), the distribution of electrons will vary continuously as a function of angle as in figure below. This distribution will vary only slightly with changes in electron energy.

Figure: Particle Model Diffraction.
Continuous distribution of electrons as a function of angle.

If electrons behave as a wave, however, a diffraction pattern will emerge. We can make an analogy with the diffraction of x-rays by a crystal.

Review of x-ray Diffraction

Crystals act as three-dimensional gratings; they scatter the wave and produce observable interference effects. In fact, x-rays scatter from the lattice planes (or Bragg planes) of the crystal as shown in Figure below.

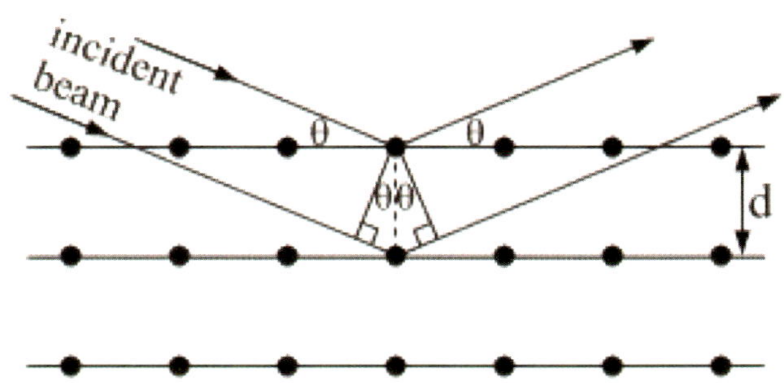

Figure: Electron Waves Reflected from Atomic Planes.

There are two conditions for constructive interference of waves:

1. The angle of incidence must equal the angle of reflection
2. The difference in path length must be an integral number of wavelengths

$n\lambda = 2\,d \sin\theta.$ Bragg's Law

where n is an integer and is the order of diffraction.

By looking at the diffraction pattern and calculating the angle of diffraction, Bragg's Law allows us to determine either:

- The wavelength of the x-rays if the crystal structure is known, or
- The interatomic spacing of the crystal if the x-ray wavelength is known.

The second method is often used to determine the structure of an unknown crystal by performing an x-ray diffraction experiment.

If electrons act like waves, we should be able to apply Bragg's Law to the diffraction of electrons. In that case the beam would appear as concentric rings around a bright center. This pattern can be thought of as a one-dimensional diffraction pattern of bright spots rotated about its center.

Figure: Wave Model Diffraction.
Constructive interference of an electron beam through a polycrystal.

Electron Diffraction in a TEM

Electron diffraction of solids is usually performed in a Transmission Electron Microscope (TEM) where the electrons pass through a thin film of the material to be studied. The resulting diffraction pattern is then observed on a fluorescent screen, recorded on photographic film, on imaging plates or using a CCD camera.

Benefits

The wavelength of an electron accelerated in a TEM is much smaller than that of the radiation usually used for X-ray diffraction experiments. A consequence of this is that the radius of the Ewald sphere is much larger in electron diffraction experiments than in X-ray diffraction. This allows the diffraction experiment to reveal more of the two-dimensional distribution of reciprocal lattice points.

Furthermore, electron lenses allows the geometry of the diffraction experiment to be varied. The conceptually simplest geometry referred to as selected area electron diffraction (SAED) is that of a parallel beam of electrons incident on the specimen, with the specimen field selected using a sub-specimen image-plane aperture. However, by converging the electrons in a cone onto the specimen, one can in effect perform a diffraction experiment over several incident angles simultaneously. This technique is called Convergent Beam Electron Diffraction (CBED) and can reveal the full three-dimensional symmetry of the crystal.

In a TEM, a single crystal grain or particle may be selected for the diffraction experiments. This means that the diffraction experiments can be performed on single crystals of nanometer size, whereas other diffraction techniques would be limited to studying the diffraction from a multicrystalline or powder sample. Furthermore, electron diffraction in TEM can be combined with direct imaging of the sample, including high resolution imaging of the crystal lattice, and a range of other techniques. These include solving and refining crystal structures by electron crystallography, chemical analysis of the sample composition through energy-dispersive X-ray spectroscopy, investigations of electronic structure and bonding through electron energy loss spectroscopy, and studies of the mean inner potential through electron holography.

Practical Aspects

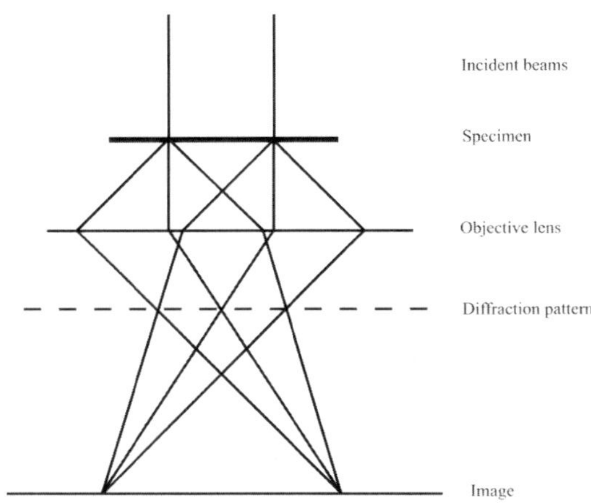

Sketch of the electron beam-path in a TEM.

The figure above is a simple sketch of the path of a parallel beam of electrons in a TEM from just above the sample and down the column to the fluorescent screen. As the electrons pass through the sample, they are scattered by the electrostatic potential set up by the constituent elements. After the electrons have left the sample they pass through the electromagnetic objective lens. This lens acts to collect all electrons scattered from one point of the sample in one point on the fluorescent screen, causing an image of the sample to be formed. We note that at the dashed line in the figure, electrons scattered in the same direction by the sample are collected into a single point. This is the back focal plane of the microscope, and is where the diffraction pattern is formed. By manipulating the magnetic lenses of the microscope, the diffraction pattern may be observed by projecting it onto the screen instead of the image.

If the sample is tilted with respect to the incident electron beam, one can obtain diffraction patterns from several crystal orientations. In this way, the reciprocal lattice of the crystal can be mapped in three dimensions. By studying the systematic absence of diffraction spots the Bravais lattice and any screw axes and glide planes present in the crystal structure may be determined.

Limitations

Electron diffraction in TEM is subject to several important limitations. First, the sample to be studied must be electron transparent, meaning the sample thickness must be of the order of 100 nm or less. Careful and time-consuming sample preparation may therefore be needed. Furthermore, many samples are vulnerable to radiation damage caused by the incident electrons.

The study of magnetic materials is complicated by the fact that electrons are deflected in magnetic fields by the Lorentz force. Although this phenomenon may be exploited to study the magnetic domains of materials by *Lorentz force microscopy*, it may make crystal structure determination virtually impossible.

Furthermore, electron diffraction is often regarded as a *qualitative* technique suitable for symmetry determination, but too inaccurate for determination of lattice parameters and atomic positions. But there are also several examples where unknown crystal structures (inorganic, organic and biological) have been solved by electron crystallography. Lattice parameters of high accuracy can in fact be obtained from electron diffraction, relative errors less than 0.1% have been demonstrated. However, the right experimental conditions may be difficult to obtain, and these procedures are often viewed as too time-consuming and the data too difficult to interpret. X-ray or neutron diffraction are therefore often the preferred methods for determining lattice parameters and atomic positions.

However, the main limitation of electron diffraction in TEM remains the comparatively high level of user interaction needed. Whereas both the execution of powder X-ray (and neutron) diffraction experiments and the data analysis are highly automated and routinely performed, electron diffraction requires a much higher level of user input.

Classical Electron Radius

The classical electron radius is a combination of fundamental physical quantities that define a length scale for problems involving electrons interacting with electromagnetic radiation. According to modern understanding, the electron is a point particle with a point charge and no spatial extent. Attempts to model the electron as a non-point particle are considered ill-conceived and counter-pedagogic. Nevertheless, it is useful to define a length that arises in electron interactions in atomic-scale problems. The classical electron radius is given as (in SI units)

$$r_e = \frac{1}{4\pi\varepsilon_0}\frac{e^2}{m_e c^2} = 2.8179403227(19)\times10^{-15}\ m,$$

where e and m_e are the electric charge and the mass of the electron, c is the speed of light, and ε_0 is the permittivity of free space. This numerical value is several times larger than the radius of the proton.

In cgs units, the permittivity factor does not enter, but the classical electron radius has the same value.

The classical electron radius is sometimes known as the Lorentz radius or the Thomson scattering length. It is one of a trio of related scales of length, the other two being the Bohr radius a_0 and the Compton wavelength of the electron λ_e. The classical electron radius is built from the electron mass m_e, the speed of light c and the electron charge e. The Bohr radius is built from m_e, e and Planck's constant h. The Compton wavelength is built from m_e, h and c. Any one of these three length scales can be written in terms of any other using the fine structure constant α:

$$r_e = \alpha \frac{\hbar c}{m_e c^2} = \alpha \frac{\lambda_e}{2\pi} = \alpha^2 a_0.$$

Derivation

The classical electron radius length scale can be motivated by considering the energy necessary to assemble an amount of charge q into a sphere of a given radius r. The electrostatic potential at a distance r from a charge q is

$$V(r) = \frac{1}{4\pi\varepsilon_0}\frac{q}{r}.$$

To bring an additional amount of change dq from infinity necessitates putting energy into the system, dU, by an amount

$$dU = V(r)dq.$$

If the sphere is assumed to have constant charge density, ρ, then

$$q = \rho \frac{4}{3}\pi r^3 \text{ and } dq = \rho 4\pi r^2 dr.$$

Doing the integration for r starting at zero up to a final radius r leads to the expression for the total energy, U, necessary to assemble total charge q into a uniform sphere of radius r:

$$U = \frac{1}{4\pi\varepsilon_0}\frac{3}{5}\frac{q^2}{r}.$$

This is called the electrostatic self-energy of the object. The charge q is now interpreted as the electron charge, e, and the energy U is set equal to the relativistic mass-energy of the electron, mc^2, and the numerical factor 3/5 is ignored as being specific to the special case of a uniform charge density. The radius r is then defined to be the classical electron radius, r_e, and one arrives at the expression given above.

Note that this derivation does not say that r_e is the actual radius of an electron. It only establishes a dimensional link between electrostatic self energy and the mass-energy scale of the electron.

Electron Excitation

Electrons can absorb energy from external sources, such as lasers, arc-discharge lamps, and tungsten-halogen bulbs, and be promoted to higher energy levels. This topic explores how photon energy is absorbed by an electron to elevate it into a higher energy level and how the energy can subsequently be released, in the form of a lower energy photon, when the electron falls back to the original ground state.

Visible light displays classical wave-like properties, but it also exhibits properties reminiscent of particles, which are manifested through entities that possess energy and momentum (but no mass), and are referred to as photons. The atom is the source of all forms of electromagnetic radiation, whether visible or invisible. Higher-energy forms of radiation, such as gamma waves and X-rays, are produced by events that occur to disrupt the nuclear stability of the atom. Radiation having lower energy, such as ultraviolet, visible, and infrared light, as well as radio and microwaves, originate from the electron clouds that surround the nucleus or the interaction of one atom with another. These forms of radiation occur due to fact that electrons moving in orbits around the nucleus of an atom are arranged in different energy levels within their probability distribution functions. Many of the electrons can absorb additional energy from external sources of electromagnetic radiation, which results in their promotion to an inherently unstable higher energy level.

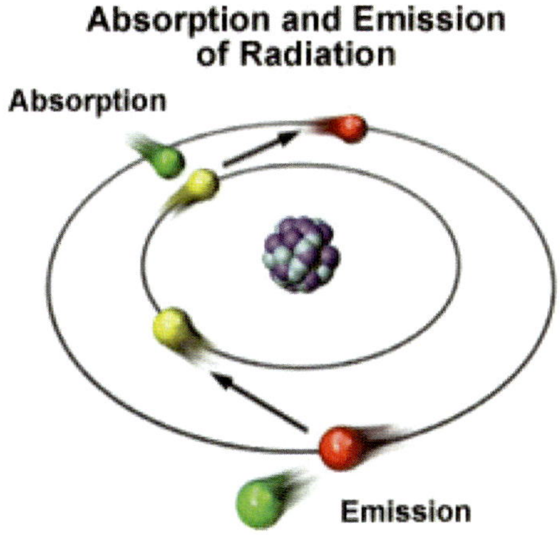

Eventually, the "excited" electron loses the extra energy by emitting electromagnetic radiation of lower energy and, in doing so, falls back into its original and stable energy level. The energy of the emitted radiation equals the energy that was originally absorbed by the electron minus other small quantities of energy lost through a number of secondary processes.

Electromagnetic radiation energy levels can vary to a significant degree depending upon the energy of source electrons or nuclei. For example, radio waves possess significantly less energy than do microwaves, infrared rays, or visible light, and all of these waves contain far less energy than ultraviolet light, X-rays, and gamma waves. As a rule, higher electromagnetic radiation energies are associated with shorter wavelengths than similar forms of radiation having lower energy. The relationship between the energy of an electromagnetic wave and its frequency is expressed by the equation:

$$E = h\nu = hc/\lambda$$

where E is the energy in kilojoules per mole, *h* is Planck's constant, and the other variables are defined as discussed previously. Based on this equation, the energy of an electromagnetic wave is directly proportional to its frequency and inversely proportional to the wavelength. Thus, as frequency increases (with a corresponding decrease in wavelength), the electromagnetic wave energy increases, and vice versa.

Proton

A proton is a subatomic particle with a positive electric charge that lies within the atomic nucleus of atoms. The number of protons in the atomic nucleus is that which determines the atomic number of an element, as indicated in the periodic table of the elements.

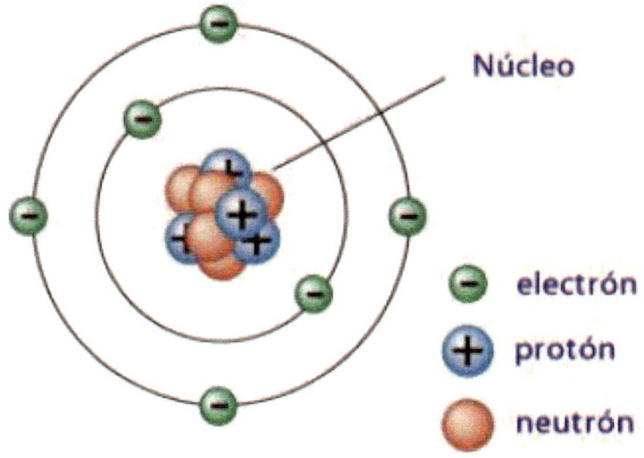

The proton has +1 charge (or, alternatively, 1,602 x 10^{-19} coulombs), exactly the opposite of the charge -1 containing the electron. In mass, however, there is no competition because the mass of the proton is approximately 1,836 times greater than that of an electron.

The proton is classified as a baryon, and is composed of three quarks (uud). The corresponding antiparticle, the antiproton, has the same characteristics as the proton but with negative electric charge.

The proton is stable by itself. In some rare types of radioactive decay emit free protons, and the result of the decomposition of free neutrons in other disintegrations. As a free proton, it has the facility to pick up an electron and become a neutral hydrogen isotope, which can react chemically very easily. Free protons may exist in plasmas, cosmic rays or in the solar wind.

Description of Protons

Protons are composed of three quarks of 1/2 spin. Protons are classified as baryons which are a subtype of hadrons. The two top quarks and a quark below the protons are held together by strong nuclear interaction. The proton has a positive charge distribution and decays exponentially.

Protons and neutrons are nucleons. Both are linked in the nucleus by a strong nuclear force. The most common hydrogen isotope is a nucleus with a proton. The nuclei of the heavy hydrogen isotopes (deuterium and tritium) contain one proton and one or two neutrons, respectively. These two hydrogen isotopes are used as nuclear fuel in nuclear fusion reactions. All other types of atoms are composed of two or more protons and a different number of neutrons.

The number of protons in the nucleus of an atom determines its chemical properties and, therefore, the chemical element is represented by the number of protons in a nucleus (Z). To determine the isotopes of an element, the number of neutrons (N) is also used by summing all the nucleons, and is known as the mass number (A).

According to the current of proton particle physics experiments, the proton is a stable particle, which means that it does not disintegrate into other particles and, therefore, within the experimental limits, its life is eternal. This point is summarized in the conservation of the number of baryons in the processes between elementary particles. In fact, the lightest baryon is precisely the proton and, if the baryon number is to be stored, it can not decay in any lighter particle.

Stability

The free proton (a proton not bound to nucleons or electrons) is a stable particle that has not been observed to break down spontaneously to other particles. Free protons are found naturally in a number of situations in which energies or temperatures are high enough to separate them from electrons, for which they have some affinity. Free protons exist in plasmas in which temperatures are too high to allow them to combine with electrons. Free protons of high energy and velocity make up 90% of cosmic rays, which propagate in vacuum for interstellar distances. Free protons are emitted directly from atomic nuclei in some rare types of radioactive decay. Protons also result (along with electrons and antineutrinos) from the radioactive decay of free neutrons, which are unstable.

The spontaneous decay of free protons has never been observed, and protons are therefore considered stable particles according to the Standard Model. However, some grand unified theories (GUTs) of particle physics predict that proton decay should take place with lifetimes between 10^{31} to 10^{36} years and experimental searches have established lower bounds on the mean lifetime of a proton for various assumed decay products.

Experiments at the Super-Kamiokande detector in Japan gave lower limits for proton mean lifetime of 6.6×10^{33} years for decay to an antimuon and a neutral pion, and 8.2×10^{33} years for decay to a positron and a neutral pion. Another experiment at the Sudbury Neutrino Observatory in Canada searched for gamma rays resulting from residual nuclei resulting from the decay of a proton from oxygen-16. This experiment was designed to detect decay to any product, and established a lower limit to a proton lifetime of 2.1×10^{29} years.

However, protons are known to transform into neutrons through the process of electron capture (also called inverse beta decay). For free protons, this process does not occur spontaneously but only when energy is supplied. The equation is:

$$p^{+} + e^{-} \rightarrow n + v_{e}$$

The process is reversible; neutrons can convert back to protons through beta decay, a common form of radioactive decay. In fact, a free neutron decays this way, with a mean lifetime of about 15 minutes.

Quarks and the Mass of a Proton

In quantum chromodynamics, the modern theory of the nuclear force, most of the mass of protons and neutrons is explained by special relativity. The mass of a proton is about 80–100 times greater than the sum of the rest masses of the quarks that make it up, while the gluons have zero rest mass. The extra energy of the quarks and gluons in a region within a proton, as compared to the rest energy of the quarks alone in the QCD vacuum, accounts for almost 99% of the mass. The rest mass of a proton is, thus, the invariant mass of the system of moving quarks and gluons that make up the particle, and, in such systems, even the energy of massless particles is still measured as part of the rest mass of the system.

Two terms are used in referring to the mass of the quarks that make up protons: *current quark mass* refers to the mass of a quark by itself, while *constituent quark mass* refers to the current quark mass plus the mass of the gluon particle field surrounding the quark. These masses typically have very different values. As noted, most of a proton's mass comes from the gluons that bind the current quarks together, rather than from the quarks themselves. While gluons are inherently massless, they possess energy—to be more specific, quantum chromodynamics binding energy (QCBE)—and it is this that contributes so greatly to the overall mass of protons. A proton has a mass of approximately 938 MeV/c^2, of which the rest mass of its three valence quarks contributes only about 9.4 MeV/c^2; much of the remainder can be attributed to the gluons' QCBE.

The internal dynamics of protons are complicated, because they are determined by the quarks' exchanging gluons, and interacting with various vacuum condensates. Lattice QCD provides a way of calculating the mass of a proton directly from the theory to any accuracy, in principle. The most recent calculations claim that the mass is determined to better than 4% accuracy, even to 1% accuracy. These claims are still controversial, because the calculations cannot yet be done with quarks as light as they are in the real world. This means that the predictions are found by a process of extrapolation, which can introduce systematic errors. It is hard to tell whether these errors are controlled properly, because the quantities that are compared to experiment are the masses of the hadrons, which are known in advance.

These recent calculations are performed by massive supercomputers, and, as noted by Boffi and Pasquini: "a detailed description of the nucleon structure is still missing because ... long-distance behavior requires a nonperturbative and/or numerical treatment..." More conceptual approaches to the structure of protons are: the topological soliton approach originally due to Tony Skyrme and the more accurate AdS/QCD approach that extends it to include a string theory of gluons, various QCD-inspired models like the bag model and the constituent quark model, which were popular in the 1980s, and the SVZ sum rules, which allow for rough approximate mass calculations. These methods do not have the same accuracy as the more brute-force lattice QCD methods, at least not yet.

Charge Radius

The problem of defining a radius for an atomic nucleus (proton) is similar to the problem of atomic radius, in that neither atoms nor their nuclei have definite boundaries. However, the nucleus can be modeled as a sphere of positive charge for the interpretation of electron scattering experiments: because there is no definite boundary to the nucleus, the electrons "see" a range of cross-sections, for which a mean can be taken. The qualification of "rms" (for "root mean square") arises because it is the nuclear cross-section, proportional to the square of the radius, which is determining for electron scattering.

The internationally accepted value of a proton's charge radius is 0.8768 fm. This value is based on measurements involving a proton and an electron (namely, electron scattering measurements and complex calculation involving scattering cross section based on Rosenbluth equation for momentum-transfer cross section), and studies of the atomic energy levels of hydrogen and deuterium.

However, in 2010 an international research team published a proton charge radius measurement via the Lamb shift in muonic hydrogen (an exotic atom made of a proton and a negatively charged muon). As a muon is 200 times heavier than an electron, its de Broglie wavelength is correspondingly shorter. This smaller atomic orbital is much more sensitive to the proton's charge radius, so allows more precise measurement. Their measurement of the root-mean-square charge radius of a proton is "0.84184(67) fm, which differs by 5.0 standard deviations from the CODATA value of 0.8768(69) fm". In January 2013, an updated value for the charge radius of a proton—0.84087(39) fm—was published. The precision was improved by 1.7 times, increasing the significance of the discrepancy to 7σ. The 2014 CODATA adjustment slightly reduced the recommended value for the proton radius (computed using electron measurements only) to 0.8751(61) fm, but this leaves the discrepancy at 5.6σ.

The international research team that obtained this result at the Paul Scherrer Institut in Villigen includes scientists from the Max Planck Institute of Quantum Optics, Ludwig-Maximilians-Universität, the Institut für Strahlwerkzeuge of Universität Stuttgart, and the University of Coimbra, Portugal. The team is now attempting to explain the discrepancy, and re-examining the results of both previous high-precision measurements and complex calculations involving scattering cross section. If no errors are found in the measurements or calculations, it could be necessary to re-examine the world's most precise and best-tested fundamental theory: quantum electrodynamics. The proton radius remains a puzzle as of 2017. Perhaps the discrepancy is due to new physics, or the explanation may be an ordinary physics effect that has been missed.

The radius is linked to the form factor and momentum transfer cross section. The atomic form factor G modifies the cross section corresponding to point-like proton.

$$R_e^2 = -6\frac{dG_e}{dq^2}\bigg|_{q^2=0}$$

$$\frac{d\sigma}{d\Omega} = \frac{d\sigma}{d\Omega}\bigg|_{point} G^2(q^2)$$

The atomic form factor is related to the wave function density of the target:

$$G(q^2) = \int e^{iqr}\psi(r)^2\,dr^3$$

The form factor can be split in electric and magnetic form factors. These can be further written as linear combinations of Dirac and Pauli form factors.

$$G_m = F_D + F_P$$

$$G_e = F_D - \tau F_P$$

$$\frac{d\sigma}{d\Omega} = \frac{d\sigma}{d\Omega}\bigg|_{NS}\frac{1}{1+\tau}\left(G_e^2\left(q^2\right) + \frac{\tau}{\epsilon}G_m^2\left(q^2\right)\right)$$

Interaction of Free Protons with Ordinary Matter

Although protons have affinity for oppositely charged electrons, this is a relatively low-energy interaction and so free protons must lose sufficient velocity (and kinetic energy) in order to become closely associated and bound to electrons. High energy protons, in traversing ordinary matter, lose energy by collisions with atomic nuclei, and by ionization of atoms (removing electrons) until they are slowed sufficiently to be captured by the electron cloud in a normal atom.

However, in such an association with an electron, the character of the bound proton is not changed, and it remains a proton. The attraction of low-energy free protons to any electrons present in normal matter (such as the electrons in normal atoms) causes free protons to stop and to form a new chemical bond with an atom. Such a bond happens at any sufficiently "cold" temperature (i.e., comparable to temperatures at the surface of the Sun) and with any type of atom. Thus, in interaction with any type of normal (non-plasma) matter, low-velocity free protons are attracted to electrons in any atom or molecule with which they come in contact, causing the proton and molecule to combine. Such molecules are then said to be "protonated", and chemically they often, as a result, become so-called Brønsted acids.

Proton in Chemistry

Atomic Number

In chemistry, the number of protons in the nucleus of an atom is known as the atomic number, which determines the chemical element to which the atom belongs. For example, the atomic number of chlorine is 17; this means that each chlorine atom has 17 protons and that all atoms with 17 protons are chlorine atoms. The chemical properties of each atom are determined by the number of (negatively charged) electrons, which for neutral atoms is equal to the number of (positive) protons so that the total charge is zero. For example, a neutral chlorine atom has 17 protons and 17 electrons, whereas a Cl^- anion has 17 protons and 18 electrons for a total charge of −1.

All atoms of a given element are not necessarily identical, however, as the number of neutrons may vary to form different isotopes, and energy levels may differ forming different nuclear isomers. For example, there are two stable isotopes of chlorine: ${}^{35}_{17}Cl$- with 35 − 17 = 18 neutrons and ${}^{37}_{17}Cl$ with 37 − 17 = 20 neutrons.

Hydrogen Ion

The term "hydrogen ion" (H^+) implies that that H-atom has lost its one electron, causing only a proton to remain. Thus, in chemistry, the terms "proton" and "hydrogen ion" (for the protium isotope) are used synonymously

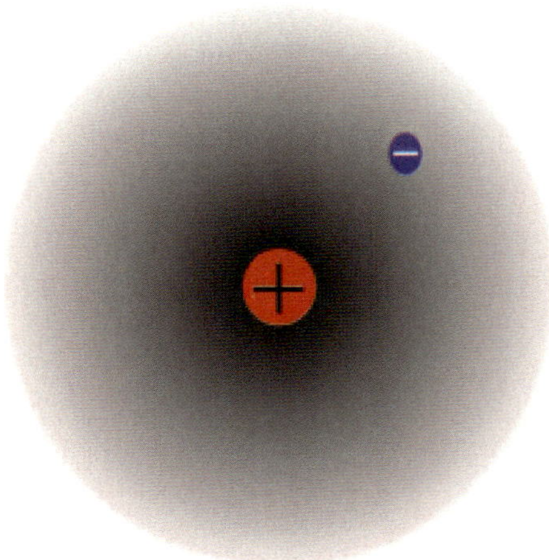

Protium, the most common isotope of hydrogen, consists of one proton and one electron (it has no neutrons).

In chemistry, the term proton refers to the hydrogen ion, H^+. Since the atomic number of hydrogen is 1, a hydrogen ion has no electrons and corresponds to a bare nucleus, consisting of a proton (and 0 neutrons for the most abundant isotope *protium* ^1_1H). The proton is a "bare charge" with only about 1/64,000 of the radius of a hydrogen atom, and so is extremely reactive chemically. The free proton, thus, has an extremely short lifetime in chemical systems such as liquids and it reacts immediately with the electron cloud of any available molecule. In aqueous solution, it forms the hydronium ion, H_3O^+, which in turn is further solvated by water molecules in clusters such as $[H_5O_2]^+$ and $[H_9O_4]^+$.

The transfer of H^+ in an acid–base reaction is usually referred to as "proton transfer". The acid is referred to as a proton donor and the base as a proton acceptor. Likewise, biochemical terms such as proton pump and proton channel refer to the movement of hydrated H^+ ions.

The ion produced by removing the electron from a deuterium atom is known as a deuteron, not a proton. Likewise, removing an electron from a tritium atom produces a triton.

Proton Nuclear Magnetic Resonance (NMR)

Also in chemistry, the term "proton NMR" refers to the observation of hydrogen-1 nuclei in (mostly organic) molecules by nuclear magnetic resonance. This method uses the spin of the proton, which has the value one-half. The name refers to examination of protons as they occur in protium (hydrogen-1 atoms) in compounds, and does not imply that free protons exist in the compound being studied.

Proton Decay

Much rests on the existence of proton decay, and yet we've never seen a proton die. The reason may simply be that protons rarely decay, a hypothesis borne out by both experiment and theory. Experiments say the proton lifetime has to be greater than about 10^{34} years: That's a 1 followed by 34 zeroes.

For reference, the universe is only 13.8 billion years old, which is roughly a 1 followed by 10 zeros. Protons on average will outlast every star, galaxy and planet, even the ones not yet born.

The key phrase in that last sentence is "on average." As Feng says, it's not like "every single proton will last for 10^{34} years and then at 10^{34} years they all boom! poof! in a puff of smoke, they all disappear."

Because of quantum physics, the time any given proton decays is random, so a tiny fraction will decay long before that 10^{34}-year lifetime. So, "what you need to do is to get a whole bunch of protons together," he says. Increasing the number of protons increases the chance that one of them will decay while you're watching.

The second essential step is to isolate the experiment from particles that could mimic proton decay, so any realistic proton decay experiment must be located deep underground to isolate it from random particle passers-by. That's the strategy pursued by the currently operating Super-Kamiokande experiment in Japan, which consists of a huge tank with 50,000 tons of water in a mine. The upcoming Deep Underground Neutrino Experiment, to be located in a former gold mine in South Dakota, will consist of 40,000 tons of liquid argon.

Because the two experiments are based on different types of atoms, they are sensitive to different ways protons might decay, which will reveal which GUT is correct ... if any of the current models is right. Both Super-Kamiokande and DUNE are neutrino experiments first, Feng says, "but we're just as interested in the proton decay possibilities of these experiments as in the neutrino aspects."

After all, proton decay follows from profound concepts of how the cosmos fundamentally operates. If protons do decay, it's so rare that human bodies would be unaffected, but not our understanding. The impact of that knowledge would be immense, and worth a tiny bit of instability.

Baryogenesis

One of the outstanding problems in modern physics is the predominance of matter over antimatter in the universe. The universe, as a whole, seems to have a nonzero positive baryon number density – that is, matter exists. Since it is assumed in cosmology that the particles we see were created using the same physics we measure today, it would normally be expected that the overall baryon number should be zero, as matter and antimatter should have been created in equal amounts. This has led to a number of proposed mechanisms for symmetry breaking that favour the creation of normal matter (as opposed to antimatter) under certain conditions. This imbalance would have been exceptionally small, on the order of 1 in every 10000000000 (10^{10}) particles a small fraction of a second after the Big Bang, but after most of the matter and antimatter annihilated, what was left over was all the baryonic matter in the current universe, along with a much greater number of bosons. Experiments reported in 2010 at Fermilab, however, seem to show that this imbalance is much greater than previously assumed. In an experiment involving a series of particle collisions, the amount of generated matter was approximately 1% larger than the amount of generated antimatter. The reason for this discrepancy is yet unknown.

Most grand unified theories explicitly break the baryon number symmetry, which would account for this discrepancy, typically invoking reactions mediated by very massive X bosons (X) or massive Higgs bosons (H^o). The rate at which these events occur is governed largely by the mass of the intermediate X or H^o particles, so by assuming these reactions are responsible for the majority of the baryon number seen today, a maximum mass can be calculated above which the rate would be too slow to explain the presence of matter today. These estimates predict that a large volume of material will occasionally exhibit a spontaneous proton decay.

Experimental Evidence

Proton decay is one of the key predictions of the various grand unified theories (GUTs) proposed in the 1970s, another major one being the existence of magnetic monopoles. Both concepts have been the focus of major experimental physics efforts since the early 1980s. To date, all attempts to observe these events have failed; however, these experiments have been able to establish lower bounds on the half-life of the proton. Currently the most precise results come from the Super-Kamiokande water Cherenkov radiation detector in Japan: a 2015 analysis placed a lower bound on the proton's half-life of 1.67×10^{34} years via positron decay, and similarly, a 2012 analysis gave a lower bound to the proton's half-life of 1.08×10^{34} years via antimuon decay. close to a supersymmetry (SUSY) prediction of 10^{34}–10^{36} years. An upgraded version, Hyper-Kamiokande, probably will have sensitivity 5–10 times better than Super-Kamiokande.

Theoretical Motivation

Despite the lack of observational evidence for proton decay, some grand unification theories, such as the SU(5) Georgi–Glashow model and SO(10), along with their supersymmetric variants, require it. According to such theories, the proton has a half-life of about 10^{31} to 10^{36} years and decays into a positron and a neutral pion that itself immediately decays into 2 gamma ray photons:

$$p^+ \rightarrow e^+ + \pi^0$$

$$\pi^0 \rightarrow 2\gamma$$

Since a positron is an antilepton this decay preserves $B - L$ number, which is conserved in most GUTs.

Additional decay modes are available (e.g.: $p^+ \rightarrow \mu^+ + \pi^0$), both directly and when catalyzed via interaction with GUT-predicted magnetic monopoles. Though this process has not been observed experimentally, it is within the realm of experimental testability for future planned very large-scale detectors on the megaton scale. Such detectors include the Hyper-Kamiokande.

Early grand unification theories (GUTs) such as the Georgi–Glashow model, which were the first consistent theories to suggest proton decay, postulated that the proton's half-life would be at least 10^{31} years. As further experiments and calculations were performed in the 1990s, it became clear that the proton half-life could not lie below 10^{32} years. Many books from that period refer to this figure for the possible decay time for baryonic matter. More recent findings have pushed the minimum proton half-life to at least 10^{34}-10^{35} years, ruling out the simpler GUTs (including minimal SU(5)/Georgi–Glashow) and most non-SUSY models. The maximum upper limit on proton lifetime (if unstable), is calculated at 6×10^{39} years, a bound applicable to SUSY models, with a maximum for (minimal) non-SUSY GUTs at 1.4×10^{36} years.

Although the phenomenon is referred to as "proton decay", the effect would also be seen in neutrons bound inside atomic nuclei. Free neutrons—those not inside an atomic nucleus—are already known to decay into protons (and an electron and an antineutrino) in a process called beta decay. Free neutrons have a half-life of about 10 minutes (610.2±0.8 s) due to the weak interaction. Neutrons bound inside a nucleus have an immensely longer half-life—apparently as great as that of the proton.

Projected Proton Lifetimes

Theory class	Proton lifetime (years)
Minimal SU(5) (Georgi–Glashow)	$10^{30}...10^{31}$
Minimal SUSY SU(5)	$10^{28}...10^{32}$
SUGRA SU(5)	$10^{32}...10^{34}$
SUSY SU(5)(MSSM)	$\sim 10^{34}$
Minimal (Basic) SO(10) – Non SUSY	$< \sim 10^{35}$ (maximum range)
SUSY SO(10)	$10^{32}...10^{35}$
SUSY SO(10) MSSM G(224)	$2 \cdot 10^{34}$
Flipped SU(5) (MSSM)	$10^{35}...10^{36}$
SUSY SU(5) – 5 dimensions	$10^{34}...10^{35}$

Decay Operators

Dimension-6 Proton Decay Operators

The dimension-6 proton decay operators are $\frac{qqql}{\Lambda^2}$, $\frac{d^c u^c u^c e^c}{\Lambda^2}$, $\frac{\overline{e^c u^c} qq}{\Lambda^2}$ and $\frac{\overline{d^c u^c} ql}{\Lambda^2}$ where Λ is the cutoff scale for the Standard Model. All of these operators violate both baryon number (B) and lepton number (L) conservation but not the combination $B - L$.

In GUT models, the exchange of an X or Y boson with the mass Λ_{GUT} can lead to the last two operators suppressed by $\frac{1}{\Lambda_{GUT}^2}$. The exchange of a triplet Higgs with mass M can lead to all of the operators suppressed by $\frac{1}{M^2}$. See doublet–triplet splitting problem.

Proton decay. These graphics refer to the X bosons and Higgs bosons.

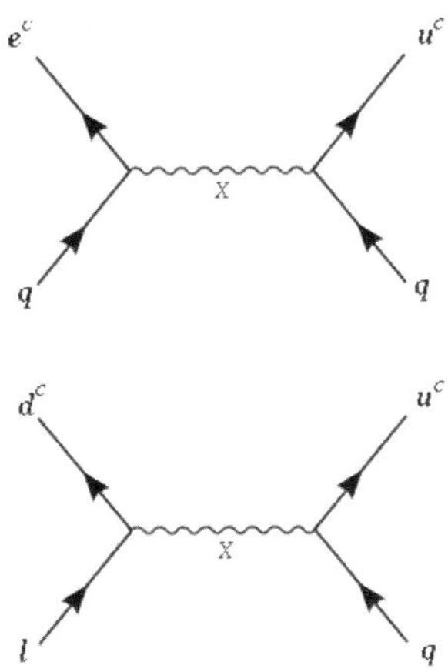

Dimension 6 proton decay mediated by the X boson $(3,2)_{-5/6}$ in SU(5) GUT

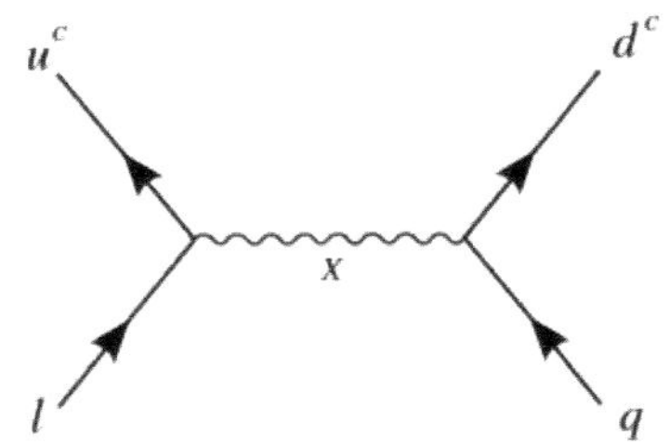

Dimension 6 proton decay mediated by the X boson $(3,2)_{1/6}$ in flipped SU(5) GUT

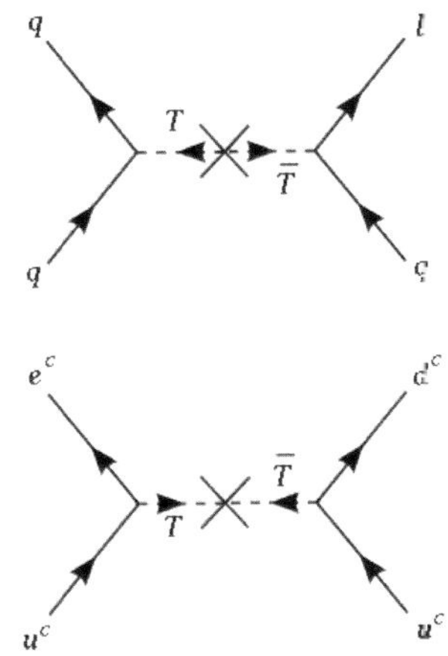

Dimension-6 proton decay mediated by the triplet Higgs T $(3,1)_{-1/3}$ and the anti-triplet Higgs $\bar{T}\,(3,1)_{1/3}$ in SU(5) GUT

Dimension-5 Proton Decay Operators

In supersymmetric extensions (such as the MSSM), we can also have dimension-5 operators involving two fermions and two sfermions caused by the exchange of a tripletino of mass M. The sfermions will then exchange a gaugino or Higgsino or gravitino leaving two fermions. The overall Feynman diagram has a loop (and other complications due to strong interaction physics). This decay rate is suppressed by $\frac{1}{MM_{SUSY}}$ where M_{SUSY} is the mass scale of the superpartners.

Dimension-4 Proton Decay Operators

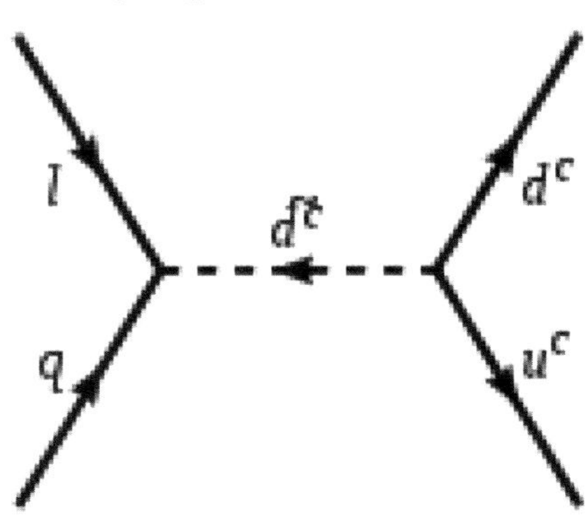

In the absence of matter parity, supersymmetric extensions of the Standard Model can

give rise to the last operator suppressed by the inverse square of sdown quark mass. This is due to the dimension-4 operators $q \ell \tilde{d}^c$ and $u^c d^c \tilde{d}^c$.

Life-span of Proton

The proton decay rate is only suppressed by $\frac{1}{M_{SUSY}^2}$ which is far too fast unless the couplings are very small.

There are several models for GUTs, but mostly they predict lifetimes longer than 10^{30} years! This is unimaginably long, especially when compared to the age of the universe, which is estimated to be approximately 138x10^8 years. Of course, we cannot keep on observing for such long period. However, the lifetime of particles is defined as the time in which the number of particles decreases to 1/2.72 of the initial number. So it may be possible to measure proton lifetime if we prepare a huge number of protons, even though the observation period is minute compared to the lifetime of a proton. This is the reason why we need a large detector in order to measure proton lifetime.

Currently, the most sensitive detector in the world used to examine proton decay is Super-Kamiokande (SK), which contains 7.5x10^{33} protons. SK has been performing its observations for more than 12 years, but still proton decay has not been observed and 10^{34} years has been obtained as the lower limit of proton lifetime.

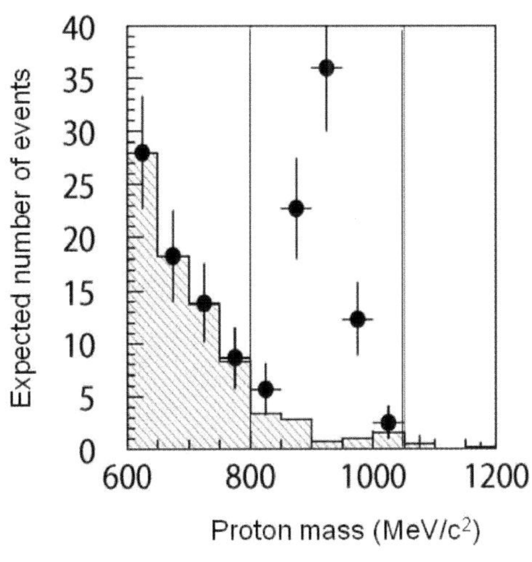

Reconstructed proton mass distribution after 10 years run of Hyper-Kamiokande assuming the current lower limit as proton lifetime. Upper figure corresponds to lower reconstructed proton momentum case (< 100 MeV/c) and lower shows higher momentum case (100 ~ 250 MeV/c). Dots shows sum of signal and background, Hatched histogram shows only background.

Hyper-Kamiokande is about 10 times larger than SK and it can overtake the current reach by SK within two years. Hyper-Kamiokande can detect all final particles and mass (938MeV/c^2) and momentum of proton can be reconstructed from decayed particles. If proton lifetime is assumed as the current lower limit obtained by SK, we will see clear peak above background as seen in figure above. Especially, the lower proton momentum region (< 100 MeV/c, above figure) is expected to be almost background free and only a few events in this region could be an evidence of the proton decay. Furthermore, Hyper-Kamiokande has sensitivity up to more than one order longer than the current lower lifetime of proton and most of GUT models can be examined (Figure below). Hyper-Kamiokande will discover proton decays and we will challenge the root of materials and mysteries in the genesis of the universe beyond the Standard Model.

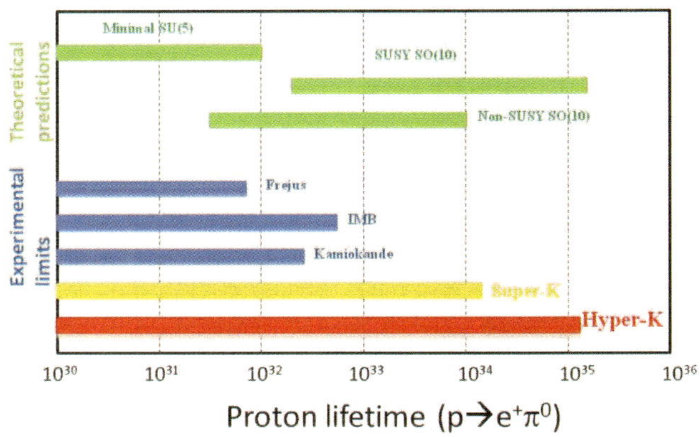

Figure: Predictions of proton lifetime and sensitivity of Hyper-Kamiokande. The sensitivity can be jumped up more than 10 times after 10 years run of Hyper-Kamiokande and can cover most of predictions.

Proton Emission

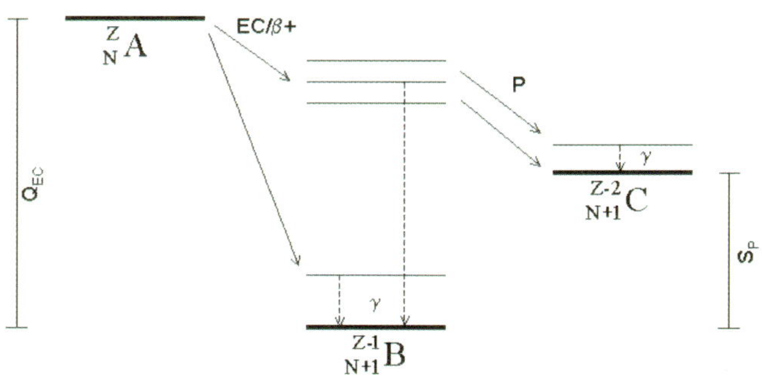

The decay of a proton rich nucleus A populates excited states of a daughter nucleus B by β+ emission or electron capture (EC).

Those excited states that lie below the separation energy for protons (Sp) decay by γ emission towards the groundstate of daughter B. For the higher excited states a competitive decay channel of proton emission to the granddaughter C exists, called β-delayed proton emission.

Proton emission (also known as proton radioactivity) is a rare type of radioactive decay in which a proton is ejected from a nucleus. Proton emission can occur from high-lying excited states in a nucleus following a beta decay, in which case the process is known as beta-delayed proton emission, or can occur from the ground state (or a low-lying isomer) of very proton-rich nuclei, in which case the process is very similar to alpha decay. For a proton to escape a nucleus, the proton separation energy must be negative - the proton is therefore unbound, and tunnels out of the nucleus in a finite time. Proton emission is not seen in naturally occurring isotopes; proton emitters can be produced via nuclear reactions, usually using linear particle accelerators.

Although prompt (i.e. not beta-delayed) proton emission was observed from an isomer in cobalt-53 as early as 1969, no other proton-emitting states were found until 1981, when the proton radioactive ground states of lutetium-151 and thulium-147 were observed at experiments at the GSI in West Germany. Research in the field flourished after this breakthrough, and to date more than 25 isotopes have been found to exhibit proton emission. The study of proton emission has aided the understanding of nuclear deformation, masses, and structure, and it is a pure example of quantum tunneling.

In 2002, the simultaneous emission of two protons was observed from the nucleus iron-45 in experiments at GSI and GANIL (Grand Accélérateur National d'Ions Lourds at Caen). In 2005 it was experimentally determined (at the same facility) that zinc-54 can also undergo double proton decay.

Proton Magnetic Moment

The proton magnetic moment is the magnetic dipole moment of the proton, symbol μ_p. Protons and neutrons, both nucleons, comprise the nucleus of an atom, and both nucleons act as small magnets whose strength is measured by their magnetic moments. The magnitude of the proton's magnetic moment indicates that the proton is not an elementary particle.

Description

CODATA's recommended value for the magnetic moment of the proton is μ_p = 2.7928473508(85) μ_N. A more precise measurement has since been claimed, with a result of μ_p = 2.79284734462(82) μ_N, for an 11-fold improvement in precision. In these values, μ_N is the nuclear magneton, a physical constant and standard unit for the magnetic moments of nuclear components. In SI units, the CODATA value of μ_p is $1.4106067873(97)\times10^{-26}$ J·T^{-1}. A magnetic moment is a vector quantity, and the direction of the proton's magnetic moment is defined by its spin. The torque on the proton resulting from an external magnetic field is towards aligning the proton's spin vector in the same direction as the magnetic field vector.

The nuclear magneton is the spin magnetic moment of a Dirac particle, a charged, spin 1/2 elementary particle, with a proton's mass m_p. In SI units, the nuclear magneton is

$$\mu_N = \frac{e\hbar}{2m_p}$$

where e is the elementary charge and $\hbar$ is the reduced Planck constant. The magnetic moment of this particle is parallel to its spin. Since the proton has charge +1 e, it should have magnetic moment equal to 1 μ_N by this expression. The larger magnetic moment of the proton indicates that it is not an elementary particle. The sign of the proton's magnetic moment is that of a positively charged particle. Similarly, the fact that the magnetic moment of the neutron, $\mu_n = -1.913\ \mu_N$, is finite and negative indicates that it too is not an elementary particle. Protons and neutrons are composed of quarks, and the magnetic moments of the quarks can be used to compute the magnetic moments of the nucleons.

The magnetic moment of the antiproton is the same magnitude, but is of opposite sign, as that of the proton.

Measurement

The anomalously large magnetic moment of the proton was discovered in 1933 by Otto Stern in Hamburg. Stern won the Nobel Prize in 1943 for this discovery.

By 1934 groups led by Stern, now in Pittsburgh, and I. I. Rabi in New York had independently measured the magnetic moments of the proton and deuteron. While the measured values for these particles were only in rough agreement between the groups, the Rabi group confirmed the earlier Stern measurements that the magnetic moment for the proton was unexpectedly large. Since a deuteron is composed of a proton and a neutron with aligned spins, the neutron's magnetic moment could be inferred by subtracting the deuteron and proton magnetic moments. The resulting value was not zero and had sign opposite to that of the proton. By the late 1930s accurate values for the magnetic moment of the proton had been measured by the Rabi group using newly developed nuclear magnetic resonance techniques. The large value for the proton's magnetic moment and the inferred negative value for the neutron's magnetic moment were unexpected and raised many questions. The anomalous values for the magnetic moments of the nucleons would remain a puzzle until the quark model was developed in the 1960s.

Proton g-factor and Gyromagnetic Ratio

The magnetic moment of a nucleon is sometimes expressed in terms of its g-factor, a dimensionless scalar. The conventional formula is

$$\mu = \frac{g\mu_N}{\hbar} \mathrm{I}$$

where μ is the intrinsic magnetic moment of the nucleon, I is the nuclear spin angular momentum, and g is the effective g-factor. For the proton, I is $1/2\ \hbar$, so the proton's g-factor, symbol g_p, is 5.585694713(46).

The gyromagnetic ratio, symbol γ, of a particle or system is the ratio of its magnetic moment to its spin angular momentum, or

$$\mu = \gamma I$$

For nucleons, the ratio is conventionally written in terms of the proton mass and charge, by the formula

$$\gamma = \frac{g\mu_N}{\hbar} = g\frac{e}{2m_p}$$

The proton's gyromagnetic ratio, symbol γ_p, is 2.675222005(63)×10⁸ rad·s⁻¹·T⁻¹. The gyromagnetic ratio is also the ratio between the observed angular frequency of Larmor precession (in rad s^{-1}) and the strength of the magnetic field in proton NMR applications, such as in MRI imaging or proton magnetometers. For this reason, the value of γ_p is often given in units of MHz/T. The quantity $\gamma_p/2\pi$ ("gamma bar") is therefore convenient, which has the value 42.5774806(10) MHz·T^{-1}.

Physical Significance

When a proton is put into a magnetic field produced by an external source, it is subject to a torque tending to orient its magnetic moment parallel to the field (hence its spin also parallel to the field). Like any magnet, the amount of this torque is proportional both to the magnetic moment and the external magnetic field. Since the proton has spin angular momentum, this torque will cause the proton to precess with a well-defined frequency, called the Larmor frequency. It is this phenomenon that enables the measurement of nuclear properties through nuclear magnetic resonance. The Larmor frequency can be determined by the product of the gyromagnetic ratio with the magnetic field strength. Since the sign of γ_p is positive, the proton's spin angular momentum precesses clockwise about the direction of the external magnetic field.

Since an atomic nucleus consists of a bound state of protons and neutrons, the magnetic moments of the nucleons contribute to the nuclear magnetic moment, or the magnetic moment for the nucleus as a whole. The nuclear magnetic moment also includes contributions from the orbital motion of the nucleons. The deuteron has the simplest example of a nuclear magnetic moment, with measured value 0.857 μ_N. This value is within 3% of the sum of the moments of the proton and neutron, which gives 0.879 μ_N. In this calculation, the spins of the nucleons are aligned, but their magnetic moments offset because of the neutron's negative magnetic moment.

Magnetic Moment, Quarks and the Standard Model

Within the quark model for hadrons, such as the neutron, the proton is composed of one down quark (charge -1/3 e) and two up quarks (charge +2/3 e). The magnetic moment of the proton can be modeled as a sum of the magnetic moments of the constituent quarks, although this simple model belies the complexities of the Standard Model of particle physics.

In one of the early successes of the Standard Model (SU(6) theory), in 1964 Mirza A. B. Beg, Benjamin W. Lee, and Abraham Pais theoretically calculated the ratio of proton to neutron magnetic moments to be −3/2, which agrees with the experimental value to within 3%. The measured value for this ratio is −1.45989806(34). A contradiction of the quantum mechanical basis of this calculation with the Pauli exclusion principle, led to the discovery of the color charge for quarks by Oscar W. Greenberg in 1964.

From the nonrelativistic, quantum mechanical wavefunction for baryons composed of three quarks, a straightforward calculation gives fairly accurate estimates for the magnetic moments of protons, neutrons, and other baryons. The calculation assumes that the quarks behave like pointlike Dirac particles, each having their own magnetic moment, as computed using an expression similar to the one above for the nuclear magneton. For a proton, the end result of this calculation is that the magnetic moment of the neutron is given by $\mu_p = 4/3\,\mu_u - 1/3\,\mu_d$, where μ_u and μ_d are the magnetic moments for the up and down quarks, respectively. This result combines the intrinsic magnetic moments of the quarks with their orbital magnetic moments.

Baryon	**Magnetic moment of quark model**	**Computed (μ_N)**	**Observed (μ_N)**
p	$4/3\,\mu_u - 1/3\,\mu_d$	2.79	2.793
n	$4/3\,\mu_d - 1/3\,\mu_u$	−1.86	−1.913

While the results of this calculation are encouraging, the masses of the up or down quarks were assumed to be 1/3 the mass of a nucleon, whereas the masses of these quarks are only about 1% that of a nucleon. The discrepancy stems from the complexity of the Standard Model for nucleons, where most of their mass originates in the gluon fields and virtual particles that are essential aspects of the strong force. Further, the complex system of quarks and gluons that constitute a neutron requires a relativistic treatment. A calculation of nucleon magnetic moments from first principles is not yet available.

Proton-to-electron Mass Ratio

In physics, the proton-to-electron mass ratio, μ or β, is simply the rest mass of the proton divided by that of the electron. Because this is a ratio of like-dimensioned physical quantities, it is a dimensionless quantity, a function of the dimensionless physical constants, and has numerical value independent of the system of units, namely:

$$\mu = m_p/m_e = 1836.15267389(17).$$

The number enclosed in parentheses is the measurement uncertainty on the last two digits. The value of μ is known to about 0.1 parts per billion.

Discussion

μ is an important fundamental physical constant because:

- Nearly all of science deals with baryonic matter and how the fundamental interactions affect such matter. Baryonic matter consists of quarks and particles made from quarks, like protons and neutrons. Free neutrons have a half life of 613.9 seconds. Electrons and protons appear to be stable, to the best of current knowledge. (Theories of proton decay predict that the proton has a half life on the order of at least 10^{32} years. To date, there is no experimental evidence of proton decay.);
- The proton is the most important baryon, while the electron is the most important lepton;
- μ and the fine structure constant α are the two dimensionless quantities emerging in elementary physics, and two of the three dimensionless quantities discussed in Barrow (2002);
- The proton mass m_p is composed primarily of gluons, and of the quarks (the up quark and down quark) making up the proton. Hence m_p, and therefore the ratio μ, are easily measurable consequences of the strong force. In fact, in the chiral limit, m_p is proportional to the QCD energy scale, Λ_{QCD}. At a given energy scale, the strong coupling constant α_s is related to the QCD scale (and thus μ) as

$$\alpha_s = -\frac{2\pi}{\beta_0 \ln(E / \Lambda_{QCD})}$$

where $\beta_0 = -11 + 2n/3$, with n being the number of flavors of quarks.

Variation of μ Over Time

Astrophysicists have tried to find evidence that μ has changed over the history of the universe. (The same question has also been asked of the fine structure constant.) One interesting cause of such change would be change over time in the strength of the strong force.

Astronomical searches for time-varying μ have typically examined the Lyman series and Werner transitions of molecular hydrogen which, given a sufficiently large redshift, occur in the optical region and so can be observed with ground-based spectrographs.

If μ were to change, then the change in the wavelength λ_i of each rest frame wavelength can be parameterised as:

$$\lambda_i = \lambda_0[1 + K_i(\Delta\mu / \mu)],$$

where $\Delta\mu/\mu$ is the proportional change in μ and K_i is a constant which must be calculated within a theoretical (or semi-empirical) framework.

Reinhold et al. (2006) reported a potential 4 standard deviation variation in μ by analysing the molecular hydrogen absorption spectra of quasars Q0405-443 and Q0347-373. They found that $\Delta\mu/\mu = (2.4 \pm 0.6)\times10^{-5}$. King et al. (2008) reanalysed the spectral data of Reinhold et al. and collected new data on another quasar, Q0528-250. They estimated that $\Delta\mu/\mu = (2.6 \pm 3.0)\times10^{-6}$, different from the estimates of Reinhold et al. (2006).

Murphy et al. (2008) used the inversion transition of ammonia to conclude that $|\Delta\mu/\mu| < 1.8\times10^{-6}$ at redshift $z = 0.68$.

Bagdonaite et al. (2013) used methanol transitions in the spiral lens galaxy PKS 1830-211 to find $\Delta\mu/\mu = (0.0 \pm 1.0) \times 10^{-7}$ at $z = 0.89$, a stringent limit at this redshift.

Note that any comparison between values of $\Delta\mu/\mu$ at substantially different redshifts will need a particular model to govern the evolution of $\Delta\mu/\mu$. That is, results consistent with zero change at lower redshifts do not rule out significant change at higher redshifts.

Proton Radius Puzzle

The proton radius puzzle (PRP) concerns the difference between the radius of the proton as measured with electron scattering and atomic hydrogen spectroscopy, and that measured in muonic hydrogen. In 2010, the CREMA Collaboration published their measurement of the proton radius Rp = 0.8409(4) fm, which was made by studying the Lamb shift in muonic hydrogen. This result, although ten times more precise than, is completely incompatible with the CODATA value of Rp = 0.8775(51) fm. Until that point, the CODATA value was in good agreement with both scattering and spectroscopy results. Since 2010, various attempts have been made to resolve the PRP by looking closely at the experimental data and its interpretation: looking for errors in the theory of the radius extraction from the spectroscopy data, looking for errors in the data itself, and re-analyzing and refitting the electron scattering data, but all to no avail.

There were initially many suggestions as to what could theoretically cause this discrepancy in radius measurements, but, to date, none of the postulated solutions have been generally accepted as resolving the puzzle. Many have been ruled out. Remaining explanations include possible physics beyond the Standard Model (BSM), whereby the electron is measuring an electromagnetic radius, while the muon is measuring an electromagnetic radius modified by BSM effects. There are also suggestions that there might be a currently unconstrained correction in the proton polarizability. Neither is universally accepted but neither has been conclusively ruled out.

The puzzle, if anything, has become more puzzling since its inception in 2010. Now the main focus for resolution of the PRP is on new measurements: atomic hydrogen spectroscopy measurements are being repeated; the CREMA collaboration is extending the range of muonic measurements to include D, ^{3}He and ^{4}He; new, lower Q^2, electron scattering measurements are planned and underway JLab and MAMI; and a simultaneous measurement of muon and electron scattering, of both charge states, will be performed at PSI by the MUSE collaboration. By measuring elastic scattering cross sections down to low Q^2, it is possible to test for the new particles postulated as part of the BSM explanation. By measuring in both charge states, MUSE can access two-photon exchange effects, postulated to be responsible for the breakdown of the Rosenbluth separation and which would be enhanced by the enhanced proton polarizability proposition.

Radius Measurements

Electron Scattering Measurements

The "radius" of the proton is defined as the slope of the electric form factor at $Q^2 = 0$. To date, the most comprehensive elastic electron scattering cross section measurement was performed at MAMI by Bernauer *et al.* They fit the data with a variety of functional forms and determined the proton electric radius to be $< r2E >1/2= 0.879(5)_{stat}(4)_{syst}(2)_{model}(4)_{group}$ fm. The data agree with the recent JLab polarized measurement and the CODATA value. Other fits including the world data set, but deliberately excluding the Mainz data, agree with the CODATA value, strengthening the puzzle.

The Muonic Spectroscopy Measurements

The muonic hydrogen measurements were made as the knowledge of the proton radius limits precision in many electromagnetic physics measurements. Due to the finite size of the proton, the electron in atomic hydrogen actually has a finite probability to be found inside the proton, when occupying S-states. While inside the proton, some of its attractive potential is screened, thereby reducing the average attractive potential seen by the electron and perturbing the energy levels of the S-states. This perturbs the S-P transitions, so by measuring them very precisely, and applying some theoretical calculations, one can extract the radius of the proton.

The probability that the electron is found inside the proton is simply proportional to the volume of the proton relative to the volume of the atom. By replacing the electron with a muon, which is ~205 times as massive, one shrinks the size of the atom and thereby increases the strength of the perturbation, vastly improving the sensitivity of the measurement to the proton radius, and enabling a much more precise radius extraction.

The measurement by Pohl *et al.* of $rp = 0.84184 \pm 0.00067$ fm was highly unexpected. The result was confirmed in 2013 by the analysis of Antognini *et al.*, who measured the proton electric radius to be $rp = 0.84087 \pm 0.00039$ fm. Interestingly, Antognini *et al.* produced a magnetic radius which agrees with that obtained from electron scattering, while still supporting the discrepant electric radius of Pohl *et al.*

The CREMA collaboration has extended their measurements beyond hydrogen to deuterium, ^3He and ^4He. Preliminary results from muonic deuterium are consistent with the CODATA, electronscattering and spectroscopy measured values for the deuterium radius. The ^{3}He measurements are also underway, and the ^{4}He data is under analysis.

New Measurements

It is commonly agreed that the only way to solve the proton radius puzzle is by new measurements. Some of these plan to confirm the old results with new, more precise measurements, some intend to extend their kinematic range, and some to measure completely new things, such as precise muon scattering or spectroscopy of new muonic ions.

Spectroscopy and Electron Scattering Measurements

There are current efforts underway to remeasure the proton radius via atomic hydrogen spectroscopy. In addition to this, as described above, the CREMA collaboration are extending their measurement series to a number of light nuclei. There are also new elastic electron scattering experiments, planned and underway, to extend the measurements at JLab and Mainz down to lower Q^2. To date, there have been experiments in spectroscopy involving electrons and muons and elastic scattering experiments involving electrons, but no elastic muon scattering experiment at a level of precision able to answer the proton radius puzzle. The MUSE experiment will change that.

The MUon Proton Scattering Experiment (MUSE)

MUSE entails a simultaneous measurement of elastic scattering of muons and electrons on the proton, investigating both positive and negative charge states. The experiment will reach as low a Q^2 as the existing electron scattering measurements, with sub-percent precision. The experiment will use beam line Cherenkov, scintillating fiber and GEM detectors to determine the trajectory and species of each incoming beam particle. The scattered particles will be detected in Straw Tube Trackers and a double scintillator wall. This will allow a direct comparison of cross sections and form factors, and, ultimately, the proton radius, as extracted from muon and electron elastic scattering. The simultaneous measurement of electrons and muons allows very good systematic control and measurement of both charge states also allows a test of two-photon-exchange effects.

The MUSE experiment will take place at the πM1 beam line of the Paul Scherrer Institute (PSI). The πM1 beam line is a secondary beam line and the beam consists of a mixture of pions, electrons and muons with a 50 MHz time structure. We will measure at three different beam momenta: p_{in} _ 115, 153 and 210 MeV/c2. These momenta are chosen to allow good separation of the different particle types in time and give overlapping kinematic points for good systematic control while providing data in the range of kinematic interest for measurements sensitive to the proton radius.

Due to the large dispersion and the mixed particle nature of the beam, it is necessary to monitor beam timing and angle accurately. Precise timing will be achieved by tagging beam particles as they enter the experimental area at the two higher momenta using a Cherenkov detector, following the idea prototyped by Albrow *et al.* This will provide accurate timing of beam particles relative to the beam RF signals and time of flight. The Cherenkov detector will be the first detector in the beamline in the target area. At the lowest beam momentum (115 MeV), timing constraints are not as strict and multiple scattering is a much larger issue, so the Cherenkov will be replaced by a thin plastic scintillator.

The position and timing of beam particles for trigger purposes will then be measured using a scintillating fiber (SciFi) detector in the target area. This consists of four layers of forty fibres each, with double-ended maPMT readout. The SciFi detector is followed by a stack of three Gas Electron Multiplier (GEM) chambers, repurposed from OLYMPUS at DESY. These provide accurate tracking for offline scattering angle reconstruction, but the faster SciFi signal is necessary for trigger-purposes. While all the other equipment will be custom-made for the experiment, the GEM chambers already exist, are already present in PSI and have been used for test beam times. The GEM chambers are, however, being reconfigured to shorten their stack length in order to minimize their impact on multiple scattering in the experiment. The SciFi and GEM chambers will be used in the offline analysis to determine the exact angle of the particles entering the target to enable accurate scattering angle definition.

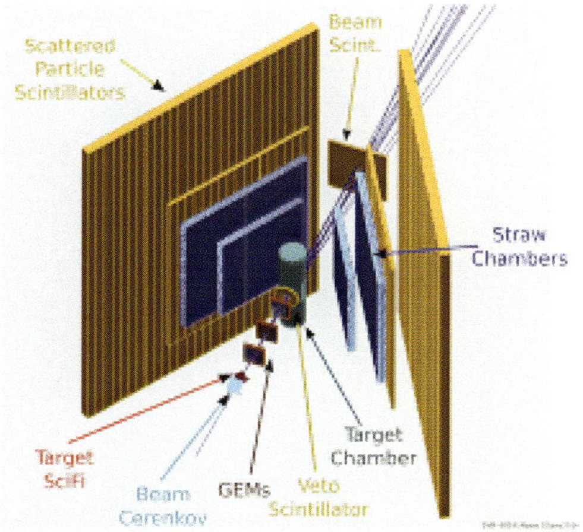

Figure: The MUSE setup: showing Cherenkov detector, SciFi, and GEM beamline detectors, and the scattered particle detector arms, with STTs and triggering scintillator walls.

In order to compensate for the low beam flux of the πM1 beam line, the scattered particle detector system employed by the MUSE experiment will take the form of a non-magnetic, large-acceptance tracking detector. There will be two identical sectors, one on each side of the beamline. Each side will consist of two ten-layer straw tube tracking chambers (STTs) with 1425 straws per side followed by two layers of fast scintillators (92 bars in total) for triggering and timing.

The STTs will be based on the PANDA design and will be read out for timing only using

PADIWA discriminators and TRB3 boards which provide timing, scalars and trigger logic. The scintillation counter signals will be sent into PADIWA discriminators and TRB3 boards for timing etc., but also into QDCs in order to allow for offline pulse-height timing corrections.

The experimental trigger will be based on the requirement of a particle firing the Cherenkov / beam line scintillation detector, the SciFi detector and a scattered particle with appropriate timing and position reaching the scattered particle scintillators. The trigger algorithm will be processed in the specially programmed FPGAs (Field-Programmable Gate Arrays) of the TRB3 boards used to

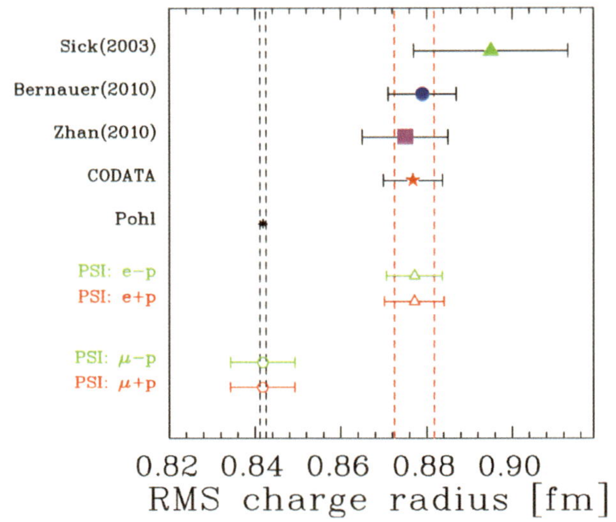

Figure: Current proton radius values and the anticipated MUSE measurement precision.

provide timing and scalar information for each chanel, with one TRB3 trigger master which will make the final trigger decision and distribute the trigger and trigger number to the TRB3s and non-TRB3 parts of the detector system.

In order to suppress backgrounds caused by muon decays in-flight, and scattering in the beamline, there will be an annular veto detector, comprising of eight scintillators, surrounding the entrance to the liquid hydrogen target scattering chamber. The signals from these detectors will be incorporated in the trigger system as a veto. These backgrounds will be further reduced in the offline analysis by studying the time of flight from the initial target Cherenkov or, at low energy, plastic scintillation detector to the scattered particle scintillator wall. In order to monitor the beam and ensure stability throughout the experiment, there will be twelve beam-monitor scintillators which will be used to monitor beam flux and position.

It is expected that MUSE will produce cross section measurements for the elastic scattering of $\mu^{+/-}$ and $e^{+/-}$ with sub-1% relative precision over a Q^2 range from 0.002 to 0.07 GeV^2. Using these data, form factors will be extracted and the resulting radii from muon and electron scattering, and from both charge states, will be compared.

Neutron

A neutron is one of the subatomic particles that make up matter. In the universe, neutrons are abundant, making up more than half of all visible matter. It has no electric charge and a rest mass equal to $1.67493 \times 10-27$ kg—marginally greater than that of the proton but nearly 1839 times greater than that of the electron. The neutron has a mean square radius of about $0.8 \times 10-15$ m, or 0.8 fm, and it is a spin-½ fermion.

The neutrons exist in the nuclei of typical atoms, along with their positively charged counterparts, the protons. Neutrons and protons, commonly called nucleons, are bound together in the atomic nucleus, where they account for 99.9 percent of the atom's mass. Research in high-energy particle physics in the 20th century revealed that neither the neutron nor the proton is not the smallest building block of matter. Protons and neutrons have also their structure. Inside the protons and neutrons, we find true elementary particles called quarks. Within the nucleus, protons and neutrons are bound together through the strong force, a fundamental interaction that governs the behaviour of the quarks that make up the individual protons and neutrons.

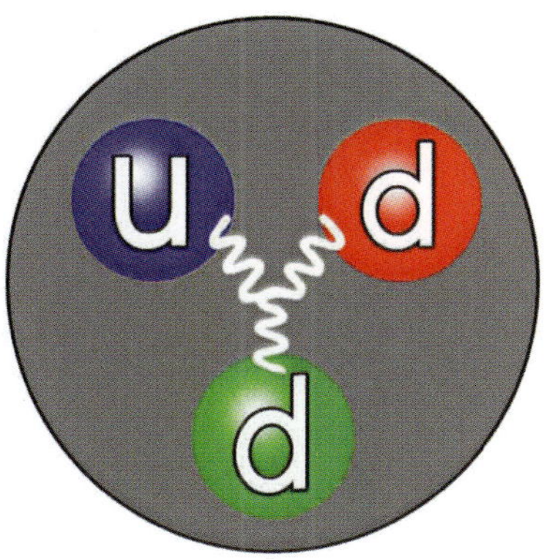

The quark structure of the neutron. The color assignment of individual quarks is arbitrary, but all three colors must be present. Forces between quarks are mediated by gluons.

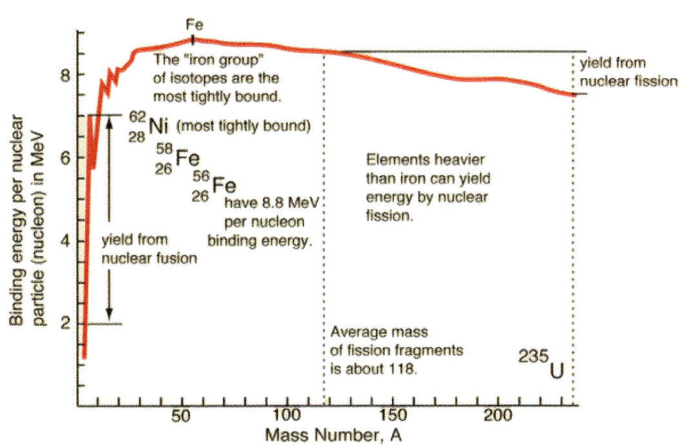

Nuclear binding energy curve.

A nuclear stability is determined by the competition between two fundamental interactions. Protons and neutrons are attracted each other via strong force. On the other hand protons repel each other via the electric force due to their positive charge. Therefore neutrons within the nucleus act somewhat like nuclear glue, neutrons attract each other and protons , which helps offset the electrical repulsion between protons. There are only certain combinations of neutrons and protons, which forms stable nuclei. For example, the most common nuclide of the common chemical element lead (Pb) has 82 protons and 126 neutrons.

Because of the strength of the nuclear force at short distances, the nuclear binding energy (the energy required to disassemble a nucleus of an atom into its component parts) of nucleons is more than seven orders of magnitude larger than the electromagnetic energy binding electrons in atoms. Nuclear reactions (such as nuclear fission or nuclear fusion) therefore have an energy density that is more than 10 000 000x that of chemical reactions. Knowledge of the behaviour and properties of neutrons is essential to the production of nuclear power. Shortly after the neutron was discovered in 1932, it was quickly realized that neutrons might act to form a nuclear chain reaction. When nuclear fission was discovered in 1938, it became clear that, if a fission reaction produced free neutrons, each of these neutrons might cause further fission reaction in a cascade known as a chain reaction. Knowledge of cross-sections (the key parameter representing probability of interaction between a neutron and a nucleus) became crucial for design of reactor cores and the first nuclear weapon (Trinity, 1945).

Structure of the Neutron

The quark structure of the neutron. The color assignment of individual quarks is arbitrary, but all three colors must be present. Forces between quarks are mediated by gluons.

Neutrons and protons are classified as hadrons, subatomic particles that are subject to the strong force and as baryons since they are composed of three quarks. The neutron is a composite particle made of two down quarks with charge $-1/2$ e and one up quark with charge $+2/3$ e. Since the neutron has no net electric charge, it is not affected by eletric forces, but the neutron does have a slight distribution of electric charge within it. This results in non-zero magnetic moment (dipole moment) of the neutron. Therefore the neutron interacts also via electromagnetic interaction, but much weaker than the proton.

The mass of the neutron is 939.565 MeV/c^2, whereas the mass of the three quarks is only about 12 MeV/c^2 (only about 1% of the mass-energy of the neutron). Like the proton, most of mass (energy) of the neutron is in the form of the strong nuclear force energy (gluons). The quarks of the neutron are held together by gluons, the exchange particles for the strong nuclear force. Gluons carry the color charge of the strong nuclear force.

Interactions of Neutrons with Matter

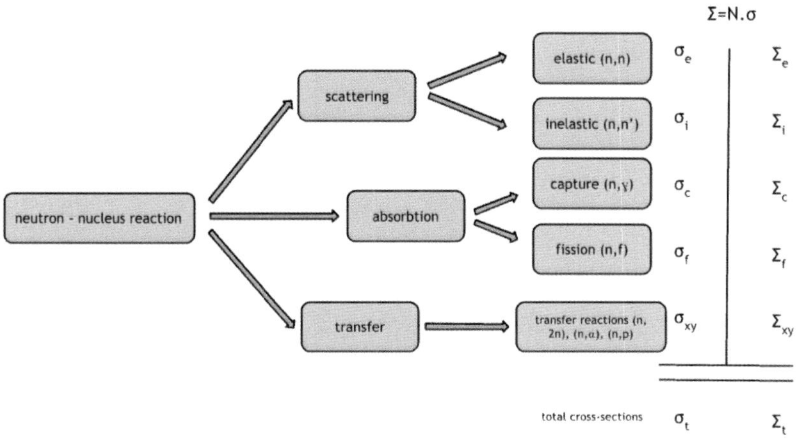

Neutrons are neutral particles, therefore they travel in straight lines, deviating from their path only when they actually collide with a nucleus to be scattered into a new direction or absorbed. Neither the electrons surrounding (atomic electron cloud) a nucleus nor the electric field caused by a positively charged nucleus affect a neutron's flight. In short, neutrons collide with nuclei, not with atoms. A very descriptive feature of the transmission of neutrons through bulk matter is the mean free path length (λ – lambda), which is the mean distance a neutron travels between interactions. It can be calculated from following equation:

$$\lambda=1/\Sigma$$

Types of neutron-nuclear reactions:

- Elastic Scattering Reaction
- Inelastic Scattering Reaction
- Neutron Absorption
- Radiative Capture
- Nuclear Fission
- Neutron Emission
- Charged Particle Ejection

Detection of Neutrons

Since the neutrons are electrically neutral particles, they are mainly subject to strong nuclear forces but not to electric forces. Therefore neutrons are not directly ionizing and they have usually to be converted into charged particles before they can be detected. Generally every type of neutron detector must be equipped with converter (to convert neutron radiation to common detectable radiation) and one of the conventional radiation detectors (scintillation detector, gaseous detector, semiconductor detector, etc.).

Neutron Converters

Two basic types of neutron interactions with matter are for this purpose available:

- Elastic scattering. The free neutron can be scattered by a nucleus, transferring some of its kinetic energy to the nucleus. If the neutron has enough energy to scatter off nuclei the recoiling nucleus ionizes the material surrounding the converter. In fact, only hydrogen and helium nuclei are light enough for practical application. Charge produced in this way can be collected by the conventional detector to produce a detected signal. Neutrons can transfer more energy to light nuclei. This method is appropriate for detecting fast neutrons (fast neutrons do not have high cross-section for absorption) allowing detection of fast neutrons without a moderator.
- Neutron absorption. This is a common method allowing detection of neutrons of entire energy spectrum. This method is is based on variety of absorption reactions (radiative capture, nuclear fission, rearrangement reactions, etc.). The neutron is here absorbed by target material (converter) emitting secondary particles such as protons, alpha particles, beta particles, photons (gamma rays) or fission fragments. Some reactions are threshold reactions (requiring a minimum energy of neutrons), but most of reactions occurs at epithermal and thermal energies. That means the moderation of fast neutrons is required leading in poor energy information of the neutrons. Most common nuclei for the neutron converter material are:
    - $^{10}B(n,\alpha)$. Where the neutron capture cross-section for thermal neutrons is $\sigma = 3820$ barns and the natural boron has abundance of ^{10}B 19,8%.
    - $^{3}He(n,p)$. Where the neutron capture cross-section for thermal neutrons is $\sigma = 5350$ barns and the natural helium has abundance of ^{3}He 0.014%.
    - $^{6}Li(n,\alpha)$. Where the neutron capture cross-section for thermal neutrons is $\sigma = 925$ barns and the natural lithium has abundance of ^{6}Li 7,4%.
    - $^{113}Cd(n,\gamma)$. Where the neutron capture cross-section for thermal neutrons is $\sigma = 20820$ barns and the natural cadmium has abundance of ^{113}Cd 12,2%.

- o ^{235}U(n,fission). Where the fission cross-section for thermal neutrons is σ = 585 barns and the natural uranium has abundance of ^{235}U 0.711%. Uranium as a converter produces fission fragments which are heavy charged particles. This have significant advantage. The heavy charged particles (fission fragments) create a high output signal, because the fragments deposit a large amount of energy in a detector sensitive volume. This allows an easy discrimination of the background radiation (i.e. gamma radiation). This important feature can be used for example in a nuclear reactor power measurement, where the neutron field is accompanied by a significant gamma background.

Neutron Sources

A neutron source is any device that emits neutrons. Neutron sources have many applications, they can be used in research, engineering, medicine, petroleum exploration, biology, chemistry and nuclear power. A neutron source is characterized by a number of factors:

- Significance of the source
- Intensity. The rate of neutrons emitted by the source.
- Energy distribution of emitted neutrons.
- Angular distribution of emitted neutrons.
- Mode of emission. Continuous or pulsed operation.

Classification by Significance of the Source

- Large (Significant) neutron sources
  - o Nuclear Reactors: There are nuclei that can undergo fission on their own spontaneously, but only certain nuclei, like uranium-235, uranium-233 and plutonium-239, can sustain a fission chain reaction. This is because these nuclei release neutrons when they break apart, and these neutrons can induce fission of other nuclei. Uranium-235 which exists as 0.7% of naturally occurring uranium undergoes nuclear fission with thermal neutrons with the production of, on average, 2.4 fast neutrons and the release of ~ 180 MeV of energy per fission. Free neutrons released by each fission play very important role as a trigger of the reaction, but they can be also used fo another purpose. For example: One neutron is required to trigger a further fission. Part of free neutrons (let say 0.5 neutrons/fission) is absorbed in other material, but an excess of neutrons (0.9 neutrons/fission) is able to leave the surface of the reactor core and can be used as a neutron source.
- Fusion Systems: Nuclear fusion is a nuclear reaction in which two or more atomic nuclei (e.g. D+T) collide at a very high energy and fuse together. Thy

byproduct of DT fusion is a free neutron, therefore also nuclear fusion reaction has the potential to produces large quantities of neutrons.

- Spallation Sources: A spallation source is a high-flux neutron source in which protons that have been accelerated to high energies hit a heavy target material, causing the emission of neutrons. The reaction occurs above a certain energy threshold for the incident particle, which is typically 5 – 15 MeV.
- Medium neutron sources
  - Bremssstrahlung from Electron Accelerators/Photofission: Energetic electrons when slowed down rapidly in a heavy target emit intense gamma radiation during the deceleration process. This is known as Bremsstrahlung or braking radiation. The interaction of the gamma radiation with the target produces neutrons via the (γ,n) reaction, or the (γ,fission) reaction when a fissile target is used. e-→Pb → γ→ Pb →(γ,n) and (γ,fission). The Bremsstrahlung γ energy exceeds the binding energy of the "last" neutron in the target. A source strength of 10^{13} neutrons/second produced in short (i.e. < 5 μs) pulses can be readily realised.
  - Dense plasme focus: The dense plasma focus (DPF) is a device that is known as an efficient source of neutrons from fusion reactions. Mechanism of dense plasma focus (DPF) is based on nuclear fusionof short-lived plasma of deuterium and/or tritium. This device produces a short-lived plasma by electromagnetic compression and acceleration that is called a pinch. This plasma is during the pinch hot and dense enough to cause nuclear fusion and the emission of neutrons.
  - Light ion accelerators: Neutrons can be also produced by particle accelerators using targets of deuterium, tritium, lithium, beryllium, and other low-Z materials. In this case the target must be bombarded with accelerated hydrogen (H), deuterium (D), or tritium (T) nuclei.
- Small neutron sources
  - Neutron Generators: Neutrons are produced in the fusion of deuterium and tritium in the following exothermic reaction. $^{2}D + {}^{3}T \rightarrow {}^{4}He + n + 17.6$ MeV. The neutron is produced with a kinetic energy of 14.1 MeV. This can be achieved on a small scale in the laboratory with a modest 100 kV accelerator for deuterium atoms bombarding a tritium target. Continuous neutron sources of ~10^{11} neutrons/second can be achieved relatively simply.
  - Radioisotope source – (α,n) reactions: In certain light isotopes the 'last' neutron in the nucleus is weakly bound and is released when the compound nucleus formed following α-particle bombardment decays. The bombardment of beryllium by α-particles leads to the production of

neutrons by the following exothermic reaction: $^{4}He + ^{9}Be \rightarrow ^{12}C + n + 5.7$ MeV. This reaction yields a weak source of neutrons with an energy spectrum resembling that from a fission source and is used nowadays in portable neutron sources. Radium, plutonium or americium can be used as an α-emitter.

- Radioisotope source – (γ,n) reactions: (γ,n) reactions can also be used for the same purpose. In this type of source, because of the greater range of the γ-ray, the two physical components of the source can be separated making it possible to 'switch off' the reaction if so required by removing the radioactive source from the beryllium. (γ,n) sources produce a monoenergetic neutrons unlike (α,n) sources. The (γ,n) source uses antimony-124 as the gamma emitter in the following endothermic reaction.

$$^{124}Sb \rightarrow ^{124}Te + \beta^- + \gamma$$

$$\gamma + ^{9}Be \rightarrow ^{8}Be + n - 1.66\ \text{MeV}$$

- Radioisotope source – spontaneous fission: Certain isotopes undergo spontaneous fission with emission of neutrons. The most commonly used spontaneous fission source is the radioactive isotope californium-252. Cf-252 and all other spontaneous fission neutron sources are produced by irradiating uranium or another transuranic element in a nuclear reactor, where neutrons are absorbed in the starting material and its subsequent reaction products, transmuting the starting material into the SF isotope.

Description

Atomic nuclei are formed by a number of protons, Z the atomic number, and a number of neutrons, N the neutron number, bound together by the nuclear force. The atomic number defines the chemical properties of the atom, and the neutron number determines the isotope or nuclide. The terms isotope and nuclide are often used synonymously, but they refer to chemical and nuclear properties, respectively. Strictly speaking, isotopes are two or more nuclides with the same number of protons; nuclides with the same number of neutrons are called isotones. The atomic mass number, symbol A, equals Z+N. Nuclides with the same atomic mass number are called isobars. The nucleus of the most common isotope of the hydrogen atom (with the chemical symbol ^{1}H) is a lone proton. The nuclei of the heavy hydrogen isotopes deuterium (D or ^{2}H) and tritium (T or ^{3}H) contain one proton bound to one and two neutrons, respectively. All other types of atomic nuclei are composed of two or more protons and various numbers of neutrons. The most common nuclide of the common chemical element lead, ^{208}Pb, has 82 protons and 126 neutrons, for example. The table of nuclides comprises all the known nuclides. Even though it is not a chemical element, the neutron is included in this table.

The free neutron has a mass of 939,565,413.3 eV/c^2, or $1.674927471 \times 10^{-27}$ kg, or 1.00866491588 u. The neutron has a mean square radius of about 0.8×10^{-15} m, or

0.8 fm, and it is a spin-½ fermion. The neutron has no measurable electric charge. With its positive electric charge, the proton is directly influenced by electric fields, whereas the neutron is unaffected by electric fields. The neutron has a magnetic moment, however, so the neutron is influenced by magnetic fields. The neutron's magnetic moment has a negative value, because its orientation is opposite to the neutron's spin.

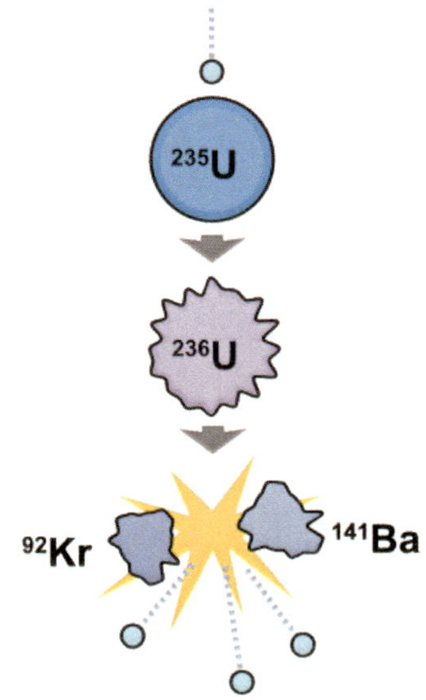

Nuclear fission caused by absorption of a neutron by uranium-235. The heavy nuclide fragments into lighter components and additional neutrons.

A free neutron is unstable, decaying to a proton, electron and antineutrino with a mean lifetime of just under 15 minutes (881.5±1.5 s). This radioactive decay, known as beta decay, is possible because the mass of the neutron is slightly greater than the proton. The free proton is stable. Neutrons or protons bound in a nucleus can be stable or unstable, however, depending on the nuclide. Beta decay, in which neutrons decay to protons, or vice versa, is governed by the weak force, and it requires the emission or absorption of electrons and neutrinos, or their antiparticles.

Protons and neutrons behave almost identically under the influence of the nuclear force within the nucleus. The concept of isospin, in which the proton and neutron are viewed as two quantum states of the same particle, is used to model the interactions of nucleons by the nuclear or weak forces. Because of the strength of the nuclear force at short distances, the binding energy of nucleons is more than seven orders of magnitude larger than the electromagnetic energy binding electrons in atoms. Nuclear reactions (such as nuclear fission) therefore have an energy density that is more than ten million times that of chemical reactions. Because of the mass–energy equivalence, nuclear binding energies add or subtract from the mass of nuclei. Ultimately, the ability of the nuclear force to store energy arising from the electromagnetic repulsion of nuclear components is the basis for most of the energy that makes nuclear reactors or bombs possible. In nuclear fission, the absorption of a neutron by a heavy nuclide (e.g., uranium-235) causes the nuclide to become unstable and break into light nuclides and additional neutrons. The positively charged light nuclides then repel, releasing electromagnetic potential energy.

The neutron is classified as a *hadron*, because it is a composite particle made of quarks. The neutron is also classified as a *baryon*, because it is composed of three valence quarks. The finite size of the neutron and its magnetic moment indicates that the neutron is a composite particle, as opposed to being an elementary particle. A neutron contains two down quarks with charge $-\frac{1}{3}\,e$ and one up quark with charge $+\frac{2}{3}\,e$.

Like protons, the quarks of the neutron are held together by the strong force, mediated by gluons. The nuclear force results from secondary effects of the more fundamental strong force.

Discovery

The story of the discovery of the neutron and its properties is central to the extraordinary developments in atomic physics that occurred in the first half of the 20th century, leading ultimately to the atomic bomb in 1945. In the 1911 Rutherford model, the atom consisted of a small positively charged massive nucleus surrounded by a much larger cloud of negatively charged electrons. In 1920, Rutherford suggested that the nucleus consisted of positive protons and neutrally-charged particles, suggested to be a proton and an electron bound in some way. Electrons were assumed to reside within the nucleus because it was known that beta radiation consisted of electrons emitted from the nucleus. Rutherford called these uncharged particles *neutrons*, by the Latin root for *neutralis* (neuter) and the Greek suffix *-on* (a suffix used in the names of subatomic particles, i.e. *electron* and *proton*). References to the word *neutron* in connection with the atom can be found in the literature as early as 1899, however.

Throughout the 1920s, physicists assumed that the atomic nucleus was composed of protons and "nuclear electrons" but there were obvious problems. It was difficult to reconcile the proton–electron model for nuclei with the Heisenberg uncertainty relation of quantum mechanics. The Klein paradox, discovered by Oskar Klein in 1928, presented further quantum mechanical objections to the notion of an electron confined within a nucleus. Observed properties of atoms and molecules were inconsistent with the nuclear spin expected from the proton–electron hypothesis. Since both protons and electrons carry an intrinsic spin of $\frac{1}{2}\,\hbar$, there is no way to arrange an odd number of spins $\pm\frac{1}{2}\,\hbar$ to give a spin integer multiple of $\hbar$. Nuclei with integer spin are common, e.g., ^{14}N.

In 1931, Walther Bothe and Herbert Becker found that if alpha particle radiation from polonium fell on beryllium, boron, or lithium, an unusually penetrating radiation was produced. The radiation was not influenced by an electric field, so Bothe and Becker assumed it was gamma radiation. The following year Irène Joliot-Curie and Frédéric Joliot-Curie in Paris showed that if this "gamma" radiation fell on paraffin, or any other hydrogen-containing compound, it ejected protons of very high energy. Neither Rutherford nor James Chadwick at the Cavendish Laboratory in Cambridge were convinced by the gamma ray interpretation. Chadwick quickly performed a series of experiments that showed that the new radiation consisted of uncharged particles with about the same mass as the proton. These particles were neutrons. Chadwick won the Nobel Prize in Physics for this discovery in 1935.

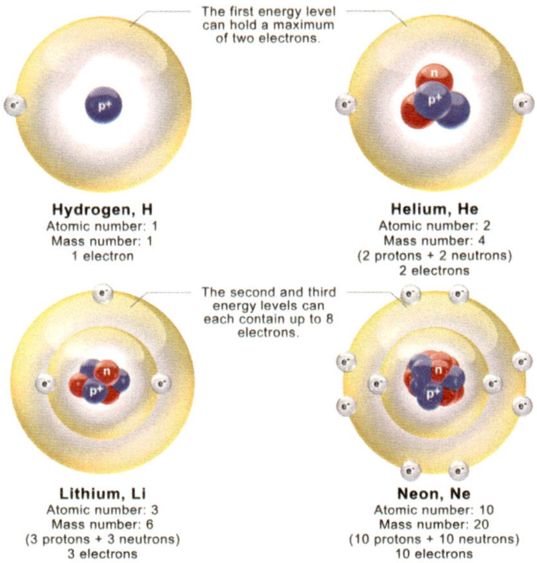

Models depicting the nucleus and electron energy levels in hydrogen, helium, lithium, and neon atoms. In reality, the diameter of the nucleus is about 100,000 times smaller than the diameter of the atom.

Models for atomic nucleus consisting of protons and neutrons were quickly developed by Werner Heisenberg and others. The proton–neutron model explained the puzzle of nuclear spins. The origins of beta radiation were explained by Enrico Fermi in 1934 by the process of beta decay, in which the neutron decays to a proton by *creating* an electron and a (as yet undiscovered) neutrino. In 1935 Chadwick and his doctoral student Maurice Goldhaber, reported the first accurate measurement of the mass of the neutron.

By 1934, Fermi had bombarded heavier elements with neutrons to induce radioactivity in elements of high atomic number. In 1938, Fermi received the Nobel Prize in Physics *"for his demonstrations of the existence of new radioactive elements produced by neutron irradiation, and for his related discovery of nuclear reactions brought about by slow neutrons"*. In 1938 Otto Hahn, Lise Meitner, and Fritz Strassmann discovered nuclear fission, or the fractionation of uranium nuclei into light elements, induced by neutron bombardment. In 1945 Hahn received the 1944 Nobel Prize in Chemistry *"for his discovery of the fission of heavy atomic nuclei."* The discovery of nuclear fission would lead to the development of nuclear power and the atomic bomb by the end of World War II.

Beta Decay and the Stability of the Nucleus

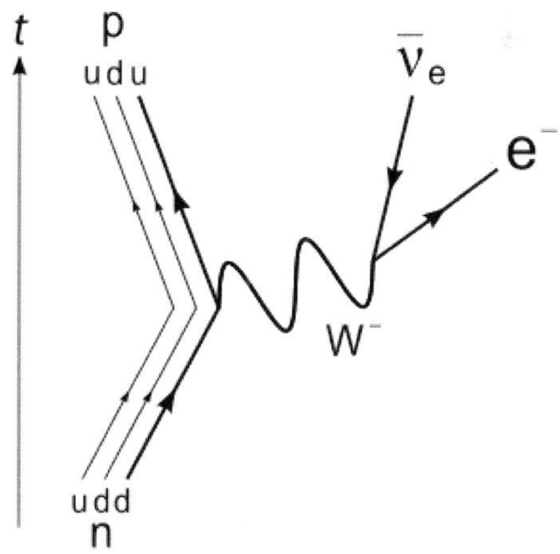

The Feynman diagram for beta decay of a neutron into a proton, electron, and electron antineutrino via an intermediate heavy W boson

Under the Standard Model of particle physics, the only possible decay mode for the neutron that conserves baryon number is for one of the neutron's quarks to change flavour via the weak interaction. The decay of one of the neutron's down quarks into a lighter up quark can be achieved by the emission of a W boson. By this process, the Standard Model description of beta decay, the neutron decays into a proton (which contains one down and two up quarks), an electron, and an electron antineutrino.

Since interacting protons have a mutual electromagnetic repulsion that is stronger than their attractive nuclear interaction, neutrons are a necessary constituent of any atomic nucleus that contains more than one proton. Neutrons bind with protons and one another in the nucleus via the nuclear force, effectively moderating the repulsive forces between the protons and stabilizing the nucleus.

Free Neutron Decay

Outside the nucleus, free neutrons are unstable and have a mean lifetime of 881.5±1.5 s (about 14 minutes, 42 seconds); therefore the half-life for this process (which differs from the mean lifetime by a factor of ln(2) = 0.693) is 611.0±1.0 s (about 10 minutes, 11 seconds). Beta decay of the neutron, described above, can be denoted by the radioactive decay:

$$n^0 \rightarrow p^+ + e^- + \bar{\nu}_e$$

where p^+, e^-, and ν_e denote the proton, electron and electron antineutrino, respectively. For the free neutron the decay energy for this process (based on the masses of the neutron, proton, and electron) is 0.782343 MeV. The maximal energy of the beta decay electron (in the process wherein the neutrino receives a vanishingly small amount of kinetic energy) has been measured at 0.782 ± 0.013 MeV. The latter number is not well-enough measured to determine the comparatively tiny rest mass of the neutrino (which must in theory be subtracted from the maximal electron kinetic energy) as well as neutrino mass is constrained by many other methods.

A small fraction (about one in 1000) of free neutrons decay with the same products, but add an extra particle in the form of an emitted gamma ray:

$$n^0 \rightarrow p^+ + e^- + \overline{\nu}_e + \gamma$$

This gamma ray may be thought of as a sort of "internal bremsstrahlung" that arises as the emitted beta particle interacts with the charge of the proton in an electromagnetic way. Internal bremsstrahlung gamma ray production is also a minor feature of beta decays of bound neutrons.

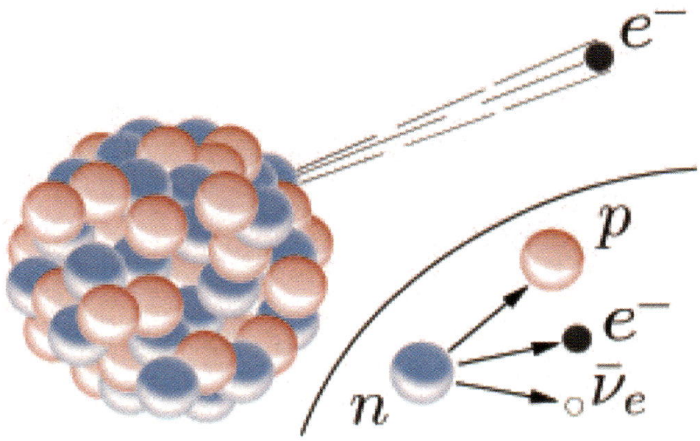

A schematic of the nucleus of an atom indicating β^- radiation, the emission of a fast electron from the nucleus (the accompanying antineutrino is omitted). In the Rutherford model for the nucleus, red spheres were protons with positive charge and blue spheres were protons tightly bound to an electron with no net charge. The inset shows beta decay of a free neutron as it is understood today; an electron and antineutrino are created in this process.

A very small minority of neutron decays (about four per million) are so-called "two-body (neutron) decays", in which a proton, electron and antineutrino are produced as usual, but the electron fails to gain the 13.6 eV necessary energy to escape the proton (the ionization energy of hydrogen), and therefore simply remains bound to it, as a neutral hydrogen atom (one of the "two bodies"). In this type of free neutron decay, almost all of the neutron decay energy is carried off by the antineutrino (the other "body"). (The hydrogen atom recoils with a speed of only about (decay energy)/(hydrogen rest energy) times the speed of light, or 250 km/s.)

The transformation of a free proton to a neutron (plus a positron and a neutrino) is energetically impossible, since a free neutron has a greater mass than a free proton. But a high-energy collision of a proton and an electron or neutrino can result in a neutron.

Bound Neutron Decay

While a free neutron has a half life of about 10.2 min, most neutrons within nuclei are stable. According to the nuclear shell model, the protons and neutrons of a nuclide are a quantum mechanical system organized into discrete energy levels with unique quantum numbers. For a neutron to decay, the resulting proton requires an available state at lower energy than the initial neutron state. In stable nuclei the possible lower energy states are all filled, meaning they are each occupied by two protons with spin up and spin down. The Pauli exclusion principle therefore disallows the decay of a neutron to a proton within stable nuclei. The situation is similar to electrons of an atom, where electrons have distinct atomic orbitals and are prevented from decaying to lower energy states, with the emission of a photon, by the exclusion principle.

Neutrons in unstable nuclei can decay by beta decay as described above. In this case, an energetically allowed quantum state is available for the proton resulting from the decay. One example of this decay is carbon-14 (6 protons, 8 neutrons) that decays to nitrogen-14 (7 protons, 7 neutrons) with a half-life of about 5,730 years.

Inside a nucleus, a proton can transform into a neutron via inverse beta decay, if an energetically allowed quantum state is available for the neutron. This transformation occurs by emission of a positron and an electron neutrino:

$$p^+ \rightarrow n^0 + e^+ + v_e$$

The transformation of a proton to a neutron inside of a nucleus is also possible through electron capture:

$$p^+ + e^- \rightarrow n^0 + v_e$$

Positron capture by neutrons in nuclei that contain an excess of neutrons is also possible, but is hindered because positrons are repelled by the positive nucleus, and quickly annihilate when they encounter electrons.

Competition of Beta Decay Types

Three types of beta decay in competition are illustrated by the single isotope copper-64 (29 protons, 35 neutrons), which has a half-life of about 12.7 hours. This isotope has one unpaired proton and one unpaired neutron, so either the proton or the neutron can decay. This particular nuclide is almost equally likely to undergo proton decay (by positron emission, 18% or by electron capture, 43%) or neutron decay (by electron emission, 39%).

Intrinsic Properties

Mass

The mass of a neutron cannot be directly determined by mass spectrometry due to lack of electric charge. However, since the masses of a proton and of a deuteron can be measured with a mass spectrometer, the mass of a neutron can be deduced by subtracting proton mass from deuteron mass, with the difference being the mass of the neutron plus the binding energy of deuterium (expressed as a positive emitted energy). The latter can be directly measured by measuring the energy (B_d) of the single 0.7822 MeV gamma photon emitted when neutrons are captured by protons (this is exothermic and happens with zero-energy neutrons), plus the small recoil kinetic energy (E_{rd}) of the deuteron (about 0.06% of the total energy).

$$m_n = m_d - m_p + B_d - E_{rd}$$

The energy of the gamma ray can be measured to high precision by X-ray diffraction techniques, as was first done by Bell and Elliot in 1948. The best modern (1986) values for neutron mass by this technique are provided by Greene, et al. These give a neutron mass of:

$$m_{\text{neutron}} = 1.008644904(14)\ \text{u}$$

The value for the neutron mass in MeV is less accurately known, due to less accuracy in the known conversion of u to MeV:

$$m_{\text{neutron}} = 939.56563(28)\ \text{MeV}/c^2.$$

Another method to determine the mass of a neutron starts from the beta decay of the neutron, when the momenta of the resulting proton and electron are measured.

Electric Charge

The total electric charge of the neutron is 0 *e*. This zero value has been tested experimentally, and the present experimental limit for the charge of the neutron is $-2(8)\times10^{-22}$ *e*, or $-3(13)\times10^{-41}$ C. This value is consistent with zero, given the experimental uncertainties (indicated in parentheses). By comparison, the charge of the proton is +1 *e*.

Magnetic Moment

Even though the neutron is a neutral particle, the magnetic moment of a neutron is not zero. The neutron is not affected by electric fields, but it is affected by magnetic fields. The magnetic moment of the neutron is an indication of its quark substructure and internal charge distribution. The value for the neutron's magnetic moment was first directly measured by Luis Alvarez and Felix Bloch at Berkeley, California, in 1940, using an extension of the magnetic resonance methods developed by Rabi. Alvarez and Bloch determined the magnetic moment of the neutron to be $\mu_n = -1.93(2)\ \mu_N$, where μ_N is the nuclear magneton.

In the quark model for hadrons, the neutron is composed of one up quark (charge +2/3 *e*) and two down quarks (charge −1/3 *e*). The magnetic moment of the neutron can be modeled as a sum of the magnetic moments of the constituent quarks. The calculation assumes that the quarks behave like pointlike Dirac particles, each having their own magnetic moment. Simplistically, the magnetic moment of the neutron can be viewed as resulting from the vector sum of the three quark magnetic moments, plus the orbital magnetic moments caused by the movement of the three charged quarks within the neutron.

In one of the early successes of the Standard Model (SU(6) theory, now understood in terms of quark behavior), in 1964 Mirza A. B. Beg, Benjamin W. Lee, and Abraham Pais theoretically calculated the ratio of proton to neutron magnetic moments to be −3/2, which agrees with the experimental value to within 3%. The measured value for this ratio is −1.45989805(34). A contradiction of the quantum mechanical basis of this calculation with the Pauli exclusion principle, led to the discovery of the color charge for quarks by Oscar W. Greenberg in 1964.

The above treatment compares neutrons with protons, allowing the complex behavior of quarks to be subtracted out between models, and merely exploring what the effects would be of differing quark charges (or quark type). Such calculations are enough to show that the interior of neutrons is very much like that of protons, save for the difference in quark composition with a down quark in the neutron replacing an up quark in the proton.

Attempts have been made to quantitatively recover the neutron magnetic moment from first principles. From the nonrelativistic, quantum mechanical wavefunction for baryons composed of three quarks, a straightforward calculation gives fairly accurate estimates for the magnetic moments of neutrons, protons, and other baryons. For a neutron, the end result of this calculation is that the magnetic moment of the neutron is given by $\mu_n = 4/3\,\mu_d - 1/3\,\mu_u$, where μ_d and μ_u are the magnetic moments for the down and up quarks, respectively. This result combines the intrinsic magnetic moments of the quarks with their orbital magnetic moments, and assumes the three quarks are in a particular, dominant quantum state.

Baryon	**Magnetic moment of quark model**	**Computed (μ_N)**	**Observed (μ_N)**
p	$4/3\,\mu_u - 1/3\,\mu_d$	2.79	2.793
n	$4/3\,\mu_d - 1/3\,\mu_u$	−1.86	−1.913

The results of this calculation are encouraging, but the masses of the up or down quarks were assumed to be 1/3 the mass of a nucleon. The masses of the quarks are actually only about 1% that of a nucleon. The discrepancy stems from the complexity of the Standard Model for nucleons, where most of their mass originates in the gluon fields, virtual particles, and their associated energy that are essential aspects of the strong force. Furthermore, the complex system of quarks and gluons that constitute a neutron requires a relativistic treatment. The nucleon magnetic moment has been successfully computed numerically from first principles, however, including all the effects mentioned and using more realistic values for the quark masses. The calculation gave results that were in fair agreement with measurement, but it required significant computing resources.

Spin

The neutron is a spin 1/2 particle, that is, it is a fermion with intrinsic angular momentum equal to 1/2 $\hbar$, where $\hbar$ is the reduced Planck constant. For many years after the discovery of the neutron, its exact spin was ambiguous. Although it was assumed to be a spin 1/2 Dirac particle, the possibility that the neutron was a spin 3/2 particle lingered. The interactions of the neutron's magnetic moment with an external magnetic field were exploited to finally determine the spin of the neutron. In 1949, Hughes and Burgy measured neutrons reflected from a ferromagnetic mirror and found that the angular distribution of the reflections was consistent with spin 1/2. In 1954, Sherwood, Stephenson, and Bernstein employed neutrons in a Stern–Gerlach experiment that used a magnetic field to separate the neutron spin states. They recorded two such spin states, consistent with a spin 1/2 particle.

As a fermion, the neutron is subject to the Pauli exclusion principle; two neutrons cannot have the same quantum numbers. This is the source of the degeneracy pressure which makes neutron stars possible.

Structure and Geometry of Charge Distribution

An article published in 2007 featuring a model-independent analysis concluded that the neutron has a negatively charged exterior, a positively charged middle, and a negative core. In a simplified classical view, the negative "skin" of the neutron assists it to be attracted to the protons with which it interacts in the nucleus. (However, the main attraction between neutrons and protons is via the nuclear force, which does not involve charge.)

The simplified classical view of the neutron's charge distribution also "explains" the fact that the neutron magnetic dipole points in the opposite direction from its spin angular momentum vector (as compared to the proton). This gives the neutron, in effect, a magnetic moment which resembles a negatively charged particle. This can be reconciled classically with a neutral neutron composed of a charge distribution in which the negative sub-parts of the neutron have a larger average radius of distribution, and therefore contribute more to the particle's magnetic dipole moment, than do the positive parts that are, on average, nearer the core.

Electric Dipole Moment

The Standard Model of particle physics predicts a tiny separation of positive and negative charge within the neutron leading to a permanent electric dipole moment. The predicted value is, however, well below the current sensitivity of experiments. From several unsolved puzzles in particle physics, it is clear that the Standard Model is not the final and full description of all particles and their interactions. New theories going beyond the Standard Model generally lead to much larger predictions for the electric dipole moment of the neutron. Currently, there are at least four experiments trying to measure for the first time a finite neutron electric dipole moment, including:

- Cryogenic neutron EDM experiment being set up at the Institut Laue–Langevin
- nEDM experiment under construction at the new UCN source at the Paul Scherrer Institute
- nEDM experiment being envisaged at the Spallation Neutron Source
- nEDM experiment being built at the Institut Laue–Langevin

Anti-neutron

The antineutron is the antiparticle of the neutron. It was discovered by Bruce Cork in the year 1956, a year after the antiproton was discovered. CPT-symmetry puts strong constraints on the relative properties of particles and antiparticles, so studying antineutrons yields provide stringent tests on CPT-symmetry. The fractional difference in the masses of the neutron and antineutron is $(9\pm6)\times10^{-5}$. Since the difference is only about two standard deviations away from zero, this does not give any convincing evidence of CPT-violation.

Neutron Compounds

Dineutrons and Tetraneutrons

The existence of stable clusters of 4 neutrons, or tetraneutrons, has been hypothesised by a team led by Francisco-Miguel Marqués at the CNRS Laboratory for Nuclear Physics based on observations of the disintegration of beryllium-14 nuclei. This is particularly interesting because current theory suggests that these clusters should not be stable.

In February 2016, Japanese physicist Susumu Shimoura of the University of Tokyo and co-workers reported they had observed the purported tetraneutrons for the first time experimentally. Nuclear physicists around the world say this discovery, if confirmed, would be a milestone in the field of nuclear physics and certainly would deepen our understanding of the nuclear forces.

The dineutron is another hypothetical particle. In 2012, Artemis Spyrou from Michigan State University and coworkers reported that they observed, for the first time, the dineutron emission in the decay of ^{16}Be. The dineutron character is evidenced by a small emission angle between the two neutrons. The authors measured the two-neutron separation energy to be 1.35(10) MeV, in good agreement with shell model calculations, using standard interactions for this mass region.

Neutronium and Neutron Stars

At extremely high pressures and temperatures, nucleons and electrons are believed to collapse into bulk neutronic matter, called neutronium. This is presumed to happen in neutron stars.

The extreme pressure inside a neutron star may deform the neutrons into a cubic symmetry, allowing tighter packing of neutrons.

Detection

The common means of detecting a charged particle by looking for a track of ionization (such as in a cloud chamber) does not work for neutrons directly. Neutrons that elastically scatter off atoms can create an ionization track that is detectable, but the experiments are not as simple to carry out; other means for detecting neutrons, consisting of allowing them to interact with atomic nuclei, are more commonly used. The commonly used methods to detect neutrons can therefore be categorized according to the nuclear processes relied upon, mainly neutron capture or elastic scattering.

Neutron Detection by Neutron Capture

A common method for detecting neutrons involves converting the energy released from neutron capture reactions into electrical signals. Certain nuclides have a high neutron capture cross section, which is the probability of absorbing a neutron. Upon neutron capture, the compound nucleus emits more easily detectable radiation, for example an alpha particle, which is then detected. The nuclides ^{3}He, ^{6}Li, ^{10}B, ^{233}U, ^{235}U, ^{237}Np, and ^{239}Pu are useful for this purpose.

Neutron Detection by Elastic Scattering

Neutrons can elastically scatter off nuclei, causing the struck nucleus to recoil. Kinematically, a neutron can transfer more energy to a light nucleus such as hydrogen or helium than to a heavier nucleus. Detectors relying on elastic scattering are called fast neutron detectors. Recoiling nuclei can ionize and excite further atoms through collisions. Charge and/or scintillation light produced in this way can be collected to produce a detected signal. A major challenge in fast neutron detection is discerning such signals from erroneous signals produced by gamma radiation in the same detector.

Fast neutron detectors have the advantage of not requiring a moderator, and are therefore capable of measuring the neutron's energy, time of arrival, and in certain cases direction of incidence.

Sources and Production

Free neutrons are unstable, although they have the longest half-life of any unstable subatomic particle by several orders of magnitude. Their half-life is still only about 10 minutes, however, so they can be obtained only from sources that produce them continuously.

Natural neutron background. A small natural background flux of free neutrons exists everywhere on Earth. In the atmosphere and deep into the ocean, the "neutron background" is caused by muons produced by cosmic ray interaction with the atmosphere. These high-energy muons are capable of penetration to considerable depths in water and soil. There, in striking atomic nuclei, among other reactions they induce spallation reactions in which a neutron is liberated from the nucleus. Within the Earth's crust a second source is neutrons produced primarily by spontaneous fission of uranium and thorium present in crustal minerals. The neutron background is not strong enough to be a biological hazard, but it is of importance to very high resolution particle detectors that are looking for very rare events, such as (hypothesized) interactions that might be caused by particles of dark matter. Recent research has shown that even thunderstorms can produce neutrons with energies of up to several tens of MeV. Recent research has shown that the fluence of these neutrons lies between 10^{-9} and 10^{-13} per ms and per m^2 depending on the detection altitude. The energy of most of these neutrons, even with initial energies of 20 MeV, decreases down to the keV range within 1 ms.

Even stronger neutron background radiation is produced at the surface of Mars, where the atmosphere is thick enough to generate neutrons from cosmic ray muon production and neutron-spallation, but not thick enough to provide significant protection from the neutrons produced. These neutrons not only produce a Martian surface neutron radiation hazard from direct downward-going neutron radiation but may also produce a significant hazard from reflection of neutrons from the Martian surface, which will produce reflected neutron radiation penetrating upward into a Martian craft or habitat from the floor.

Sources of neutrons for research. These include certain types of radioactive decay (spontaneous fission and neutron emission), and from certain nuclear reactions. Convenient nuclear reactions include tabletop reactions such as natural alpha and gamma bombardment of certain nuclides, often beryllium or deuterium, and induced nuclear fission, such as occurs in nuclear reactors. In addition, high-energy nuclear reactions (such as occur in cosmic radiation showers or accelerator collisions) also produce neutrons from disintegration of target nuclei. Small (tabletop) particle accelerators optimized to produce free neutrons in this way, are called neutron generators.

In practice, the most commonly used small laboratory sources of neutrons use radioactive decay to power neutron production. One noted neutron-producing radioisotope, californium-252 decays (half-life 2.65 years) by spontaneous fission 3% of the time with production of 3.7 neutrons per fission, and is used alone as a neutron source from this process. Nuclear reaction sources (that involve two materials) powered by radioisotopes use an alpha decay source plus a beryllium target, or else a source of high-energy gamma radiation from a source that undergoes beta decay followed by gamma decay, which produces photoneutrons on interaction of the high-energy gamma ray with ordinary stable beryllium, or else with the deuterium in heavy water. A popular source of the latter type is radioactive antimony-124 plus beryllium, a system with a half-life of 60.9 days, which can be constructed from natural antimony (which is 42.8% stable antimony-123) by activating it with neutrons in a nuclear reactor, then transported to where the neutron source is needed.

Institut Laue–Langevin (ILL) in Grenoble, France – a major neutron research facility.

Nuclear fission reactors naturally produce free neutrons; their role is to sustain the energy-producing chain reaction. The intense neutron radiation can also be used to produce various radioisotopes through the process of neutron activation, which is a type of neutron capture.

Experimental nuclear fusion reactors produce free neutrons as a waste product. However, it is these neutrons that possess most of the energy, and converting that energy to a useful form has proved a difficult engineering challenge. Fusion reactors that generate neutrons are likely to create radioactive waste, but the waste is composed of neutron-activated lighter isotopes, which have relatively short (50–100 years) decay periods as compared to typical half-lives of 10,000 years for fission waste, which is long due primarily to the long half-life of alpha-emitting transuranic actinides.

Neutron Beams and Modification of Beams After Production

Free neutron beams are obtained from neutron sources by neutron transport. For access to intense neutron sources, researchers must go to a specialized neutron facility that operates a research reactor or a spallation source.

The neutron's lack of total electric charge makes it difficult to steer or accelerate them. Charged particles can be accelerated, decelerated, or deflected by electric or magnetic fields. These methods have little effect on neutrons. However, some effects may be attained by use of inhomogeneous magnetic fields because of the neutron's magnetic moment. Neutrons can be controlled by methods that include moderation, reflection, and velocity selection. Thermal neutrons can be polarized by transmission through magnetic materials in a method analogous to the Faraday effect for photons. Cold neutrons of wavelengths of 6–7 angstroms can be produced in beams of a high degree of polarization, by use of magnetic mirrors and magnetized interference filters.

Applications

The neutron plays an important role in many nuclear reactions. For example, neutron capture often results in neutron activation, inducing radioactivity. In particular, knowledge of neutrons and their behavior has been important in the development of nuclear reactors and nuclear weapons. The fissioning of elements like uranium-235 and plutonium-239 is caused by their absorption of neutrons.

Cold, *thermal*, and *hot* neutron radiation is commonly employed in neutron scattering facilities, where the radiation is used in a similar way one uses X-rays for the analysis of condensed matter. Neutrons are complementary to the latter in terms of atomic contrasts by different scattering cross sections; sensitivity to magnetism; energy range for inelastic neutron spectroscopy; and deep penetration into matter.

The development of "neutron lenses" based on total internal reflection within hollow glass capillary tubes or by reflection from dimpled aluminum plates has driven ongoing research into neutron microscopy and neutron/gamma ray tomography.

A major use of neutrons is to excite delayed and prompt gamma rays from elements in materials. This forms the basis of neutron activation analysis (NAA) and prompt gamma neutron activation analysis (PGNAA). NAA is most often used to analyze small samples of materials in a nuclear reactor whilst PGNAA is most often used to analyze subterranean rocks around bore holes and industrial bulk materials on conveyor belts.

Another use of neutron emitters is the detection of light nuclei, in particular the hydrogen found in water molecules. When a fast neutron collides with a light nucleus, it loses a large fraction of its energy. By measuring the rate at which slow neutrons return to the probe after reflecting off of hydrogen nuclei, a neutron probe may determine the water content in soil.

Neutron Temperature

Thermal Neutrons

A *thermal neutron* is a free neutron that is Boltzmann distributed with kT= 0.0253 eV (4.0×10^{-21} J) at room temperature. This gives characteristic (not average, or median) speed of 2.2 km/s. The name 'thermal' comes from their energy being that of the room temperature gas or material they are permeating. After a number of collisions (often in the range of 10–20) with nuclei, neutrons arrive at this energy level, provided that they are not absorbed.

In many substances, thermal neutron reactions show a much larger effective cross-section than reactions involving faster neutrons, and thermal neutrons can therefore be absorbed more readily (i.e., with higher probability) by any atomic nuclei that they collide with, creating a heavier — and often unstable — isotope of the chemical element as a result.

Most fission reactors use a neutron moderator to slow down, or *thermalize* the neutrons that are emitted by nuclear fission so that they are more easily captured, causing further fission. Others, called fast breeder reactors, use fission energy neutrons directly.

Cold Neutrons

Cold neutrons are thermal neutrons that have been equilibrated in a very cold substance such as liquid deuterium. Such a *cold source* is placed in the moderator of a research reactor or spallation source. Cold neutrons are particularly valuable for neutron scattering experiments.

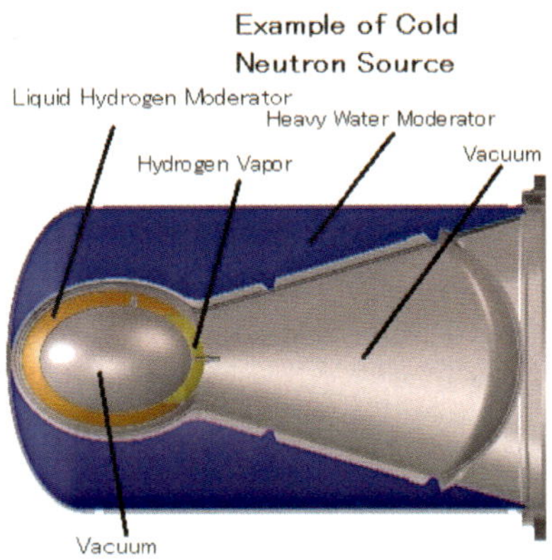

Cold neutron source providing neutrons at about the temperature of liquid hydrogen

Ultracold Neutrons

Ultracold neutrons are produced by elastically scattering cold neutrons in substances with a temperature of a few kelvins, such as solid deuterium or superfluid helium. An alternative production method is the mechanical deceleration of cold neutrons.

Fission energy Neutrons

A *fast neutron* is a free neutron with a kinetic energy level close to 1 MeV (1.6×10^{-13} J), hence a speed of ~14000 km/s (~5% of the speed of light). They are named *fission energy* or *fast* neutrons to distinguish them from lower-energy thermal neutrons, and high-energy neutrons produced in cosmic showers or accelerators. Fast neutrons are produced by nuclear processes such as nuclear fission. Neutrons produced in fission, as noted above, have a Maxwell–Boltzmann distribution of kinetic energies from 0 to ~14 MeV, a mean energy of 2 MeV (for U-235 fission neutrons), and a mode of only 0.75

MeV, which means that more than half of them do not qualify as fast (and thus have almost no chance of initiating fission in fertile materials, such as U-238 and Th-232).

Fast neutrons can be made into thermal neutrons via a process called moderation. This is done with a neutron moderator. In reactors, typically heavy water, light water, or graphite are used to moderate neutrons.

Fusion Neutrons

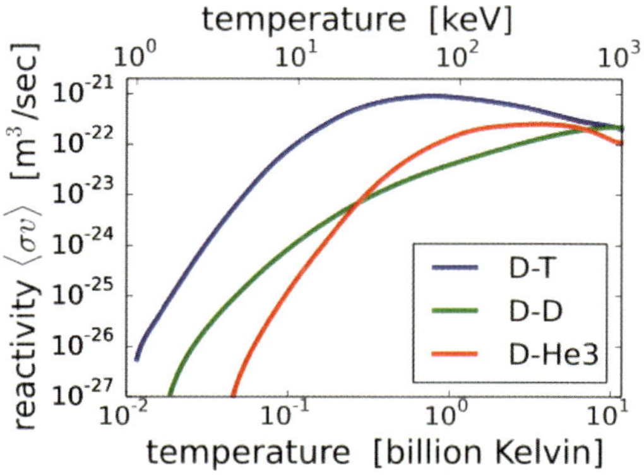

The fusion reaction rate increases rapidly with temperature until it maximizes and then gradually drops off. The DT rate peaks at a lower temperature (about 70 keV, or 800 million kelvins) and at a higher value than other reactions commonly considered for fusion energy.

D–T (deuterium–tritium) fusion is the fusion reaction that produces the most energetic neutrons, with 14.1 MeV of kinetic energy and traveling at 17% of the speed of light. D–T fusion is also the easiest fusion reaction to ignite, reaching near-peak rates even when the deuterium and tritium nuclei have only a thousandth as much kinetic energy as the 14.1 MeV that will be produced.

14.1 MeV neutrons have about 10 times as much energy as fission neutrons, and are very effective at fissioning even non-fissile heavy nuclei, and these high-energy fissions produce more neutrons on average than fissions by lower-energy neutrons. This makes D–T fusion neutron sources such as proposed tokamak power reactors useful for transmutation of transuranic waste. 14.1 MeV neutrons can also produce neutrons by knocking them loose from nuclei.

On the other hand, these very high-energy neutrons are less likely to simply be captured without causing fission or spallation. For these reasons, nuclear weapon design extensively utilizes D–T fusion 14.1 MeV neutrons to cause more fission. Fusion neutrons are able to cause fission in ordinarily non-fissile materials, such as depleted uranium (uranium-238), and these materials have been used in the jackets of thermonuclear weapons. Fusion neutrons also can cause fission in substances that are unsuitable or difficult to make into primary fission bombs, such as reactor grade plutonium. This physical fact thus causes ordinary non-weapons grade materials to become of concern in certain nuclear proliferation discussions and treaties.

Other fusion reactions produce much less energetic neutrons. D–D fusion produces a 2.45 MeV neutron and helium-3 half of the time, and produces tritium and a proton but no neutron the rest of the time. D–^{3}He fusion produces no neutron.

Intermediate-energy Neutrons

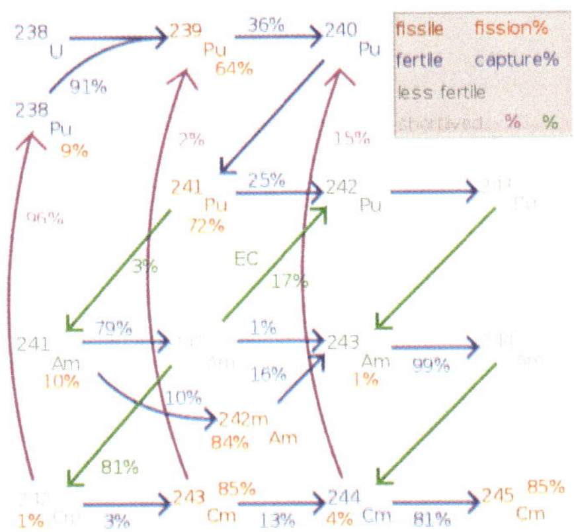

Transmutation flow in light water reactor, which is a thermal-spectrum reactor

A fission energy neutron that has slowed down but not yet reached thermal energies is called an epithermal neutron.

Cross sections for both capture and fission reactions often have multiple resonance peaks at specific energies in the epithermal energy range. These are of less significance in a fast neutron reactor, where most neutrons are absorbed before slowing down to this range, or in a well-moderated thermal reactor, where epithermal neutrons interact mostly with moderator nuclei, not with either fissile or fertile actinide nuclides. However, in a partially moderated reactor with more interactions of epithermal neutrons with heavy metal nuclei, there are greater possibilities for transient changes in reactivity that might make reactor control more difficult.

Ratios of capture reactions to fission reactions are also worse (more captures without fission) in most nuclear fuels such as plutonium-239, making epithermal-spectrum reactors using these fuels less desirable, as captures not only waste the one neutron captured but also usually result in a nuclide that is not fissile with thermal or epithermal neutrons, though still fissionable with fast neutrons. The exception is uranium-233 of the thorium cycle, which has good capture-fission ratios at all neutron energies.

High-energy Neutrons

High-energy neutrons have much more energy than fission energy neutrons and are generated as secondary particles by particle accelerators or in the atmosphere from cosmic rays. These high-energy neutrons are extremely efficient at ionization and far more likely to cause cell death than X-rays or protons.

Neutron Detection

The method of detection relies on the absorption of the neutron by an atom with the simultaneous emission of a γ-ray photon, often referred to as an (n,γ) reaction. Since the absorbing material must absorb neutrons and be capable of existing in gaseous form, then the choice of materials is very limited. The most common is ^3He gas, which relies on the reaction:

$$^3He + {^1n} \rightarrow {^4He} + \gamma$$

This produces the stable isotope of helium together with a γ-ray of a specific energy. Another suitable gas is BF_3, which uses the absorption properties of the isotope ^{10}B. For thermal wavelength neutrons the ^3He gas detector is considered best, but for long wavelength neutrons BF_3 is actually better, but is rarely used now due to the technical difficulties of handling a corrosive and toxic gas. Because the capture cross-section of gases is very small, ^3He detectors are usually filled above atmospheric pressure, e.g. 5 to 10 bar. In size, they are typically 10 to 15 cm long and 2 to 5 cm in diameter.

Recent developments by the detector group at the ILL, Grenoble, have led to the production of single-tube He-gas detectors with positional sensitivity. This is achieved by measuring the charge developed at both ends of a resistive anode wire, a larger magnitude of charge detected being related to a shorter distance travelled by the pulse along the wire. These will have huge potential for high-resolution angle-dispersive powder diffractometers since they will allow a greater solid angle to be measured without comprising instrumental resolution, the asymmetry due to the curved Debye-Scherrer diffractions being removed via software binning of the 2-dimensional data to the conventional 1-dimensional powder pattern.

Neutron multidetectors are also ^3He gas based, though in addition they are also filled with a quenching gas such as a mixture of xenon (Xe) and methane (CH_4). The original multidetectors consisted of an array of vertical wires at fixed intervals apart, usually chosen so that the interval corresponded to a nice angle in degrees, e.g. 0.1°. They are very large, can be both one- and two-dimensional, and can cover a huge area of solid angle. A recent development to improve the spacing interval has been the development of microstrips which replace the conventional anode wires in these detectors.

Scintillation detectors have been developed using, for example, ^6Li or Gd as the absorbing atoms. The latter may be used in the mixed metal oxide-sulphide $Gd_2O_3.Gd_2S_3$, which is able to convert the γ-ray photon directly into a UV/visible photon. The latter is counted by using a conventional photomultiplier tube. With scintillation detectors, the neutron is detected in a smaller region of physical space compared to a gas detector due to the higher density of absorbing material. This has advantages for some applications. A potential disadvantage is that they may detect sample fluorescence due to (n,γ) reactions within the sample itself.

Thus one of the problems in detecting neutrons is the potential background from rogue γ-rays produced by neutrons from either the reactor or the pulsed source, plus those produced by reaction with a monochromator or even the sample. Therefore, the electronics controlling the detector must be capable of discriminating the energy (or emission time at pulsed sources) of the γ-ray so that only those produced in the detector are counted. In addition, the detector must be highly shielded from the general "cloud" of neutrons that exist around the diffractometer. The shielding is very thick: many centimetres of polythene are used to slow down any fast neutrons; the background neutrons then being absorbed by B_4C-doped plastic or rubber.

Signatures by which a Neutron May be Detected

Atomic and subatomic particles are detected by the signature they produce through interaction with their surroundings. The interactions result from the particles' fundamental characteristics.

- Charge: Neutrons are neutral particles and do not ionize directly; hence they are harder than charged particles to detect directly. Further, their paths of motion are only weakly affected by electric and magnetic fields.

- Mass: The neutron mass of 1.0086649156(6) u is not directly detectable, but does influence reactions through which it can be detected.

- Reactions: Neutrons react with a number of materials through elastic scattering producing a recoiling nucleus, inelastic scattering producing an excited nucleus, or absorption with transmutation of the resulting nucleus. Most detection approaches rely on detecting the various reaction products.

- Magnetic moment: Although neutrons have a magnetic moment of −1.9130427(5) μ_N, techniques for detection of the magnetic moment are too insensitive to use for neutron detection.

- Electric dipole moment: The neutron is predicted to have only a tiny electric dipole moment, which has not yet been detected. Hence it is not a viable detection signature.

- Decay: Outside the nucleus, free neutrons are unstable and have a mean lifetime of 885.7±0.8 s (about 14 minutes, 46 seconds). Free neutrons decay by emission of an electron and an electron antineutrino to become a proton, a process known as beta decay:

  $$n^o \rightarrow p^+ + e^- + \overline{\nu}_e.$$

  Although the p^+ and e^- produced by neutron decay are detectable, the decay rate is too low to serve as the basis for a practical detector system.

Classic Neutron Detection Options

As a result of these properties, detection of neutrons fall into several major categories:

- Absorptive reactions with prompt reactions - low energy neutrons are typically detected indirectly through absorption reactions. Typical absorber materials used have high cross sections for absorption of neutrons and include helium-3, lithium-6, boron-10, and uranium-235. Each of these reacts by emission of high energy ionized particles, the ionization track of which can be detected by a number of means. Commonly used reactions include $^{3}He(n,p)\,^{3}H$, $^{6}Li(n,t)\,^{4}He$, $^{10}B(n,\alpha)\,^{7}Li$ and the fission of uranium.
- Activation processes - Neutrons may be detected by reacting with absorbers in a radiative capture, spallation or similar reaction, producing reaction products that then decay at some later time, releasing beta particles or gammas. Selected materials (e.g., indium, gold, rhodium, iron ($^{56}Fe(n,p)\,^{56}Mn$), aluminum ($^{27}Al(n,\alpha)^{24}Na$), niobium ($^{93}Nb(n,2n)\,^{92m}Nb$), & silicon ($^{28}Si(n,p)\,^{28}Al$)) have extremely large cross sections for the capture of neutrons within a very narrow band of energy. Use of multiple absorber samples allows characterization of the neutron energy spectrum. Activation also allows recreation of an historic neutron exposure (e.g., forensic recreation of neutron exposures during an accidental criticality).
- Elastic scattering reactions (also referred to as proton-recoil) - High energy neutrons are typically detected indirectly through elastic scattering reactions. Neutron collide with the nucleus of atoms in the detector, transferring energy to that nucleus and creating an ion, which is detected. Since the maximum transfer of energy occurs when the mass of the atom with which the neutron collides is comparable to the neutron mass, hydrogenous materials are often the preferred medium for such detectors.

Types of Neutron Detectors

Gas Proportional Detectors

Gas proportional detectors can be adapted to detect neutrons. While neutrons do not typically cause ionization, the addition of a nuclide with high neutron cross-section allows the detector to respond to neutrons. Nuclides commonly used for this purpose are helium-3, lithium-6, boron-10 and uranium-235. Since these materials are most likely to react with thermal neutrons (i.e., neutrons that have slowed to equilibrium with their surroundings), they are typically surrounded by moderating materials to reduce their energy and increase the likelihood of detection.

Further refinements are usually necessary to differentiate the neutron signal from the effects of other types of radiation. Since the energy of a thermal neutron is relatively low, charged particle reactions are discrete (i.e., essentially monoenergetic and lie within a narrow bandwidth of energies) while other reactions such as gamma reactions will span a broad energy range, it is possible to discriminate among the sources.

As a class, gas ionization detectors measure the number (count rate), and not the energy of neutrons.

^3He Gas-filled Proportional Detectors

An isotope of Helium, ^3He provides for an effective neutron detector material because ^3He reacts by absorbing thermal neutrons, producing a ^1H and ^3H ion. Its sensitivity to gamma rays is negligible, providing a very useful neutron detector. Unfortunately the supply of ^3He is limited to production as a byproduct from the decay of tritium (which has a 12.3 year half-life); tritium is produced either as part of weapons programs as a booster for nuclear weapons or as a byproduct of reactor operation.

BF_3 Gas-filled Proportional Detectors

As elemental boron is not gaseous, neutron detectors containing boron may alternately use boron trifluoride (BF_3) enriched to 96% boron-10 (natural boron is 20% ^{10}B, 80% ^{11}B). It should be noted that boron trifluoride is highly toxic. The sensitivity of this detector is around 35-40 CPS/nv whereas that of Boron lined is about 4 CPS/nv. This is because in Boron lined, n reacts with Boron and hence produce ion pairs inside the layer. Hence charged particles produced (Alpha and Li) they lose some of their energy inside that layer. Low energy charged particles are able to reach the Ionization chabber's gas environment. Hence ionization produced in gas is also lower.

Whereas in BF3 gas filled, N reacts with B in gas. and fully energetic Alpha and Li are able to produce more ionization and give more pulses.

Boron Lined Proportional Detectors

Alternately, boron-lined gas-filled proportional counters react similarly to BF_3 gas-filled proportional detectors, with the exception that the walls are coated with ^{10}B. In this design, since the reaction takes place on the surface, only one of the two particles will escape into the proportional counter.

Scintillation Neutron Detectors

Scintillation neutron detectors include liquid organic scintillators, crystals, plastics, glass and scintillation fibers.

Neutron-sensitive Scintillating Glass Fiber Detectors

Scintillating ^6Li glass for neutron detection was first reported in the scientific literature in 1957 and key advances were made in the 1960s and 1970s. Scintillating fiber was demonstrated by Atkinson M. *et al.* in 1987 and major advances were made in the late 1980s and early 1990s at Pacific Northwest National Laboratory where it was developed as a classified technology. It was declassified in 1994 and first licensed by Oxford Instruments in 1997, followed by a transfer to Nucsafe in 1999. The fiber and fiber detectors are now manufactured and sold commercially by Nucsafe, Inc.

The scintillating glass fibers work by incorporating ^6Li and Ce^{3+} into the glass bulk composition. The ^6Li has a high cross-section for thermal neutron absorption through the $^6Li(n,\alpha)$ reaction. Neutron absorption produces a tritium ion, an alpha particle, and kinetic energy. The alpha particle and triton interact with the glass matrix to produce ionization, which transfers energy to Ce^{3+} ions and results in the emission of photons with wavelength 390 nm - 600 nm as the excited state Ce^{3+} ions return to the ground state. The event results in a flash of light of several thousand photons for each neutron absorbed. A portion of the scintillation light propagates through the glass fiber, which acts as a waveguide. The fibers ends are optically coupled to a pair of photomultiplier tubes (PMTs) to detect photon bursts. The detectors can be used to detect both neutrons and gamma rays, which are typically distinguished using pulse-height discrimination. Substantial effort and progress in reducing fiber detector sensitivity to gamma radiation has been made. Original detectors suffered from false neutrons in a 0.02 mR gamma field. Design, process, and algorithm improvements now enable operation in gamma fields up to 20 mR/h (^{60}Co).

The scintillating fiber detectors have excellent sensitivity, they are rugged, and have fast timing (~60 ns) so that a large dynamic range in counting rates is possible. The detectors have the advantage that they can be formed into any desired shape, and can be made very large or very small for use in a variety of applications. Further, they do not rely on ^3He or any raw material that has limited availability, nor do they contain toxic or regulated materials. Their performance matches or exceeds that of ^3He tubes for gross neutron counting due to the higher density of neutron absorbing species in the solid glass compared to high-pressure gaseous ^3He. Even though the thermal neutron cross section of ^6Li is low compared to ^3He (940 barns vs. 5330 barns), the atom density of ^6Li in the fiber is fifty times greater, resulting in an advantage in effective capture density ratio of approximately 10:1.

$LiCaAlF_6$

$LiCaAlF_6$ is a neutron sensitive inorganic scintillator crystal which like neutron-sensitive scintillating glass fiber detectors makes use of neutron capture by ^6Li. Unlike scintillating glass fiber detectors however the ^6Li is part of the crystalline structure of the scintillator giving it a naturally high ^6Li density. A doping agent is added to provide the crystal with its scintillating properties, two common doping agents are cesium and europium. Europium doped $LiCaAlF_6$ has the advantage over other materials that the number of optical photons produced per neutron capture is around 30.000 which is 5 times higher than for example in neutron-sensitive scintillating glass fibers. This property makes neutron photon discrimination easier. Due to its high ^{6}Li density this material is suitable for producing light weight compact neutron detectors, as a result $LiCaAlF_6$ has been used for neutron detection at high altitudes on balloon missions. The long decay time of Europium doped $LiCaAlF_6$ makes it less suitable for measurements in high radiation environments, the cesium doped variant has a shorter decay time but suffers from a lower light-yield.

Semiconductor Neutron Detectors

There are two basic types of semiconductor neutron detectors, the first being electron devices coated with a neutron reactive material and the second being a semiconductor being partly composed of neutron reactive material. The most successful of these configurations is the coated device type, and an example would be a common planar Si diode coated with either ^{10}B or ^{6}LiF. This type of detector was first proposed by Babcock et al. The concept is straightforward. A neutron is absorbed in the reactive film and spontaneously emits energetic reaction products. A reaction product may reach the semiconductor surface, and upon entering the semiconductor produces electron-hole pairs. Under a reverse bias voltage, these electrons and holes are drifted through the diode to produce an induced current, usually integrated in pulse mode to form a voltage output. The maximum intrinsic efficiency for single-coated devices is approximately 5% for thermal neutrons (0.0259 eV), and the design and operation are thoroughly described in the literature. The neutron detection efficiency limitation is a consequence of reaction-product self-absorption. For instance, the range in a boron film of 1.47 MeV α particles from the ^{10}B(n,α) ^{7}Li reaction is approximately 4.5 microns, and the range in LiF of 2.7 MeV tritons from the ^{10}B(n,α) ^{7}Li reaction is approximately 28 microns. Reaction products originating at distances further from the film/semiconductor interface can not reach the semiconductor surface, and consequently will not contribute to neutron detection. Devices coated with natural Gd have also been explored, mainly because of its large thermal neutron microscopic cross section of 49,000 barns. However, the Gd(n,γ) reaction products of interest are mainly low energy conversion electrons, mostly grouped around 70 keV. Consequently, discrimination between neutron induced events and gamma-ray events (mainly producing Compton scattered electrons) is difficult for Gd-coated semiconductor diodes. A compensated pixel design sought to remedy the problem. Overall, devices coated with either ^{10}B or ^{6}LiF are preferred mainly because the energetic charged-particle reaction products are much easier to discriminate from background radiations.

The low efficiency of coated planar diodes led to the development of microstructured semiconductor neutron detectors (MSND). These detectors have microscopic structures etched into a semiconductor substrate, subsequently formed into a pin style diode. The microstructures are backfilled with neutron reactive material, usually ^{6}LiF, although ^{10}B has been used. The increased semiconductor surface area adjacent to the reactive material and the increased probability that a reaction product will enter the semiconductor greatly increase the intrinsic neutron detection efficiency.

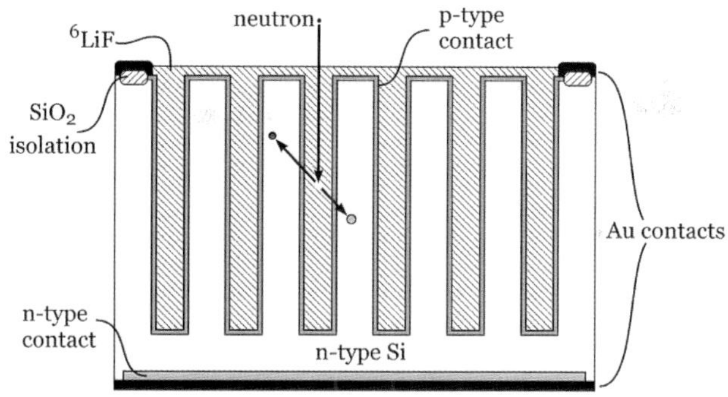

Basic design of a microstructured semiconductor neutron detector (MSND).

The MSND device configuration was first proposed by Muminov and Tsvang, and later by Schelten et al. It was years later when the first working example of a MSND was fabricated and demonstrated , then having only 3.3% thermal neutron detection efficiency. Since that initial work, MSNDs have achieved greater than 30% thermal neutron detection efficiency. Although MSNDs can operate on the built-in potential (zero applied voltage), they perform best when 2-3 volts are applied. There are several groups now working on MSND variations. The most successful types are the variety backfilled with ^6LiF material. MSNDs are now manufactured and sold commercially by Radiation Detection Technologies, Inc. Advanced experimental versions of double-sided MSNDs with opposing microstructures on both sides of a semiconductor wafer have been reported with over 50% thermal neutron detection efficiency, and are theoretically capable of over 70% efficiency.

Semiconductor detectors in which one of more constituent atoms are neutron reactive are called bulk semiconductor neutron detectors. Bulk solid-state neutron detectors can be divided into two basic categories: those that rely on the detection of charged-particle reaction products and those that rely on the detection of prompt capture gamma rays. In general, this type of neutron detector is difficult to make reliably and presently are not commercially available.

The bulk materials that rely upon charged-particle emissions are based on boron and lithium containing semiconductors. In the search for bulk semiconductor neutron detectors, the boron-based materials, such as BP, BAs, BN, and B_4C, have been investigated more than other potential materials.

Boron-based semiconductors in cubic form are difficult to grow as bulk crystals, mainly because they require high temperatures and high pressure for synthesis. BP and Bas can decompose into undesirable crystal structures (cubic to icosahedral form) unless synthesized under high pressure. B_4C also forms icosahedral units in a rhombohedral crystal structure, an undesirable transformation because the icosahedral structure has relatively poor charge collection properties which make these icosahedral forms unsuitable for neutron detection.

BN can be formed as either simple hexagonal, cubic (zincblende) or wurtzite crystals, depending on the growth temperature, and it is usually grown by thin film methods. It is the simple hexagonal form of BN that has been most studied as a neutron detector. Thin film chemical vapor deposition methods are usually employed to produce BP, BAs, BN, or B_4C. These boron-based films are often grown upon n-type Si substrates, which can form a pn junction with the Si and, therefore, produce a coated Si diode. Consequently, the neutron response from the device can be easily mistaken as a bulk response when it is actually a coated diode response. To date, there is sparse evidence of boron-based semiconductors producing intrinsic neutron signals.

Li-containing semiconductors, categorized as Nowotny-Juza compounds, have also been investigated as bulk neutron detectors. The Nowotny-Juza compound LiZnAs has been demonstrated as a neutron detector; however, the material is difficult and expensive to synthesize, and only small semiconductor crystals have been reported. Finally, traditional semiconductor materials with neutron reactive dopants have been investigated, namely, Si(Li) detectors. Neutrons interact with the lithium dopant in the material and produce energetic reaction products. However, the dopant concentration is relatively low in Li drifted Si detectors (or other doped semiconductors), typically less than 10^{19} cm^{-3}. For a degenerate concentration of Li on the order of 10^{19} cm^{-3}, a 5-cm thick block of natural Si(Li) would have less than 1% thermal-neutron detection efficiency, while a 5-cm thick block of a Si(^{6}Li) detector would have only 4.6% thermal-neutron detection efficiency.

Prompt gamma-ray emitting semiconductors, such as CdTe, and HgI_2 have been successfully used as neutron detectors. These detectors rely upon the prompt gamma-ray emissions from the ^{113}Cd(n, γ)^{114}Cd reaction (producing 558.6 keV and 651.3 keV gamma rays) and the ^{199}Hg(n, γ) ^{200}Hg reaction (producing 368.1 keV and 661.1 keV gamma rays). However, these semiconductor materials are designed for use as gamma-ray spectrometers and, hence, are intrinsically sensitive to the gamma-ray background. With adequate energy resolution, pulse height discrimination can be used to separate the prompt gamma-ray emissions from neutron interactions. However, the effective neutron detection efficiency is compromised because of the relatively small Compton ratio. In other words, the majority of events add to the Compton continuum rather than to the full energy peak, thus, making discrimination between neutrons and background gamma rays difficult. Also, both natural Cd and Hg have relatively large thermal-neutron (n,γ) cross sections of 2444 b and 369.8 b, respectively. Consequently, most thermal neutrons are absorbed near the detector surface so that nearly half of the prompt gamma rays are emitted in directions away from the detector bulk and, thus, produce poor gamma-ray reabsorption or interaction efficiency.

Neutron Activation Detectors

Activation samples may be placed in a neutron field to characterize the energy spectrum and intensity of the neutrons. Activation reactions that have differing energy thresholds can be used including $^{56}Fe(n,p)^{56}Mn$, $^{27}Al(n,\alpha)^{24}Na$, $^{93}Nb(n,2n)^{92m}Nb$, & $^{28}Si(n,p)^{28}Al$.

Fast Neutron Detectors

Fast neutrons are often detected by first moderating (slowing) them to thermal energies. However, during that process the information on the original energy of the neutron, its direction of travel, and the time of emission is lost. For many applications, the detection of "fast" neutrons that retain this information is highly desirable.

Typical fast neutron detectors are liquid scintillators, 4-He based noble gas detectors and plastic detectors. Fast neutron detectors differentiate themselves from one another by their 1.) capability of neutron/gamma discrimination (through pulse shape discrimination) and 2.) sensitivity. The capability to distinguish between neutrons and gammas is excellent in noble gas based 4-He detectors due to their low electron density and excellent pulse shape discrimination property.

Detection of fast neutrons poses a range of special problems. A directional fast-neutron detector has been developed using multiple proton recoils in separated planes of plastic scintillator material. The paths of the recoil nuclei created by neutron collision are recorded; determination of the energy and momentum of two recoil nuclei allow calculation of the direction of travel and energy of the neutron that underwent elastic scattering with them.

Applications

Neutron detection is used for varying purposes. Each application has different requirements for the detection system.

- Reactor instrumentation: Since reactor power is essentially linearly proportional to the Neutron flux, neutron detectors provide an important measure of power in nuclear power and research reactors. Boiling water reactors may have dozens of neutron detectors, one per fuel assembly. Most neutron detectors used in thermal-spectrum nuclear reactors are optimized to detect thermal neutrons.
- Particle physics: Neutron detection has been proposed as a method of enhancing neutrino detectors.
- Materials science: Elastic and inelastic neutron scattering enables experimentalists to characterize the morphology of materials from scales ranging from ångströms to about one micrometer.

- Radiation safety: Neutron radiation is a hazard associated with neutron sources, space travel, accelerators and nuclear reactors. Neutron detectors used for radiation safety must take into account the relative biological effectiveness (i.e., the way damage caused by neutrons varies with energy).
- Cosmic ray detection: Secondary neutrons are one component of particle showers produced in Earth's atmosphere by cosmic rays. Dedicated ground-level neutron detectors, namely neutron monitors, are employed to monitor variations in cosmic ray flux.
- Special nuclear material detection: Special nuclear materials (SNM) such as uranium-233 and plutonium-239 decay by spontaneous fission, yielding neutrons. Neutrons detectors can be used for monitor for SNM in commerce.

Experimental Neutron Detection

Experiments that make use of this science include scattering experiments in which neutrons directed and then scattered from a sample are to be detected. Facilities include the ISIS neutron source at the Rutherford Appleton Laboratory, the Spallation Neutron Source at the Oak Ridge National Laboratory, and the Spallation Neutron Source (SINQ) at the Paul Scherrer Institute, in which the neutrons are produced by spallation reaction, and the traditional research reactor facilities in which neutrons are produced during fission of uranium isotopes. Noteworthy among the various neutron detection experiments is the trademark experiment of the European Muon Collaboration, first performed at CERN and now termed the "EMC experiment." The same experiment is performed today with more sophisticated equipment to obtain more definite results related to the original EMC effect.

Challenges in Neutron Detection in an Experimental Environment

Neutron detection in an experimental environment is not an easy science. The major challenges faced by modern-day neutron detection include background noise, high detection rates, neutron neutrality, and low neutron energies.

Background Noise

The main components of background noise in neutron detection are high-energy photons, which aren't easily eliminated by physical barriers. The other sources of noise, such as alpha and beta particles, can be eliminated by various shielding materials, such as lead, plastic, thermo-coal, etc. Thus, photons cause major interference in neutron detection, since it is uncertain if neutrons or photons are being detected by the neutron detector. Both register similar energies after scattering into the detector from the target or ambient light, and are thus hard to distinguish. Coincidence detection can also be used to discriminate real neutron events from photons and other radiation.

High Detection Rates

If the detector lies in a region of high beam activity, it is hit continuously by neutrons and background noise at overwhelmingly high rates. This obfuscates collected data, since there is extreme overlap in measurement, and separate events are not easily distinguished from each other. Thus, part of the challenge lies in keeping detection rates as low as possible and in designing a detector that can keep up with the high rates to yield coherent data.

Neutrality of Neutrons

Neutrons are neutral and thus do not respond to electric fields. This makes it hard to direct their course towards a detector to facilitate detection. Neutrons also do not ionize atoms except by direct collision, so gaseous ionization detectors are ineffective.

Varying Behavior with Energy

Detectors relying on neutron absorption are generally more sensitive to low-energy thermal neutrons, and are orders of magnitude less sensitive to high-energy neutrons. Scintillation detectors, on the other hand, have trouble registering the impacts of low-energy neutrons.

Experimental Setup and Method

Figure shows the typical main components of the setup of a neutron detection unit. In principle, the diagram shows the setup as it would be in any modern particle physics lab, but the specifics describe the setup in Jefferson Lab (Newport News, Virginia).

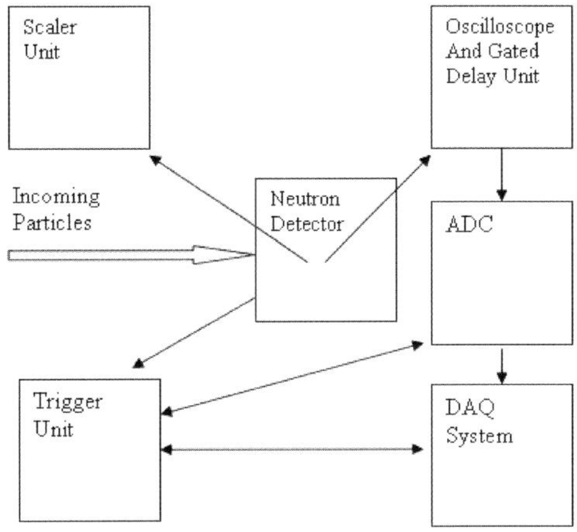

Figure: The experimental setup

In this setup, the incoming particles, comprising neutrons and photons, strike the neutron detector; this is typically a scintillation detector consisting of scintillating material, a waveguide, and a photomultiplier tube (PMT), and will be connected to a data acquisition (DAQ) system to register detection details.

The detection signal from the neutron detector is connected to the scaler unit, gated delay unit, trigger unit and the oscilloscope. The scaler unit is merely used to count the number of incoming particles or events. It does so by incrementing its tally of particles every time it detects a surge in the detector signal from the zero-point. There is very little dead time in this unit, implying that no matter how fast particles are coming in, it is very unlikely for this unit to fail to count an event (e.g. incoming particle). The low dead time is due to sophisticated electronics in this unit, which take little time to recover from the relatively easy task of registering a logical high every time an event occurs. The trigger unit coordinates all the electronics of the system and gives a logical high to these units when the whole setup is ready to record an event run.

The oscilloscope registers a current pulse with every event. The pulse is merely the ionization current in the detector caused by this event plotted against time. The total energy of the incident particle can be found by integrating this current pulse with respect to time to yield the total charge deposited at the end of the PMT. This integration is carried out in the analog-digital converter (ADC). The total deposited charge is a direct measure of the energy of the ionizing particle (neutron or photon) entering the neutron detector. This signal integration technique is an established method for measuring ionization in the detector in nuclear physics. The ADC has a higher dead time than the oscilloscope, which has limited memory and needs to transfer events quickly to the ADC. Thus, the ADC samples out approximately one in every 30 events from the oscilloscope for analysis. Since the typical event rate is around 10^6 neutrons every second, this sampling will still accumulate thousands of events every second.

Separating Neutrons from Photons

The ADC sends its data to a DAQ unit that sorts the data in presentable form for analysis. The key to further analysis lies in the difference between the shape of the photon ionization-current pulse and that of the neutron. The photon pulse is longer at the ends (or "tails") whereas the neutron pulse is well-centered. This fact can be used to identify incoming neutrons and to count the total rate of incoming neutrons. The steps leading to this separation (those that are usually performed at leading national laboratories, Jefferson Lab specifically among them) are gated pulse extraction and plotting-the-difference.

Gated Pulse Extraction

Ionization current signals are all pulses with a local peak in between. Using a logical AND gate in continuous time (having a stream of "1" and "0" pulses as one input and the current signal as the other), the tail portion of every current pulse signal is extracted. This gated discrimination method is used on a regular basis on liquid scintillators. The gated delay unit is precisely to this end, and makes a delayed copy of the original signal in such a way that its tail section is seen alongside its main section on the oscilloscope screen.

After extracting the tail, the usual current integration is carried out on both the tail section and the complete signal. This yields two ionization values for each event, which are stored in the event table in the DAQ system.

Plotting the Difference

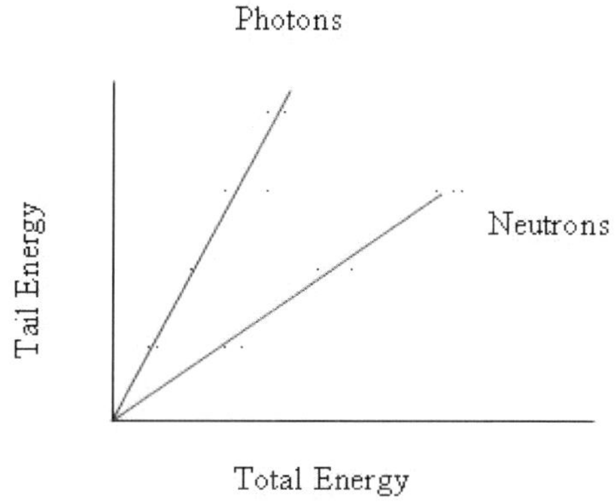

Figure: Expected plot of tail energy against energy in the complete pulse plotted for all event energies. Dots represent number densities of events.

In this step lies the crucial point of the analysis: the extracted ionization values are plotted. Specifically, the graph plots energy deposition in the tail against energy deposition in the entire signal for a range of neutron energies. Typically, for a given energy, there are many events with the same tail-energy value. In this case, plotted points are simply made denser with more overlapping dots on the two-dimensional plot, and can thus be used to eyeball the number of events corresponding to each energy-deposition. A considerable random fraction (1/30) of all events is plotted on the graph.

If the tail size extracted is a fixed proportion of the total pulse, then there will be two lines on the plot, having different slopes. The line with the greater slope will correspond to photon events and the line with the lesser slope to neutron events. This is precisely because the photon energy deposition current, plotted against time, leaves a longer "tail" than does the neutron deposition plot, giving the photon tail more proportion of the total energy than neutron tails.

The effectiveness of any detection analysis can be seen by its ability to accurately count and separate the number of neutrons and photons striking the detector. Also, the effectiveness of the second and third steps reveals whether event rates in the experiment are manageable. If clear plots can be obtained in the above steps, allowing for easy neutron-photon separation, the detection can be termed effective and the rates manageable. On the other hand, smudging and indistinguishability of data points will not allow for easy separation of events.

Rate Control

Detection rates can be kept low in many ways. Sampling of events can be used to choose only a few events for analysis. If the rates are so high that one event cannot be distinguished from another, physical experimental parameters (shielding, detector-target distance, solid-angle, etc.) can be manipulated to give the lowest rates possible and thus distinguishable events.

Finer Detection Points

It is important here to observe precisely those variables that matter, since there may be false indicators along the way. For example, ionization currents might get periodic high surges, which do not imply high rates but just high energy depositions for stray events. These surges will be tabulated and viewed with cynicism if unjustifiable, especially since there is so much background noise in the setup.

One might ask how experimenters can be sure that every current pulse in the oscilloscope corresponds to exactly one event. This is true because the pulse lasts about 50 ns, allowing for a maximum of 2×10^7 events every second. This number is much higher than the actual typical rate, which is usually an order of magnitude less, as mentioned above. This means that is it highly unlikely for there to be two particles generating one current pulse. The current pulses last 50 ns each, and start to register the next event after a gap from the previous event.

Although sometimes facilitated by higher incoming neutron energies, neutron detection is generally a difficult task, for all the reasons stated earlier. Thus, better scintillator design is also in the foreground and has been the topic of pursuit ever since the invention of scintillation detectors. Scintillation detectors were invented in 1903 by Crookes but were not very efficient until the PMT (photomultiplier tube) was developed by Curran and Baker in 1944. The PMT gives a reliable and efficient method of detection since it multiplies the detection signal tenfold. Even so, scintillation design has room for improvement as do other options for neutron detection besides scintillation.

Neutron Emission

The neutron emission is one of the radioactive decays, by which unstable nuclei may reach the stability. In general, this type of radioactive decay may occur, when nuclei contain significant excess of neutrons or excitation energy. In this type of decay a neutron is simply ejected from the nucleus.

Although the neutron emission is usually associated with nuclear decay, it must be also mentioned in connection with neutron nuclear reactions. Some neutrons interacts with a target nucleus via a compound nucleus. Among these compound nucleus reactions are also reactions, in which a neutron is ejected from nucleus and they may be referred to as neutron emission reactions. The point is that compound nuclei lose its excitation energy in a way, which is identical to the radioactive decay. Very important feature is the fact the mode of decay of compound nucleus does not depend on the way the compound nucleus was formed. The compound nucleus reactions, in which neutron emission occurs, are:

- Elastic Scattering Reaction. In some cases, if the kinetic energy of an incident neutron just right to form a resonance, the neutron may be absorbed and a compound nucleus may be formed. This interaction is more unusual (in comparison with potential scattering) and is also known as resonance elastic scattering. Due to formation of the compound nucleus, initial and final neutron are not the same and this reactions may be also referred to as one type of neutron emission reaction.
- Inelastic Scattering Reaction. In this case, the connection with neutron emission is more obvious. In an inelastic scattering reaction between a neutron and a target nucleus some energy of the incident neutron is absorbed to the recoiling nucleus and the nucleus remains in the excited state. The neutron is emitted then with a lower kinetic energy. If the kinetic energy of an incident neutron is sufficient the double, triple, or more, neutron emission may take place. These events are referred to as (n, 2n), (n, 3n) or (n, ...n) reactions. The probability of such reactions increases with increasing incident neutron energies.
- Nuclear Fission. The fission reaction is very specific reaction and is of importance in many fields of nuclear engineering. It is known the fission reaction produces fission neutrons that are of importance in any chain-reacting system. But not all neutrons are released at the same time following fission. Even the nature of creation of these neutrons is different. From this point of view we usually divide the fission neutrons into two following groups:
    - Prompt Neutrons. Prompt neutrons are emitted directly from fission and they are emitted within very short time of about 10^{-14} second.
    - Delayed Neutrons. Delayed neutrons are emitted by neutron rich fission fragments that are called the delayed neutron precursors. These precursors usually undergo beta decay but a small fraction of them are excited enough to undergo neutron emission. The fact the neutron is produced via this type of decay and this happens orders of magnitude later compared to the emission of the prompt neutrons, plays an extremely important role in the control of the reactor.

Spontaneous Neutron Emission

As a consequence of the Pauli exclusion principle nuclei with an excess of protons or neutrons have a higher average energy per nucleon. Nuclei with a sufficient excess of neutrons have a greater energy than the combination of a free neutron and a nucleus with one less neutron, and therefore can decay by neutron emission. Nuclei which can decay by this mode are described as lying beyond the neutron drip line.

Two examples of isotopes that emit neutrons are beryllium-13 (decaying to beryllium-12 with a mean life 2.7×10^{-21} s) and helium-5 (helium-4, 7×10^{-22} s).

Photoneutron Emission

Some nucleides, e.g. Be-9, can be induced to eject a neutron by gamma radiation.

Beta-delayed Neutron Emission

Neutron emission usually happens from nuclei that are in an excited state, such as the excited $^{17}O^*$ produced from the beta decay of ^{17}N. The neutron emission process itself is controlled by the nuclear force and therefore is extremely fast, sometimes referred to as "nearly instantaneous". This process allows unstable atoms to become more stable. The ejection of the neutron may be as a product of the movement of many nucleons, but it is ultimately mediated by the repulsive action of the nuclear force that exists at extremely short-range distances between nucleons. The life time of an ejected neutron inside the nucleus before it is emitted is usually comparable to the flight time of a typical neutron before it leaves the small nuclear "potential well", or about 10^{-23} seconds.

Delayed Neutrons in Reactor Control

Most neutron emission outside prompt neutron production associated with fission (either induced or spontaneous), is from neutron-heavy isotopes produced as fission products. These neutrons are sometimes emitted with a delay, giving them the term delayed neutrons, but the actual delay in their production is a delay waiting for the beta decay of fission products to produce the excited-state nuclear precursors that immediately undergo prompt neutron emission. Thus, the delay in neutron emission is not from the neutron-production process, but rather its precursor beta decay, which is controlled by the weak force, and thus requires a far longer time. The beta decay half lives for the precursors to delayed neutron-emitter radioisotopes, are typically fractions of a second to tens of seconds.

Nevertheless, the delayed neutrons emitted by neutron-rich fission products aid control of nuclear reactors by making reactivity change far more slowly than it would if it were controlled by prompt neutrons alone. About 0.65% of neutrons are released in a nuclear chain reaction in a delayed way due to the mechanism of neutron emission, and it is this fraction of neutrons that allows a nuclear reactor to be controlled on human reaction time-scales, without proceeding to a prompt critical state, and runaway melt down.

Neutron Emission in Fission

Induced Fission

A synonym for such neutron emission is "prompt neutron" production, of the type that is best known to occur simultaneously with induced nuclear fission. Induced fission happens only when a nucleus is bombarded with neutrons, gamma rays, or other carriers of energy. Many heavy isotopes, most notably californium-252, also emit prompt neutrons among the products of a similar spontaneous radioactive decay process, spontaneous fission.

Spontaneous Fission

Spontaneous fission happens when a nucleus splits into two smaller nuclei and generally one or more neutrons.

Neutron Magnetic Moment

The neutron magnetic moment is the intrinsic magnetic dipole moment of the neutron, symbol μ_n. Protons and neutrons, both nucleons, comprise the nucleus of atoms, and both nucleons behave as small magnets whose strengths are measured by their magnetic moments. The neutron interacts with normal matter through either the nuclear force or its magnetic moment. The neutron's magnetic moment is exploited to probe the atomic structure of materials using scattering methods and to manipulate the properties of neutron beams in particle accelerators. The neutron was determined to have a magnetic moment by indirect methods in the mid 1930s. Luis Alvarez and Felix Bloch made the first accurate, direct measurement of the neutron's magnetic moment in 1940. The existence of the neutron's magnetic moment indicates the neutron is not an elementary particle. For an elementary particle to have an intrinsic magnetic moment, it must have both spin and electric charge. The neutron has spin 1/2 $\hbar$, but it has no net charge. The existence of the neutron's magnetic moment was puzzling and defied a correct explanation until the quark model for particles was developed in the 1960s. The neutron is composed of three quarks, and the magnetic moments of these elementary particles combine to give the neutron its magnetic moment.

Description

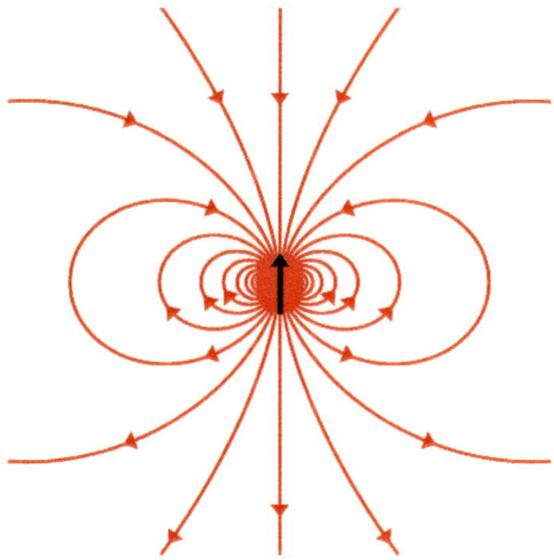

Schematic diagram depicting the spin of the neutron as the black arrow and magnetic field lines associated with the neutron's negative magnetic moment. The spin of the neutron is upward in this diagram, but the magnetic field lines at the center of the dipole are downward.

The best available measurement for the value of the magnetic moment of the neutron is $\mu_n = -1.91304272(45)\ \mu_N$. Here μ_N is the nuclear magneton, a physical constant and standard unit for the magnetic moments of nuclear components. In SI units, $\mu_n = -9.6623647(23)\times10^{-27}\ J\cdot T^{-1}$. A magnetic moment is a vector quantity, and the direction of the neutron's magnetic moment is defined by its spin. The torque on the neutron resulting from an external magnetic field is towards aligning the neutron's spin vector opposite to the magnetic field vector.

The nuclear magneton is the spin magnetic moment of a Dirac particle, a charged, spin 1/2 elementary particle, with a proton's mass m_p. In SI units, the nuclear magneton is

$$\mu_N = \frac{e\hbar}{2m_p},$$

where e is the elementary charge and $\hbar$ is the reduced Planck constant. The magnetic moment of this particle is parallel to its spin. Since the neutron has no charge, it should have no magnetic moment by this expression. The non-zero magnetic moment of the neutron indicates that it is not an elementary particle. The sign of the neutron's magnetic moment is that of a negatively charged particle. Similarly, the fact that the magnetic moment of the proton, $\mu_p = 2.793\ \mu_N$, is not equal to 1 μ_N indicates that it too is not an elementary particle. Protons and neutrons are composed of quarks, and the magnetic moments of the quarks can be used to compute the magnetic moments of the nucleons.

Although the neutron interacts with normal matter primarily through either nuclear or magnetic forces, the magnetic interactions are about seven orders of magnitude weaker than the nuclear interactions. The influence of the neutron's magnetic moment is therefore only apparent for low energy, or slow, neutrons. Because the value for the magnetic moment is inversely proportional to particle mass, the nuclear magneton is about 1/2000 as large as the Bohr magneton. The magnetic moment of the electron is therefore about 1000 times larger than that of the neutron.

The magnetic moment of the antineutron has the same magnitude as, but has the opposite sign, that of the neutron.

Measurement

Soon after the neutron was discovered in 1932, indirect evidence suggested the neutron had an unexpected non-zero value for its magnetic moment. Attempts to measure the neutron's magnetic moment originated with the discovery by Otto Stern in 1933 in Hamburg that the proton had an anomalously large magnetic moment. The proton's magnetic moment had been determined by measuring the deflection of a beam of molecular hydrogen by a magnetic field. Stern won the Nobel Prize in 1943 for this discovery.

By 1934 groups led by Stern, now in Pittsburgh, and I. I. Rabi in New York had independently measured the magnetic moments of the proton and deuteron. The measured values for these particles were only in rough agreement between the groups, but the Rabi group confirmed the earlier Stern measurements that the magnetic moment for the proton was unexpectedly large. Since a deuteron is composed of a proton and a neutron with aligned spins, the neutron's magnetic moment could be inferred by subtracting the deuteron and proton magnetic moments. The resulting value was not zero and had sign opposite to that of the proton. Values for the magnetic moment of the neutron were also determined by R. Bacher at Ann Arbor (1933) and I.Y. Tamm and S.A. Altshuler in the Soviet Union (1934) from studies of the hyperfine structure of atomic spectra. Although Tamm and Altshuler's estimate had the correct sign and order of magnitude ($\mu_n = -0.5\ \mu_N$), the result was met with skepticism. By the late 1930s, accurate values for the magnetic moment of the neutron had been deduced by the Rabi group using measurements employing newly developed nuclear magnetic resonance techniques. The large value for the proton's magnetic moment and the inferred negative value for the neutron's magnetic moment were unexpected and could not be explained. The anomalous values for the magnetic moments of the nucleons would remain a puzzle until the quark model was developed in the 1960s.

The refinement and evolution of the Rabi measurements led to the discovery in 1939 that the deuteron also possessed an electric quadrupole moment. This electrical property of the deuteron had been interfering with the measurements by the Rabi group. The discovery meant that the physical shape of the deuteron was not symmetric, which provided valuable insight into the nature of the nuclear force binding nucleons. Rabi was awarded the Nobel Prize in 1944 for his resonance method for recording the magnetic properties of atomic nuclei.

The value for the neutron's magnetic moment was first directly measured by Luis Alvarez and Felix Bloch at Berkeley, California in 1940. Using an extension of the magnetic resonance methods developed by Rabi, Alvarez and Bloch determined the magnetic moment of the neutron to be $\mu_n = -1.93(2)\ \mu_N$. By directly measuring the magnetic moment of free neutrons, or individual neutrons free of the nucleus, Alvarez and Bloch resolved all doubts and ambiguities about this anomalous property of neutrons.

Neutron g-factor and Gyromagnetic Ratio

The magnetic moment of a nucleon is sometimes expressed in terms of its g-factor, a dimensionless scalar. The convention defining the g-factor for composite particles, such as the neutron or proton, is

$$\mu = \frac{g\mu_N}{\hbar} \boldsymbol{I}$$

where μ is the intrinsic magnetic moment, I is the spin angular momentum, and g is the effective g-factor. While the g-factor is dimensionless, for composite particles it is defined relative to the natural unit of the nuclear magneton. For the neutron, I is $1/2\,\hbar$, so the neutron's g-factor, symbol g_n, is −3.82608545(90).

The gyromagnetic ratio, symbol γ, of a particle or system is the ratio of its magnetic moment to its spin angular momentum, or

$$\mu = \gamma I$$

For nucleons, the ratio is conventionally written in terms of the proton mass and charge, by the formula

$$\gamma = \frac{g\mu_N}{\hbar} = g\frac{e}{2m_p}$$

The neutron's gyromagnetic ratio, symbol γ_n, is $-1.83247179(43)\times10^8$ rad·s^{-1}·T^{-1}. The gyromagnetic ratio is also the ratio between the observed angular frequency of Larmor precession (in rad s^{-1}) and the strength of the magnetic field in nuclear magnetic resonance applications, such as in MRI imaging. For this reason, the value of γ_n is often given in units of MHz/T. The quantity $\gamma_n/2\pi$ ("gamma bar") is therefore convenient, which has the value −29.1646943(69) MHz·T^{-1}.

Physical Significance

Direction of Larmor precession for a neutron. The central arrow denotes the magnetic field, the small red arrow the spin of the neutron.

When a neutron is put into a magnetic field produced by an external source, it is subject to a torque tending to orient its magnetic moment parallel to the field (hence its spin antiparallel to the field). Like any magnet, the amount of this torque is proportional both to the magnetic moment and the external magnetic field. Since the neutron has spin angular momentum, this torque will cause the neutron to precess with a well-defined frequency, called the Larmor frequency. It is this phenomenon that enables the measurement of nuclear properties through nuclear magnetic resonance. The Larmor frequency can be determined by the product of the gyromagnetic ratio with the magnetic field strength. Since the sign of γ_n is negative, the neutron's spin angular momentum precesses counterclockwise about the direction of the external magnetic field.

The interactions of the neutron's magnetic moment with an external magnetic field were exploited to finally determine the spin of the neutron. In 1949, Hughes and Burgy measured neutrons reflected from a ferromagnetic mirror and found that the angular distribution of the reflections was consistent with spin 1/2. In 1954, Sherwood, Stephenson, and Bernstein employed neutrons in a Stern-Gerlach experiment that used a magnetic field to separate the neutron spin states. They recorded the two such spin states, consistent with a spin 1/2 particle. Until these measurements, the possibility that the neutron was a spin 3/2 particle could not have been ruled out.

Since neutrons are neutral particles, they do not have to overcome Coulomb repulsion as they approach charged targets, as experienced by protons or alpha particles. Neutrons can deeply penetrate matter. The magnetic moment of the neutron has therefore been exploited to probe the properties of matter using scattering or diffraction techniques. These methods provide information that is complementary to X-ray spectroscopy. In particular, the magnetic moment of the neutron is used to determine magnetic properties of materials at length scales of 1–100 Å using cold or thermal neutrons. Bertram Brockhouse and Clifford Shull won the Nobel Prize in physics in 1994 for developing these scattering techniques.

Without an electric charge, neutron beams cannot be controlled by the conventional electromagnetic methods employed for particle accelerators. The magnetic moment of the neutron allows some control of neutrons using magnetic fields, however, including the formation of polarized neutron beams. One technique employs the fact that cold neutrons will reflect from some magnetic materials at great efficiency when scattered at small grazing angles. The reflection preferentially selects particular spin states, thus polarizing the neutrons. Neutron magnetic mirrors and guides use this total internal reflection phenomena to control beams of slow neutrons.

Since an atomic nucleus consists of a bound state of protons and neutrons, the magnetic moments of the nucleons contribute to the nuclear magnetic moment, or the magnetic moment for the nucleus as a whole. The nuclear magnetic moment also includes contributions from the orbital motion of the nucleons. The deuteron has the simplest example of a nuclear magnetic moment, with measured value 0.857 μ_N. This value is within 3% of the sum of the moments of the proton and neutron, which gives 0.879 μ_N. In this calculation, the spins of the nucleons are aligned, but their magnetic moments offset because of the neutron's negative magnetic moment.

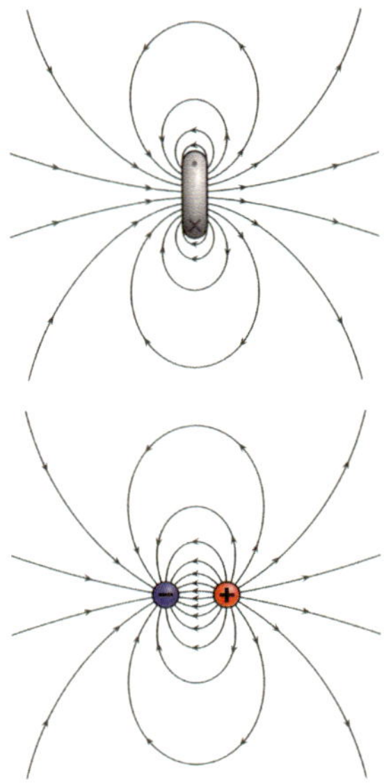

A magnetic dipole moment can be created by either a current loop (top; Ampèrian) or by two magnetic monopoles (bottom; Gilbertian). The neutron's magnetic moment is Ampèrian.

The Nature of the Neutron's Magnetic Moment

A magnetic dipole moment can be generated by two possible mechanisms. One way is by a small loop of electric current, called an "Ampèrian" magnetic dipole. Another way is by a pair of magnetic monopoles of opposite magnetic charge, bound together in some way, called a "Gilbertian" magnetic dipole. Magnetic monopoles remain hypothetical and unobserved, however. Throughout the 1930s and 1940s it was not readily apparent which of these two mechanisms caused the neutron's intrinsic magnetic moment. In 1930, E. Fermi showed that the magnetic moments of nuclei (including the proton) are Ampèrian. The two kinds of magnetic moments experience different forces in a magnetic field. Based on Fermi's arguments, the intrinsic magnetic moments of elementary particles, including the neutron, have been shown to be Ampèrian. The arguments are based on basic electromagnetism, elementary quantum mechanics, and the hyperfine structure of atomic s-state energy levels. In the case of the neutron, the theoretical possibilities were resolved by laboratory measurements of the scattering of slow neutrons from ferromagnetic materials in 1951.

Anomalous Magnetic Moments and Meson Physics

The anomalous values for the magnetic moments of the nucleons presented a theoretical quandary for the 30 years from the time of their discovery in the early 1930s to the development of the quark model in the 1960s. Considerable theoretical efforts were expended in trying to understand the origins of these magnetic moments, but the failures of these theories were glaring. Much of the theoretical focus was on developing a nuclear-force equivalence to the remarkably successful theory explaining the small anomalous magnetic moment of the electron.

The problem of the origins of the magnetic moments of nucleons was recognized as early as 1935. Gian Carlo Wick suggested the magnetic moments could be caused by the quantum mechanical fluctuations of these particles in accordance with Fermi's 1934 theory of beta decay. By this theory, a neutron is partly, regularly and briefly, disassociated into a proton, an electron, and a neutrino as a natural consequence of beta decay. By this idea, the magnetic moment of the neutron was caused by the fleeting existence of the large magnetic moment of the electron in the course of these quantum mechanical fluctuations, the value of the magnetic moment determined by the length of time the virtual electron was in existence. The theory proved to be untenable, however, when Hans Bethe and Robert Bacher showed that it predicted values for the magnetic moment that were either much too small, or much too large, depending on speculative assumptions.

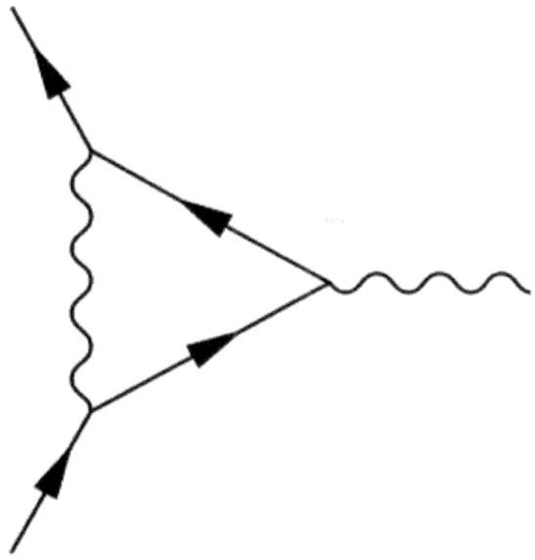

One-loop correction to a fermion's magnetic dipole moment. The solid lines at top and bottom represent the fermion (electron or nucleon), the wavey lines represent the photon mediating the electromagnetic force. The middle solid lines represent a virtual pair of particles, electron and positron for QED, pions for the nuclear force.

Similar considerations for the electron proved to be much more successful. In quantum electrodynamics (QED), the anomalous magnetic moment of a particle stems from the small contributions of quantum mechanical fluctuations to the magnetic moment of that particle. The g-factor for a "Dirac" magnetic moment is predicted to be $g = -2$ for a negatively charged, spin 1/2 particle. For particles such as the electron, this "classical" result differs from the observed value by a small fraction of a percent; the difference compared to the classical value is the anomalous magnetic moment. The actual g-factor for the electron is measured to be −2.00231930436153(53). QED results from the mediation of the electromagnetic force by photons. The physical picture is that the *effective* magnetic moment of the electron results from the contributions of the "bare" electron, which is the Dirac particle, and the cloud of "virtual," short-lived electron–positron pairs and photons that surround this particle as a consequence of QED. The small effects of these quantum mechanical fluctuations can be theoretically computed using Feynman diagrams with loops.

The one-loop contribution to the anomalous magnetic moment of the electron, corresponding to the first order and largest correction in QED, is found by calculating the vertex function shown in the diagram. The calculation was discovered by Julian Schwinger in 1948. Computed to fourth order, the QED prediction for the electron's anomalous magnetic moment agrees with the experimentally measured value to more than 10 significant figures, making the magnetic moment of the electron one of the most accurately verified predictions in the history of physics.

Compared to the electron, the anomalous magnetic moments of the nucleons are enormous. The g-factor for the proton is 5.6, and the chargeless neutron should have no magnetic moment at all. Note, however, that the anomalous magnetic moments of the nucleons, that is, their magnetic moments with the expected Dirac particle magnetic moments subtracted, are roughly equal but of opposite sign: $\mu_p - 1.00\,\mu_N = +1.79\,\mu_N$, $\mu_n - 0.00\,\mu_N = -1.91\,\mu_N$.

The Yukawa interaction for nucleons was discovered in the mid-1930s, and this nuclear force is mediated by pion mesons. In parallel with the theory for the electron, the hypothesis was that higher-order loops involving nucleons and pions may generate the anomalous magnetic moments of the nucleons. The physical picture was that the *effective* magnetic moment of the neutron arose from the combined contributions of the "bare" neutron, which is zero, and the cloud of "virtual" pions and photons that surround this particle as a consequence of the nuclear and electromagnetic forces. The Feynman diagram is roughly the first order diagram, with the role of the virtual particles played by pions. As noted by Abraham Pais, "between late 1948 and the middle of 1949 at least six papers appeared reporting on second order calculations of nucleon moments." These theories were also, as noted by Pais, "a flop" – they gave results that grossly disagreed with observation. Nevertheless, serious efforts continued along these lines for the next couple of decades, to little success. These theoretical approaches were incorrect because the nucleons are composite particles with their magnetic moments arising from their elementary components, quarks.

Quark Model for Nucleon Magnetic Moments

In the quark model for hadrons, the neutron is composed of one up quark (charge +2/3 *e*) and two down quarks (charge −1/3 *e*). The magnetic moment of the neutron can be modeled as a sum of the magnetic moments of the constituent quarks, although this simple model belies the complexities of the Standard Model of particle physics. The calculation assumes that the quarks behave like pointlike Dirac particles, each having their own magnetic moment, as computed using an expression similar to the one above for the nuclear magneton:

$$\mu_q = \frac{e_q \hbar}{2m_q},$$

where the q-subscripted variables refer to quark magnetic moment, charge, or mass. Simplistically, the magnetic moment of the neutron can be viewed as resulting from the vector sum of the three quark magnetic moments, plus the orbital magnetic moments caused by the movement of the three charged quarks within the neutron.

In one of the early successes of the Standard Model (SU(6) theory), in 1964 Mirza A. B. Beg, Benjamin W. Lee, and Abraham Pais theoretically calculated the ratio of proton to neutron magnetic moments to be −3/2, which agrees with the experimental value to within 3%. The measured value for this ratio is −1.45989806(34). A contradiction of the quantum mechanical basis of this calculation with the Pauli exclusion principle led to the discovery of the color charge for quarks by Oscar W. Greenberg in 1964.

From the nonrelativistic, quantum mechanical wavefunction for baryons composed of three quarks, a straightforward calculation gives fairly accurate estimates for the magnetic moments of neutrons, protons, and other baryons. For a neutron, the magnetic moment is given by $\mu_n = 4/3\,\mu_d - 1/3\,\mu_u$, where μ_d and μ_u are the magnetic moments for the down and up quarks, respectively. This result combines the intrinsic magnetic moments of the quarks with their orbital magnetic moments, and assumes the three quarks are in a particular, dominant quantum state.

Baryon	Magnetic moment of quark model	Computed (μ_N)	Observed (μ_N)
p	$4/3\,\mu_u - 1/3\,\mu_d$	2.79	2.793
n	$4/3\,\mu_d - 1/3\,\mu_u$	−1.86	−1.913

The results of this calculation are encouraging, but the masses of the up or down quarks were assumed to be 1/3 the mass of a nucleon. The masses of the quarks are actually only about 1% that of a nucleon. The discrepancy stems from the complexity of the Standard Model for nucleons, where most of their mass originates in the gluon fields, virtual particles, and their associated energy that are essential aspects of the strong force. Furthermore, the complex system of quarks and gluons that constitute a neutron requires a relativistic treatment. Nucleon magnetic moments have been successfully computed from first principles, requiring significant computing resources.

Free Neutron Decay

A free neutron will decay with a half-life of about 10.3 minutes but it is stable if combined into a nucleus. This decay is an example of beta decay with the emission of an electron and an electron antineutrino. The decay of the neutron involves the weak interaction as indicated in the Feynman diagram below.

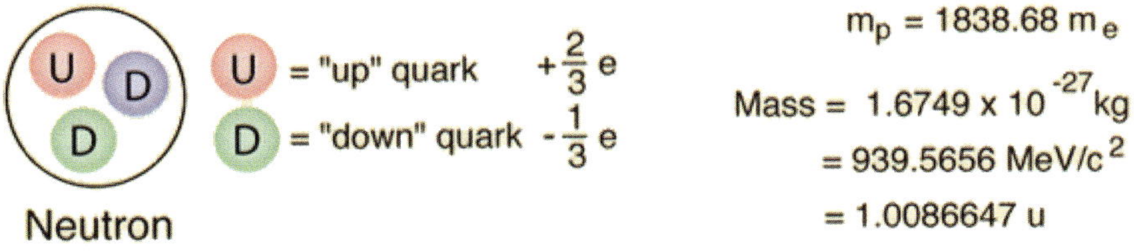

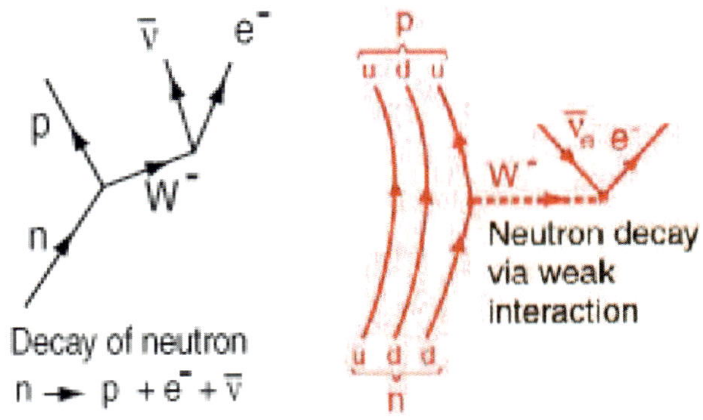

A more detailed diagram of the neutron's decay identifies it as the transformation of one of the neutron's down quarks into an up quark. It is an example of the kind of quark transformations that are involved in many nuclear processes, including beta decay.

The decay of the neutron is a good example of the observations which led to the discovery of the neutrino. An analysis of the energetics of the decay can be used to illustrate the dilemmas which faced early investigators of this process.

$$m_n = 939.5656 \ MeV$$
$$m_p = 938.2723 \ MeV$$
$$m_e = 0.510999 \ MeV$$
$$0.7823 \ MeV = Q \ \ for \ \ n \rightarrow p + e^-$$

Using the concept of binding energy, and representing the masses of the particles by their rest mass energies, the energy yield from neutron decay can be calculated from the particle masses. The energy yield is traditionally represented by the symbol Q. Because energy and momentum must be conserved in the decay, it will be shown that the lighter electron will carry away most of the kinetic energy. With a kinetic energy of this magnitude, the relativistic kinetic energy expression must be used.

For the moment we presume (incorrectly) that the decay involves just the proton and electron as products. The energy yield Q would then be divided between the proton and electron. The electron will get most of the kinetic energy and will be relativistic, but the proton is non-relativistic. The energy balance is then

$$Q = 0.7823 \ MeV = KE_e + \frac{1}{2} m_p v_p^2 = \frac{p_p^2 c^2}{2 m_p c^2}$$

In the rest frame of the neutron, conservation of momentum requires

$$pc_{electron} = - pc_{proton}$$

and $pc_{electron}$ can be expressed in terms of the electron kinetic energy

$$(pc_{electron})^2 = KE_e^2 + 2KE_e \cdot m_e c^2$$

The energy balance then becomes

$$0.7823\ MeV = KE_e + \frac{KE_e^2 + 2KE_e \cdot m_e c^2}{2m_p c^2}$$

s this value of Q, you see that the KE_e^2 term is negligible, so the required kinetic energy of the electron can be calculated.The required electron kinetic energy for this two-particle decay scheme is

$$0.7823\ MeV = KE\left[1 + \frac{m_e}{m_p}\right]; \quad KE_e = 0.7819\ MeV$$

Likewise, the momentum of the electron for this two particle decay is constrained to be

$$pc_{electron} = \sqrt{KE_e^2 + 2KE_e \cdot m_e c^2} = 1.188\ MeV$$

Momentum and energy for the two-particle decay are constrained to these values, but this is not the way nature behaves. The observed momentum and energy distributions for the electron are as shown below.

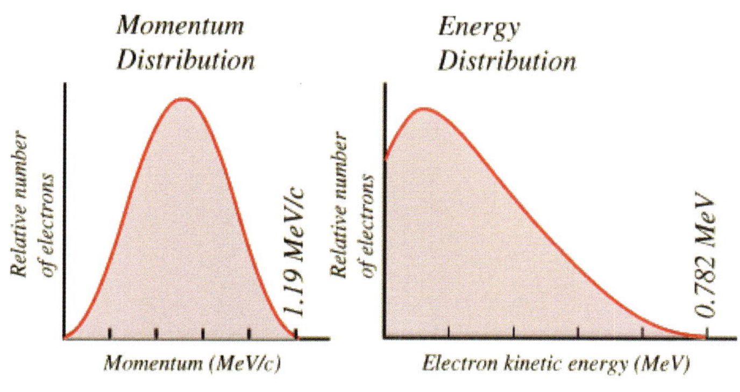

The fact that the electrons produced from the neutron decay had continuous distributions of energy and momentum was a clear indication that there was another particle emitted along with the electron and proton. It had to be a neutral particle and in certain decays carried almost all the energy and momentum of the decay. This would not have been so extraordinary except for the fact that when the electron had its maximum kinetic energy, it accounted for all the energy Q available for the decay. So there was no energy left over to account for the mass energy of the other emitted particle. The early experimenters were faced with the dilemma of a particle which could carry nearly all the energy and momentum of the decay but which had no charge and apparently no mass.

The mysterious particle was called a neutrino, but it was twenty five years before unambiguous experimental observation of the neutrino was made by Cowan and Reines. The present understanding of the decay of the neutron is

$$n \rightarrow p + e^- + \underset{\substack{electron \\ antineutrino}}{\overline{\nu}_e}$$

This decay illustrates some of the conservation laws which govern particle decays. The proton in the product satisfies the conservation of baryon number, but the emergence of the electron unaccompanied would violate conservation of lepton number. The third particle must be an electron antineutrino to allow the decay to satisfy lepton number conservation. The electron has lepton number 1, and the antineutrino has lepton number -1.

References

- Bolonkin, Alexander (2011). Universe, Human Immortality and Future Human Evaluation. Elsevier. p. 25. ISBN 9780124158016
- Heisenberg, W. (1927), "Über den anschaulichen Inhalt der quantentheoretischen Kinematik und Mechanik", Zeitschrift für Physik (in German), 43 (3–4): 172–198, Bibcode:1927ZPhy...43..172H, doi:10.1007/BF01397280
- "CODATA value: proton-electron mass ratio". 2006 CODATA recommended values. National Institute of Standards and Technology. Retrieved 2009-07-18
- Barrow, J.D. (1983). "Natural Units Before Planck". Quarterly Journal of the Royal Astronomical Society. 24: 24–26. Bibcode:1983QJRAS..24...24B
- Arabatzis, T. (2006). Representing Electrons: A Biographical Approach to Theoretical Entities. University of Chicago Press. pp. 70–74. ISBN 0-226-02421-0
- Eichten, E.J.; Peskin, M.E.; Peskin, M. (1983). "New Tests for Quark and Lepton Substructure". Physical Review Letters. 50 (11): 811–814. Bibcode:1983PhRvL..50..811E. doi:10.1103/PhysRevLett.50.811
- Main, P. (June 12, 1993). "When electrons go with the flow: Remove the obstacles that create electrical resistance, and you get ballistic electrons and a quantum surprise". New Scientist. 1887: 30. Retrieved 2008-10-09
- Gabrielse, G. (2006). "Antiproton mass measurements". International Journal of Mass Spectrometry. 251 (2–3): 273–280. Bibcode:2006IJMSp.251..273G. doi:10.1016/j.ijms.2006.02.013
- S. Hofmann (1996). "Chapter 3: Proton radioactivity". In Dorin N. Poseru. Nuclear Decay Modes. Bristol: Institute of Physics Publishing. pp. 143–203. ISBN 0-7503-0338-7
- Greenberg, O. W. (2009), "Color charge degree of freedom in particle physics", Compendium of Quantum Physics, Springer Berlin Heidelberg, pp. 109–111, arXiv:0805.0289, doi:10.1007/978-3-540-70626-7_32
- Fernandez-Alonso, Felix; Price, David (2013). Neutron Scattering Fundamentals. Amsterdam: Academic Press. p. 103. ISBN 978-0-12-398374-9. Retrieved June 30, 2016

3

Quarks and Hadrons

Quarks are the fundamental constituents of matter. These have intrinsic properties of electric charge, mass, spin and color charge. They combine to form hadrons, which are composite particles. These hadrons are either baryons, such as protons and neutrons, or mesons, such as pions. This chapter closely examines such key concepts of quarks and hadrons to provide an extensive understanding of the subject.

Quark

A quark is one of the fundamental particles in physics. They join to form hadrons, such as protons and neutrons, which are components of the nuclei of atoms. The study of quarks and the interactions between them through the strong force is called particle physics.

The antiparticle of a quark is the antiquark. Quarks and antiquarks are the only two fundamental particles that interact through all four fundamental forces of physics: gravitation, electromagnetism, and the strong and weak interactions.

Quarks and Confinement

A quark exhibits confinement, which means that the quarks are not observed independently but always in combination with other quarks. This makes determining the properties (mass, spin, and parity) impossible to measure directly; these traits must be inferred from the particles composed of them.

These measurements indicate a non-integer spin (either +1/2 or -1/2), so quarks are fermions and follow the Pauli Exclusion Principle.

In the strong interaction between quarks, they exchange gluons, which are massless vector gauge bosons that carry a pair of color and anticolor charges. When exchanging gluons, the color of the quarks change. This color force is weakest when the quarks are close together and becomes stronger as they move apart.

Quarks are so strongly bound by the color force that if there is enough energy to separate them, a quark-antiquark pair is produced and binds with any free quark to produce a hadron. As a result, free quarks are never seen alone.

Classification

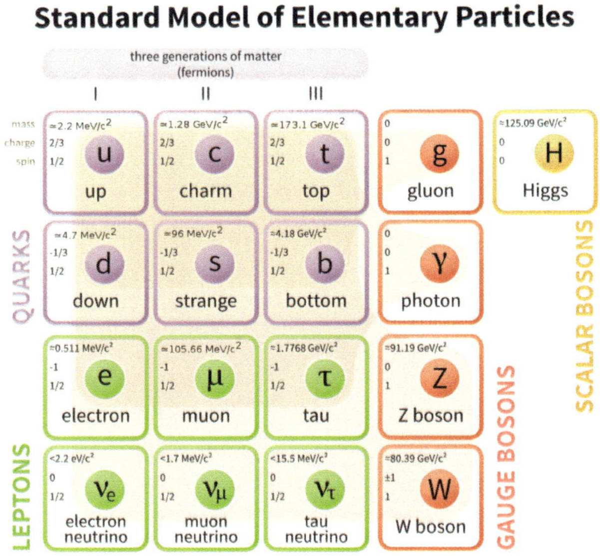

Six of the particles in the Standard Model are quarks (shown in purple).
Each of the first three columns forms a generation of matter.

The Standard Model is the theoretical framework describing all the currently known elementary particles. This model contains six flavors of quarks (q), named up (u), down (d), charm (c), strange (s), top (t), and bottom (b). Antiparticles of quarks are called antiquarks, and are denoted by a bar over the symbol for the corresponding quark, such as u for an up antiquark. As with antimatter in general, antiquarks have the same mass, mean lifetime, and spin as their respective quarks, but the electric charge and other charges have the opposite sign.

Quarks are spin-½ particles, implying that they are fermions according to the spin–statistics theorem. They are subject to the Pauli exclusion principle, which states that no two identical fermions can simultaneously occupy the same quantum state. This is in contrast to bosons (particles with integer spin), any number of which can be in the same state. Unlike leptons, quarks possess color charge, which causes them to engage in the strong interaction. The resulting attraction between different quarks causes the formation of composite particles known as hadrons.

The quarks that determine the quantum numbers of hadrons are called valence quarks; apart from these, any hadron may contain an indefinite number of virtual (or sea) quarks, antiquarks, and gluons, which do not influence its quantum numbers. There are two families of hadrons: baryons, with three valence quarks, and mesons, with a valence quark and an antiquark. The most common baryons are the proton and the neutron, the building blocks of the atomic nucleus. A great number of hadrons are known, most of them differentiated by their quark content and the properties these constituent quarks confer. The existence of "exotic" hadrons with more valence quarks, such as tetraquarks ($qq\bar{q}\bar{q}$) and pentaquarks ($qqqq\bar{q}$), were conjectured from the beginnings of the quark model but not discovered until the early 21st century.

Elementary fermions are grouped into three generations, each comprising two leptons and two quarks. The first generation includes up and down quarks, the second strange and charm quarks, and the third bottom and top quarks. All searches for a fourth generation of quarks and other elementary fermions have failed, and there is strong indirect evidence that no more than three generations exist. Particles in higher generations generally have greater mass and less stability, causing them to decay into lower-generation particles by means of weak interactions. Only first-generation (up and down) quarks occur commonly in nature. Heavier quarks can only be created in high-energy collisions (such as in those involving cosmic rays), and decay quickly; however, they are thought to have been present during the first fractions of a second after the Big Bang, when the universe was in an extremely hot and dense phase (the quark epoch). Studies of heavier quarks are conducted in artificially created conditions, such as in particle accelerators.

Having electric charge, mass, color charge, and flavor, quarks are the only known elementary particles that engage in all four fundamental interactions of contemporary physics: electromagnetism, gravitation, strong interaction, and weak interaction. Gravitation is too weak to be relevant to individual particle interactions except at extremes of energy (Planck energy) and distance scales (Planck distance). However, since no successful quantum theory of gravity exists, gravitation is not described by the Standard Model.

Flavors of Quarks

There are six flavors of quarks: up, down, strange, charm, bottom, and top. The flavor of the quark determines its properties.

Quarks with a charge of +(2/3)e are called up-type quarks and those with a charge of -(1/3)e are called down-type.

There are three generations of quarks, based on pairs of weak positive/negative, weak isospin. The first generation quarks are up and down quarks, the second generation quarks are strange, and charm quarks, the third generation quarks are top and bottom quarks.

All quarks have a baryon number (B = 1/3) and a lepton number (L = 0). The flavor determines certain other unique properties, described in individual descriptions.

The up and down quarks make up protons and neutrons, seen in the nucleus of ordinary matter. They are the lightest and most stable. The heavier quarks are produced in high-energy collisions and rapidly decay into up and down quarks. A proton is composed of two up quarks and a down quark. A neutron is composed of one up quark and two down quarks.

First-generation Quarks

Up quark (symbol *u*)

- Weak Isospin: +1/2
- Isospin (I_z): +1/2
- Charge (proportion of *e*): +2/3
- Mass (in MeV/c^2): 1.5 to 4.0

Down quark (symbol *d*)

- Weak Isospin: -1/2
- Isospin (I_z): -1/2
- Charge (proportion of *e*): -1/3
- Mass (in MeV/c^2): 4 to 8

Second-generation Quarks

Charm quark (symbol *c*)

- Weak Isospin: +1/2
- Charm (*C*): 1
- Charge (proportion of *e*): +2/3
- Mass (in MeV/c^2): 1150 to 1350

Strange quark (symbol *s*)

- Weak Isospin: -1/2
- Strangeness (*S*): -1
- Charge (proportion of *e*): -1/3
- Mass (in MeV/c^2): 80 to 130

Third-generation Quarks

Top quark (symbol *t*)

- Weak Isospin: +1/2
- Topness (*T*): 1
- Charge (proportion of *e*): +2/3
- Mass (in MeV/c^2): 170200 to 174800

Bottom quark (symbol *b*)

- Weak Isospin: -1/2
- Bottomness (*B'*): 1
- Charge (proportion of *e*): -1/3
- Mass (in MeV/c²): 4100 to 4400

Properties

Electric Charge

Quarks have fractional electric charge values – either (−⅓) or (+⅔) times the elementary charge (e), depending on flavor. Up, charm, and top quarks (collectively referred to as up-type quarks) have a charge of +⅔ e, while down, strange, and bottom quarks (down-type quarks) have −⅓ e. Antiquarks have the opposite charge to their corresponding quarks; up-type antiquarks have charges of −⅔ e and down-type antiquarks have charges of +⅓ e. Since the electric charge of a hadron is the sum of the charges of the constituent quarks, all hadrons have integer charges: the combination of three quarks (baryons), three antiquarks (antibaryons), or a quark and an antiquark (mesons) always results in integer charges. For example, the hadron constituents of atomic nuclei, neutrons and protons, have charges of 0 e and +1 e respectively; the neutron is composed of two down quarks and one up quark, and the proton of two up quarks and one down quark.

Spin

Spin is an intrinsic property of elementary particles, and its direction is an important degree of freedom. It is sometimes visualized as the rotation of an object around its own axis (hence the name "spin"), though this notion is somewhat misguided at subatomic scales because elementary particles are believed to be point-like.

Spin can be represented by a vector whose length is measured in units of the reduced Planck constant ħ (pronounced "h bar"). For quarks, a measurement of the spin vector component along any axis can only yield the values +ħ/2 or −ħ/2; for this reason quarks are classified as spin-½ particles. The component of spin along a given axis – by convention the z axis – is often denoted by an up arrow ↑ for the value +½ and down arrow ↓ for the value −½, placed after the symbol for flavor. For example, an up quark with a spin of +½ along the z axis is denoted by u↑.

Weak Interaction

A tree diagram consisting mostly of straight arrows. A down quark forks into an up quark and a wavy-arrow W^- boson, the latter forking into an electron and reversed-arrow electron antineutrino.

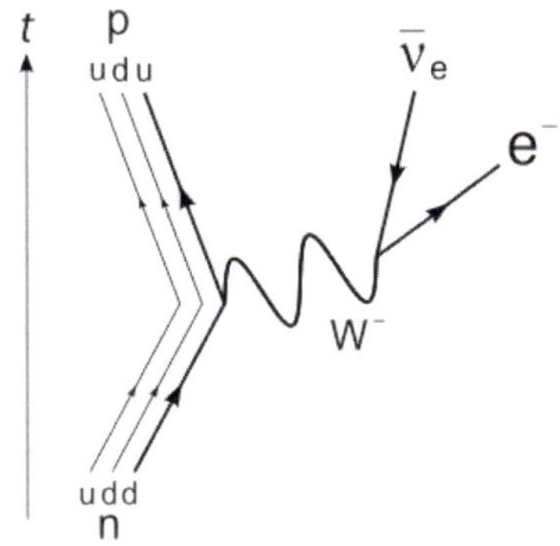

Feynman diagram of beta decay with time flowing upwards. The CKM matrix encodes the probability of this and other quark decays.

A quark of one flavor can transform into a quark of another flavor only through the weak interaction, one of the four fundamental interactions in particle physics. By absorbing or emitting a W boson, any up-type quark (up, charm, and top quarks) can change into any down-type quark (down, strange, and bottom quarks) and vice versa. This flavor transformation mechanism causes the radioactive process of beta decay, in which a neutron (n) "splits" into a proton (p), an electron (e−) and an electron antineutrino ($\bar{\nu}_e$). This occurs when one of the down quarks in the neutron (udd) decays into an up quark by emitting a virtual W^- boson, transforming the neutron into a proton (uud). The W− boson then decays into an electron and an electron antineutrino.

$$n \rightarrow p + e^- + \bar{\nu}_e \text{ (Beta decay, hadron notation)}$$

$$udd \rightarrow uud + e^- + \bar{\nu}_e \text{ (Beta decay, quark notation)}$$

Both beta decay and the inverse process of inverse beta decay are routinely used in medical applications such as positron emission tomography (PET) and in experiments involving neutrino detection.

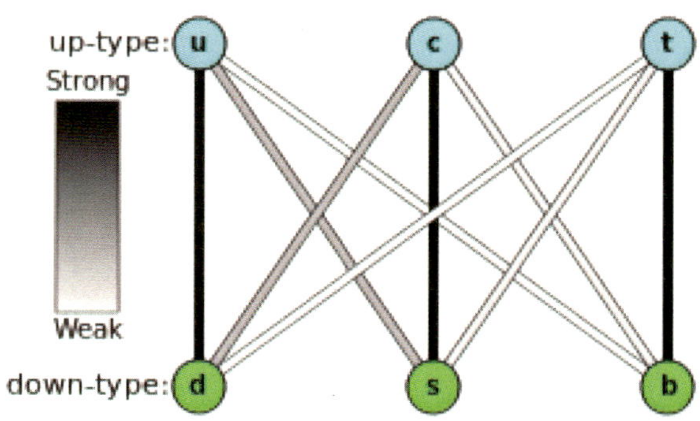

The strengths of the weak interactions between the six quarks. The "intensities" of the lines are determined by the elements of the CKM matrix.

While the process of flavor transformation is the same for all quarks, each quark has a preference to transform into the quark of its own generation. The relative tendencies of all flavor transformations are described by a mathematical table, called the Cabibbo–Kobayashi–Maskawa matrix (CKM matrix). Enforcing unitarity, the approximate magnitudes of the entries of the CKM matrix are:

$$\begin{bmatrix} |V_{ud}| & |V_{us}| & |V_{ub}| \\ |V_{cd}| & |V_{cs}| & |V_{cb}| \\ |V_{td}| & |V_{ts}| & |V_{tb}| \end{bmatrix} \approx \begin{bmatrix} 0.974 & 0.225 & 0.003 \\ 0.225 & 0.973 & 0.041 \\ 0.009 & 0.040 & 0.999 \end{bmatrix},$$

where V_{ij} represents the tendency of a quark of flavor i to change into a quark of flavor j (or vice versa).

There exists an equivalent weak interaction matrix for leptons, called the Pontecorvo–Maki–Nakagawa–Sakata matrix (PMNS matrix). Together, the CKM and PMNS matrices describe all flavor transformations, but the links between the two are not yet clear.

Strong Interaction and Color Charge

A green and a magenta ("antigreen") arrow canceling out each other out white, representing a meson; a red, a green, and a blue arrow canceling out to white, representing a baryon; a yellow ("antiblue"), a magenta, and a cyan ("antired") arrow canceling out to white, representing an antibaryon.

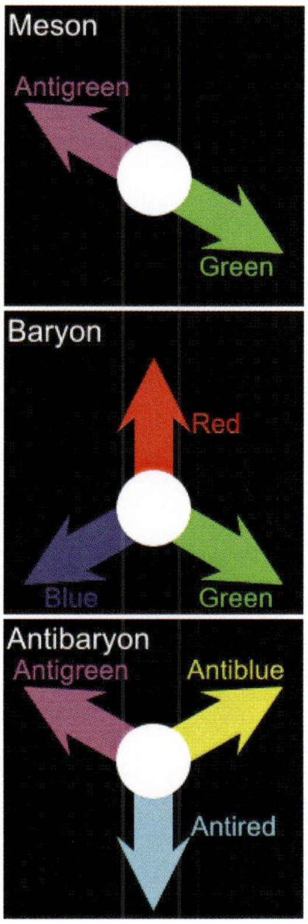

All types of hadrons have zero total color charge.

According to quantum chromodynamics (QCD), quarks possess a property called color charge. There are three types of color charge, arbitrarily labeled blue, green, and red. Each of them is complemented by an anticolor – antiblue, antigreen, and antired. Every quark carries a color, while every antiquark carries an anticolor.

The system of attraction and repulsion between quarks charged with different combinations of the three colors is called strong interaction, which is mediated by force carrying particles known as gluons; this is discussed at length below. The theory that describes strong interactions is called quantum chromodynamics (QCD). A quark, which will have a single color value, can form a bound system with an antiquark carrying the corresponding anticolor. The result of two attracting quarks will be color neutrality: a quark with color charge ξ plus an antiquark with color charge $-\xi$ will result in a color charge of 0 (or "white" color) and the formation of a meson. This is analogous to the additive color model in basic optics. Similarly, the combination of three quarks, each with different color charges, or three antiquarks, each with anticolor charges, will result in the same "white" color charge and the formation of a baryon or antibaryon.

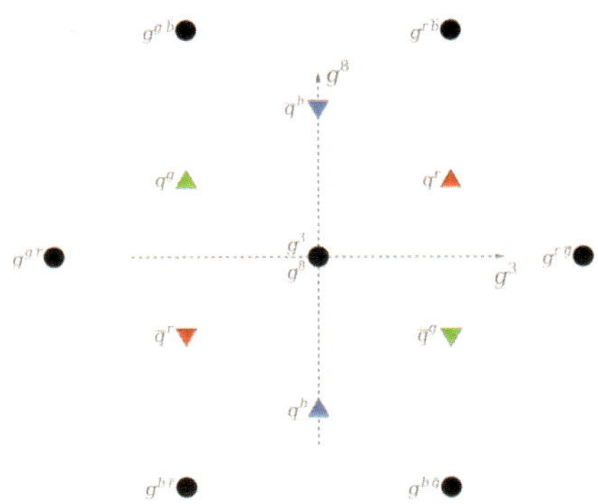

The pattern of strong charges for the three colors of quark, three antiquarks, and eight gluons (with two of zero charge overlapping).

In modern particle physics, gauge symmetries – a kind of symmetry group – relate interactions between particles. Color SU(3) (commonly abbreviated to $SU(3)_c$) is the gauge symmetry that relates the color charge in quarks and is the defining symmetry for quantum chromodynamics. Just as the laws of physics are independent of which directions in space are designated x, y, and z, and remain unchanged if the coordinate axes are rotated to a new orientation, the physics of quantum chromodynamics is independent of which directions in three-dimensional color space are identified as blue, red, and green. $SU(3)_c$ color transformations correspond to "rotations" in color space (which, mathematically speaking, is a complex space). Every quark flavor f, each with subtypes f_B, f_G, f_R corresponding to the quark colors, forms a triplet: a three-component quantum field which transforms under the fundamental representation of $SU(3)_c$. The requirement that $SU(3)_c$ should be local – that is, that its transformations be allowed to vary with space and time – determines the properties of the strong interaction. In particular, it implies the existence of eight gluon types to act as its force carriers.

Mass

Two terms are used in referring to a quark's mass: current quark mass refers to the mass of a quark by itself, while constituent quark mass refers to the current quark mass plus the mass of the gluon particle field surrounding the quark. These masses typically have very different values. Most of a hadron's mass comes from the gluons that bind the constituent quarks together, rather than from the quarks themselves. While gluons are inherently massless, they possess energy – more specifically, quantum chromodynamics binding energy (QCBE) – and it is this that contributes so greatly to the overall mass of the hadron. For example, a proton has a mass of approximately 938 MeV/c^2, of which the rest mass of its three valence quarks only contributes about 9 MeV/c^2; much of the remainder can be attributed to the field energy of the gluons.

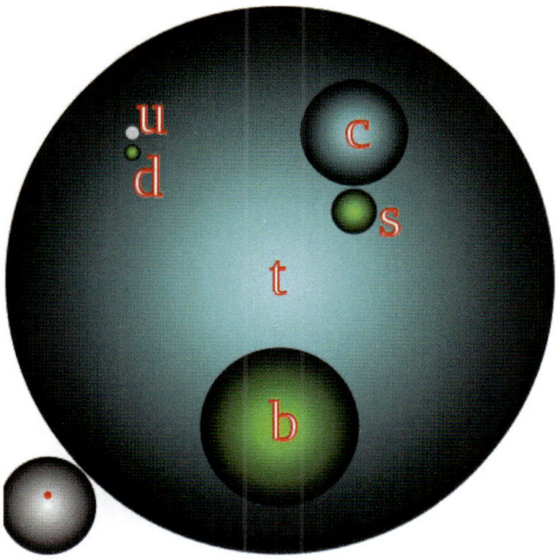

Current quark masses for all six flavors in comparison, as balls of proportional volumes. Proton (gray) and electron (red) are shown in bottom left corner for scale

The Standard Model posits that elementary particles derive their masses from the Higgs mechanism, which is associated to the Higgs boson. It is hoped that further research into the reasons for the top quark's large mass of ~173 GeV/c^2, almost the mass of a gold atom, might reveal more about the origin of the mass of quarks and other elementary particles.

Table of Properties

The following table summarizes the key properties of the six quarks. Flavor quantum numbers (isospin (I_3), charm (C), strangeness, topness (T), and bottomness (B′)) are assigned to certain quark flavors, and denote qualities of quark-based systems and hadrons. The baryon number (B) is +1⁄3 for all quarks, as baryons are made of three quarks. For antiquarks, the electric charge (Q) and all flavor quantum numbers (B, I_3, C, S, T, and B′) are of opposite sign. Mass and total angular momentum (J; equal to spin for point particles) do not change sign for the antiquarks.

Quark Flavor Properties

Name	Symbol	Mass (MeV/c²)*	J	B	Q (e)	I_3	C	S	T	B'	Antiparticle	Antiparticle symbol
First generation												
Up	u	2.3±0.7 ± 0.5	$^1/_2$	$+^1/_3$	$+^2/_3$	$+^1/_2$	0	0	0	0	Antiup	$\bar{u}$
Down	d	4.8±0.5 ± 0.3	$^1/_2$	$+^1/_3$	$-^1/_3$	$-^1/_2$	0	0	0	0	Antidown	$\bar{d}$
Second generation												
Charm	C	1275±25	½	$+^1/_3$	$+^2/_3$	0	+1	0	0	0	Antich-arm	$\bar{c}$
Strange	S	95±5	½	$+^1/_3$	$-^1/_3$	0	0	−1	0	0	An-tistrange	$\bar{s}$
Third generation												
Top	t	173210±510 ± 710	½	$+^1/_3$	$+^2/_3$	0	0	0	+1	0	Antitop	$\bar{t}$
Bottom	b	4180±30	½	$+^1/_3$	$-^1/_3$	0	0	0	0	−1	Antibot-tom	$\bar{b}$

J = total angular momentum, B = baryon number, Q = electric charge, I_3 = isospin, C = charm, S = strangeness, T = topness, B′ = bottomness.

* Notation such as 173210±510 ± 710 denotes two types of measurement uncertainty. In the case of the top quark, the first uncertainty is statistical in nature, and the second is systematic.

Interacting Quarks

As described by quantum chromodynamics, the strong interaction between quarks is mediated by gluons, massless vector gauge bosons. Each gluon carries one color charge and one anticolor charge. In the standard framework of particle interactions (part of a more general formulation known as perturbation theory), gluons are constantly exchanged between quarks through a virtual emission and absorption process. When a gluon is transferred between quarks, a color change occurs in both; for example, if a red quark emits a red–antigreen gluon, it becomes green, and if a green quark absorbs a red–antigreen gluon, it becomes red. Therefore, while each quark's color constantly changes, their strong interaction is preserved.

Since gluons carry color charge, they themselves are able to emit and absorb other gluons. This causes asymptotic freedom: as quarks come closer to each other, the chromodynamic binding force between them weakens. Conversely, as the distance between quarks increases, the binding force strengthens. The color field becomes stressed, much as an elastic band is stressed when stretched, and more gluons of appropriate color are spontaneously created to strengthen the field. Above a certain energy threshold, pairs of quarks and antiquarks are created. These pairs bind with the quarks being separated, causing new hadrons to form. This phenomenon is known as color confinement: quarks never appear in isolation. This process of hadronization occurs before quarks, formed in a high energy collision, are able to interact in any other way. The only exception is the top quark, which may decay before it hadronizes.

Sea Quarks

Hadrons contain, along with the valence quarks (q_v) that contribute to their quantum numbers, virtual quark–antiquark ($q\bar{q}$) pairs known as sea quarks (q_s). Sea quarks form when a gluon of the hadron's color field splits; this process also works in reverse in that the annihilation of two sea quarks produces a gluon. The result is a constant flux of gluon splits and creations colloquially known as "the sea". Sea quarks are much less stable than their valence counterparts, and they typically annihilate each other within the interior of the hadron. Despite this, sea quarks can hadronize into baryonic or mesonic particles under certain circumstances.

Other Phases of Quark Matter

Quark–gluon plasma exists at very high temperatures; the hadronic phase exists at lower temperatures and baryonic densities, in particular nuclear matter for relatively low temperatures and intermediate densities; color superconductivity exists at sufficiently low temperatures and high densities.

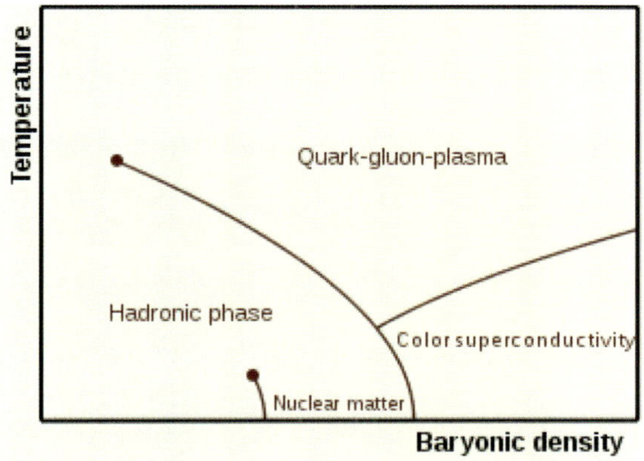

A qualitative rendering of the phase diagram of quark matter. The precise details of the diagram are the subject of ongoing research.

Under sufficiently extreme conditions, quarks may become deconfined and exist as free particles. In the course of asymptotic freedom, the strong interaction becomes weaker at higher temperatures. Eventually, color confinement would be lost and an extremely hot plasma of freely moving quarks and gluons would be formed. This theoretical phase of matter is called quark–gluon plasma. The exact conditions needed to give rise to this state are unknown and have been the subject of a great deal of speculation and experimentation. A recent estimate puts the needed temperature at $(1.90\pm0.02)\times10^{12}$ kelvin.

While a state of entirely free quarks and gluons has never been achieved (despite numerous attempts by CERN in the 1980s and 1990s), recent experiments at the Relativistic Heavy Ion Collider have yielded evidence for liquid-like quark matter exhibiting "nearly perfect" fluid motion.

The quark–gluon plasma would be characterized by a great increase in the number of heavier quark pairs in relation to the number of up and down quark pairs. It is believed that in the period prior to 10–6 seconds after the Big Bang (the quark epoch), the universe was filled with quark–gluon plasma, as the temperature was too high for hadrons to be stable.

Given sufficiently high baryon densities and relatively low temperatures – possibly comparable to those found in neutron stars – quark matter is expected to degenerate into a Fermi liquid of weakly interacting quarks. This liquid would be characterized by a condensation of colored quark Cooper pairs, thereby breaking the local $SU(3)_c$ symmetry. Because quark Cooper pairs harbor color charge, such a phase of quark matter would be color superconductive; that is, color charge would be able to pass through it with no resistance.

Top Quark

As its name states, the top quark is a quark. Quarks are one of the fundamental building blocks of which the entire universe is constructed. Scientists know of six different quarks, although only two of them (the up and down quarks) are found inside the protons and neutrons at the center of atoms. The other four quarks (strange, charm, bottom and top) are all unstable and decay in fractions of a second.

The top quark is the heaviest subatomic particle ever observed, with a mass that is about as heavy as an entire atom of gold. Top quarks are also among the most fleeting of particles, with a lifetime of about a trillionth of a trillionth of a second.

Top quarks were first observed at Fermilab in 1995 and the two experiments at Fermilab's Tevatron collider each observed about 150,000 top quarks over their lifetimes. Top quarks are now produced in even greater quantities at the Large Hadron Collider (LHC). When the LHC is running at design luminosity, it will produce a pair of top quarks every second. As of June 2012, over two million top quarks have been created in each of the big experiments at the LHC, and the rate is rapidly increasing.

The original studies concentrated on collisions in which a top quark and an antimatter top quark were simultaneously produced. Experiments at Fermilab were also the first to observe top quarks produced one at a time. This was a tricky thing to accomplish because ordinary processes can produce top quark lookalikes. Making two lookalikes at the same time is rare, which makes it easier to separate real top quark events from fakes. To find singly made top quarks required a detailed understanding of the detector.

Though 17 years have passed since the top quark was discovered, it remains a topic of intense research interest. Perhaps the most pressing mystery in contemporary particle physics is the origin of mass. While the mechanism that gives particles their mass is not yet known, a very popular theory is that the Higgs boson is responsible. Given that the top quark is the heaviest particle yet discovered, it is a natural laboratory in which to investigate how the Higgs boson interacts with matter. Further, if the Higgs boson is real, the theory makes very precise predictions on how the mass of the Higgs boson and the top quark are related. In fact, one of the ways we will be certain that an observation of a Higgs boson is the real Higgs boson (as opposed to a competitor theory) will be to measure the mass of the top quark with sufficient precision to see if the prediction is correct.

After years of intense studies, the top quark has revealed a mystery or two of its own. At Fermilab's Tevatron, measurements by both of its collider experiments showed that the top quark prefers to be created in the same direction the proton beam is travelling. This measurement disagrees with predictions and could indicate new physical phenomena or it could just mean that the regular theory needs to be tweaked.

In addition, top quarks, produced singly or in pairs, have rich panoply of decay products with signatures that can mimic collisions of even rarer phenomena. Thus it is imperative to understand in great detail how top quarks are produced.

The top quark is an intriguing particle, massive and fleeting, and one that continues to provide a broad range of interesting research topics.

Properties

- At the final Tevatron energy of 1.96 TeV, top–antitop pairs were produced with a cross section of about 7 picobarns (pb). The Standard Model prediction (at next-to-leading order with $m_t = 175$ GeV/c^2) is 6.7–7.5 pb.
- The W bosons from top quark decays carry polarization from the parent particle, hence pose themselves as a unique probe to top polarization.
- In the Standard Model, the top quark is predicted to have a spin quantum number of $\frac{1}{2}$ and electric charge $+\frac{2}{3}$. A first measurement of the top quark charge has been published, resulting in approximately 90% confidence limit that the top quark charge is indeed $+\frac{2}{3}$.

Production

Because top quarks are very massive, large amounts of energy are needed to create one. The only way to achieve such high energies is through high energy collisions. These occur naturally in the Earth's upper atmosphere as cosmic rayscollide with particles in the air, or can be created in a particle accelerator. In 2011, after the Tevatron ceased operations, the Large Hadron Collider at CERN became the only accelerator that generates a beam of sufficient energy to produce top quarks, with a center-of-mass energy of 7 TeV. There are multiple processes that can lead to the production of top quarks, but they can be conceptually divided in two categories.

Top-quark Pairs

The most common is production of a top–antitop pair via strong interactions. In a collision, a highly energetic gluon is created, which subsequently decays into a top and antitop. This process was responsible for the majority of the top events at Tevatron and was the process observed when the top was first discovered in 1995. It is also possible to produce pairs of top–antitop through the decay of an intermediate photon or Z-boson. However, these processes are predicted to be much rarer and have a virtually identical experimental signature in a hadron collider like Tevatron.

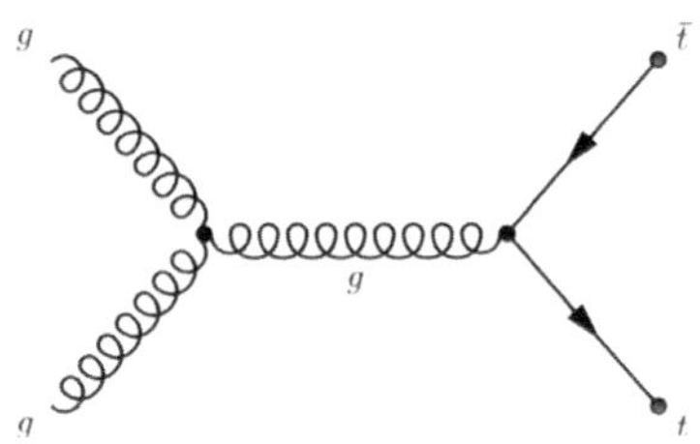

gluon-gluon fusion

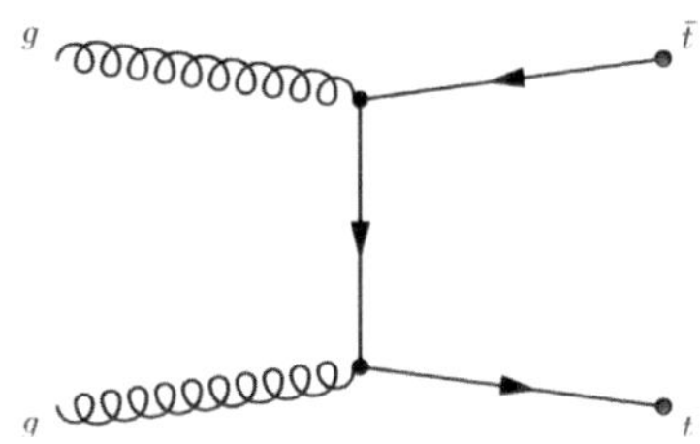

t-channel

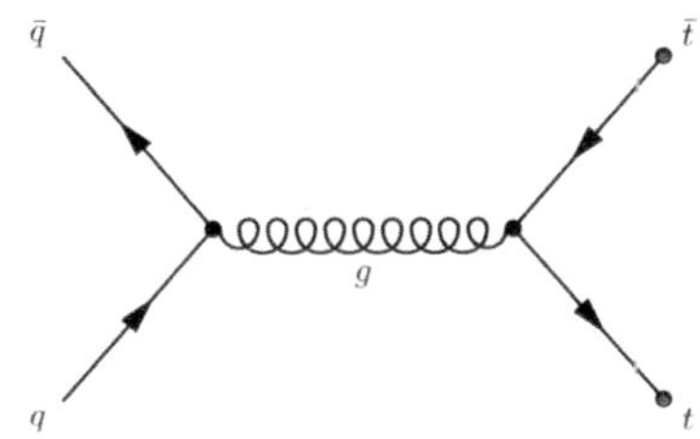

quark-antiquark annihilation

Single Top Quarks

A distinctly different process is the production of single top quarks via weak interaction. This can happen in several ways (called channels): either an intermediate W-boson decays into a top and antibottom quark ("s-channel") or a bottom quark (probably created in a pair through the decay of a gluon) transforms to a top quark by exchanging a W-boson with an up or down quark ("t-channel"). A single top quark can also be produced in association with a W boson, requiring an initial state bottom quark ("tW-channel"). The first evidence for these processes was published by the DØ collaboration in December 2006, and in March 2009 the CDF and DØ collaborations released twin papers with the definitive observation of these processes. The main significance of measuring these production processes is that their frequency is directly proportional to the $|V_{tb}|^2$ component of the CKM matrix.

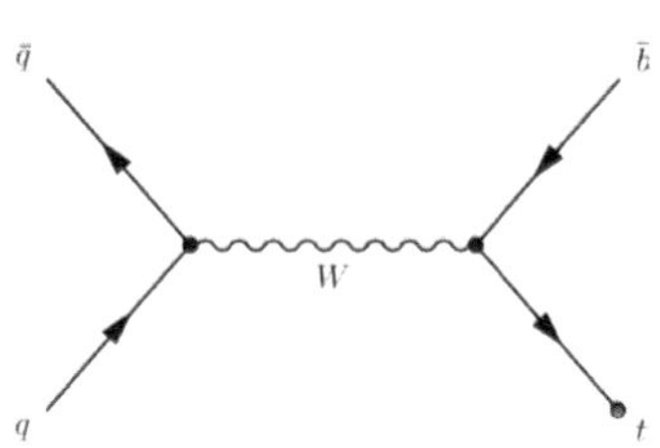

s-channel

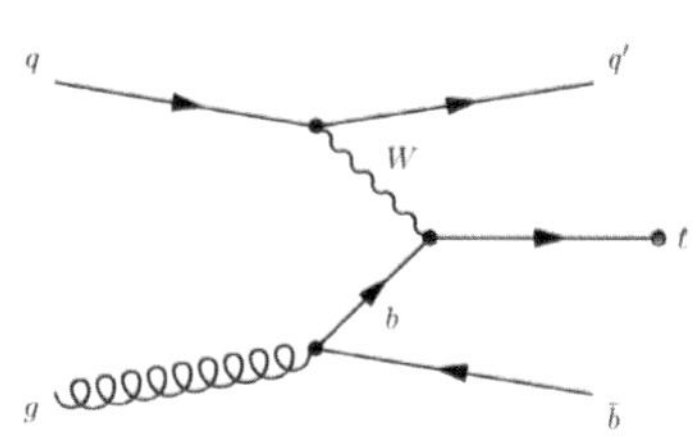

t-channel

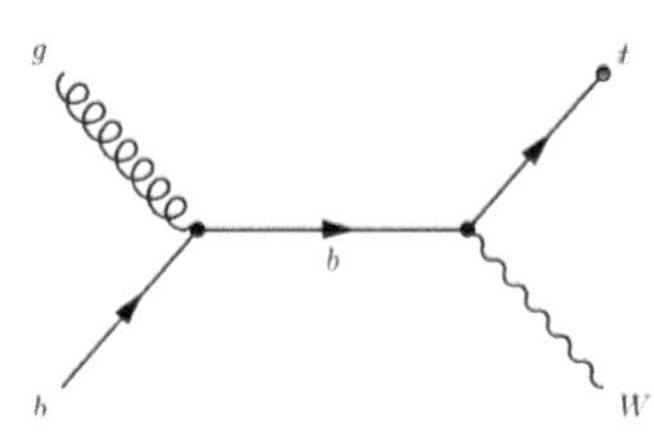

tW channel

Decay

Because of its enormous mass, the top quark is extremely short-lived with a predicted lifetime of only 5×10^{-25} s. As a result, top quarks do not have time before they decay to form hadrons as other quarks do, which provides physicists with the unique opportunity to study the behavior of a "bare" quark. The only known way the top quark can decay is through the weak interaction producing a W-boson and a down-type quark (down, strange, or bottom).

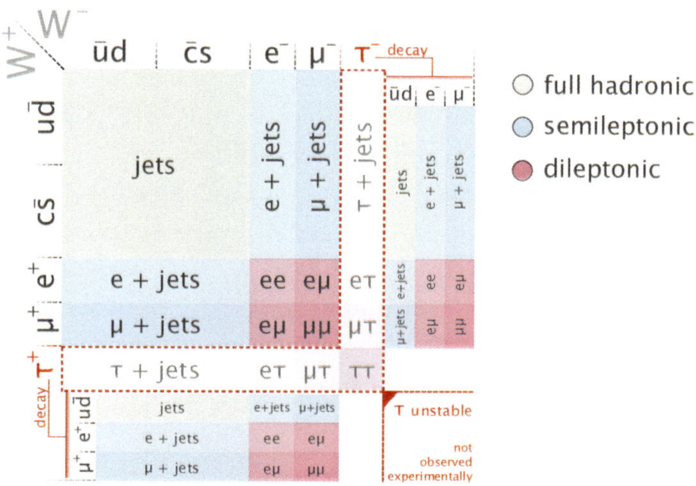

All possible final states of the decay of a top-quark pair

In particular, it is possible to directly determine the branching ratio $\Gamma(W^+b)/\Gamma(W^+q\ (q = b,s,d))$. The best current determination of this ratio is 0.91±0.04. Since this ratio is equal to $|V_{tb}|^2$ according to the Standard Model, this gives another way of determining the CKM element $|V_{tb}|$, or in combination with the determination of $|V_{tb}|$ from single top production provides tests for the assumption that the CKM matrix is unitary.

The Standard Model also allows more exotic decays, but only at one loop level, meaning that they are extremely suppressed. In particular, it is possible for a top quark to decay into another up-type quark (an up or a charm) by emitting a photon or a Z-boson. Searches for these exotic decay modes have provided no evidence for their existence in accordance with expectations from the Standard Model. The branching ratios for these decays have been determined to be less than 5.9 in 1,000 for photonic decay and less than 2.1 in 1,000 for Z-boson decay at 95% confidence.

Mass and Coupling to the Higgs Boson

The Standard Model describes fermion masses through the Higgs mechanism. The Higgs boson has a Yukawa coupling to the left- and right-handed top quarks. After electroweak symmetry breaking (when the Higgs acquires a vacuum expectation value), the left- and right-handed components mix, becoming a mass term.

$$\mathcal{L} = y_t hqu^c \rightarrow \frac{y_t v}{\sqrt{2}}(1 + h^0 / v)uu^c$$

The top quark Yukawa coupling has a value of

$$y_t = \sqrt{2}m_t / v \simeq 1$$

where v = 246 GeV is the value of the Higgs vacuum expectation value.

Yukawa Couplings

In the Standard Model, all of the quark and lepton Yukawa couplings are small compared to the top quark Yukawa coupling. Understanding this hierarchy in the fermion masses is an open problem in theoretical physics. Yukawa couplings are not constants and their values change depending on the energy scale (distance scale) at which they are measured. The dynamics of Yukawa couplings are determined by the renormalization group equation.

One of the prevailing views in particle physics is that the size of the top quark Yukawa coupling is determined by the renormalization group, leading to the "quasi-infrared fixed point."

The Yukawa couplings of the up, down, charm, strange and bottom quarks, are hypothesized to have small values at the extremely high energy scale of grand unification, 10^{15} GeV. They increase in value at lower energy scales, at which the quark masses are generated by the Higgs. The slight growth is due to corrections from the QCD coupling. The corrections from the Yukawa couplings are negligible for the lower mass quarks.

If, however, a quark Yukawa coupling has a large value at very high energies, its Yukawa corrections will evolve and cancel against the QCD corrections. This is known as a (quasi-) infrared fixed point. No matter what the initial starting value of the coupling is, if it is sufficiently large it will reach this fixed point value. The corresponding quark mass is then predicted.

The top quark Yukawa coupling lies very near the infrared fixed point of the Standard Model. The renormalization group equation is:

$$\mu \frac{\partial}{\partial \mu} y_t \approx \frac{y_t}{16\pi^2}\left(\frac{9}{2}y_t^2 - 8g_3^2 - \frac{9}{4}g_2^2 - \frac{17}{20}g_1^2\right),$$

where g_3 is the color gauge coupling, g_2 is the weak isospin gauge coupling, and g_1 is the weak hypercharge gauge coupling. This equation describes how the Yukawa coupling changes with energy scale μ. Solutions to this equation for large initial values y_t cause the right-hand side of the equation to quickly approach zero, locking y_t to the QCD coupling g_3. The value of the fixed point is fairly precisely determined in the Standard Model, leading to a top quark mass of 230 GeV. However, if there is more than one Higgs doublet, the mass value will be reduced by Higgs mixing angle effects in an unpredicted way.

In the minimal supersymmetric extension of the Standard Model (MSSM), there are two Higgs doublets and the renormalization group equation for the top quark Yukawa coupling is slightly modified:

$$\mu \frac{\partial}{\partial \mu} y_{\mathrm{t}} \approx \frac{y_{\mathrm{t}}}{16\pi^2}\left(6y_{\mathrm{t}}^2 + y_{\mathrm{b}}^2 - \frac{16}{3}g_3^2 - 3g_2^2 - \frac{13}{15}g_1^2\right),$$

where y_{b} is the bottom quark Yukawa coupling. This leads to a fixed point where the top mass is smaller, 170–200 GeV. The uncertainty in this prediction arises because the bottom quark Yukawa coupling can be amplified in the MSSM. Some theorists believe this is supporting evidence for the MSSM.

The quasi-infrared fixed point has subsequently formed the basis of top quark condensation theories of electroweak symmetry breaking in which the Higgs boson is composite at *extremely* short distance scales, composed of a pair of top and antitop quarks.

Stop Squark

In particle physics, a stop squark, symbol $\tilde{t}$, is the superpartner of the top quark as predicted by supersymmetry (SUSY). It is a sfermion, which means it is a spin-0 boson (scalar boson). While the top quark is the heaviest known quark, the stop squark is actually often the lightest squark in many supersymmetry models.

The stop squark is a key ingredient of a wide range of SUSY models that address the hierarchy problem of the Standard Model (SM) in a natural way. A boson partner to the top quark would stabilize the Higgs boson mass against quadratically divergent quantum corrections, provided its mass is close to the electroweak symmetry breaking energy scale. If this was the case then the stop squark would be accessible at the Large Hadron Collider. In the generic R-parity conserving Minimal Supersymmetric Standard Model (MSSM) the scalar partners of right-handed and left-handed top quarks mix to form two stop mass eigenstates. Depending on the specific details of the SUSY model and the mass hierarchy of the sparticles, the stop might decay into a bottom quark and a chargino, with a subsequent decay of the chargino into the lightest neutralino (which is often the lightest supersymmetric particle).

Many searches for evidence of the stop squark have been performed by both the ATLAS and CMS experiments at the LHC but so far no signal has been discovered.

Bottom Quark

Bottom quarks (also called beauty quarks) are not stable and quickly decay, mostly to charmquarks and W^- bosons. In their short lifetime, they are part of some baryons and mesons. Due to their high mass, bottom quarks are relatively easy to research in experiments, a bottom quark is slightly heavier than a helium atom.

In 1977, an experimental group at Fermilab led by Leon Lederman discovered a new resonance at 9.4 GeV/c^2 which was interpreted as a bottom-antibottom quark pair and called the Upsilon meson. From this experiment, the mass of the bottom quark is implied to be about 5 GeV/c^2. The reaction being studied was

$$p + N \rightarrow \mu^+ + \mu^- + X$$

where N was a copper or platinum nucleus. The spectrometer had a muon-pair mass resolution of about 2%, which allowed them to measure an excess of events at 9.4 GeV/c^2. This resonance has been subsequently studied at other accelerators with a detailed investigation of the bound states of the bottom-antibottom meson. Studies of the B-meson have also been productive.

Distinct Character

The bottom quark's "bare" mass is around 4.2 GeV/c^2 – a bit more than four times the mass of a proton, and many orders of magnitude larger than common "light" quarks.

Although it almost-exclusively transitions from or to a top quark, the bottom quark can decay into either an up quark or charm quark via the weak interaction. CKM matrixelements V_{ub} and V_{cb} specify the rates, where both these decays are suppressed, making lifetimes of most bottom particles (~10^{-12} s) somewhat higher than those of charmed particles (~10^{-13} s), but lower than those of strange particles (from ~10^{-10} to ~10^{-8} s).

The combination of high mass and low transition-rate gives experimental collisionby-products containing a bottom quark a distinctive signature that makes them relatively easy to identify using a technique called "B-tagging". For that reason, mesonscontaining the bottom quark are exceptionally long-lived for their mass, and are the easiest particles to use to investigate CP violation. Such experiments are being performed at th e BaBar, Belle and LHCb experiments.

Hadrons Containing Bottom Quarks

Some of the hadrons containing bottom quarks include:

- B mesons contain a bottom quark (or its antiparticle) and an up or down quark.
- B_c and B_s mesons contain a bottom quark along with a charm quark or strange quark respectively.

- There are many bottomonium states, for example the ϒ meson and $\chi_b(3P)$, the first particle discovered in LHC. These consist of a bottom quark and its antiparticle.
- Bottom baryons have been observed, and are named in analogy with strange baryons (e.g. Λ_b^0).

Up and Down Quark

The up and down quarks are the most common and least massive quarks, being the constituents of protons and neutrons and thus of most ordinary matter.

The fact that the free neutron decays

$$n -> p + e^- + \overline{v_e}$$

and nuclei decay by beta decay in processes like

$$p^{32} \; -> \; S^{32} + e^- + \overline{v_e}$$

is thought to be the result of a more fundamental quark process

$$d \; -> \; u + e^- + \overline{v_e}$$

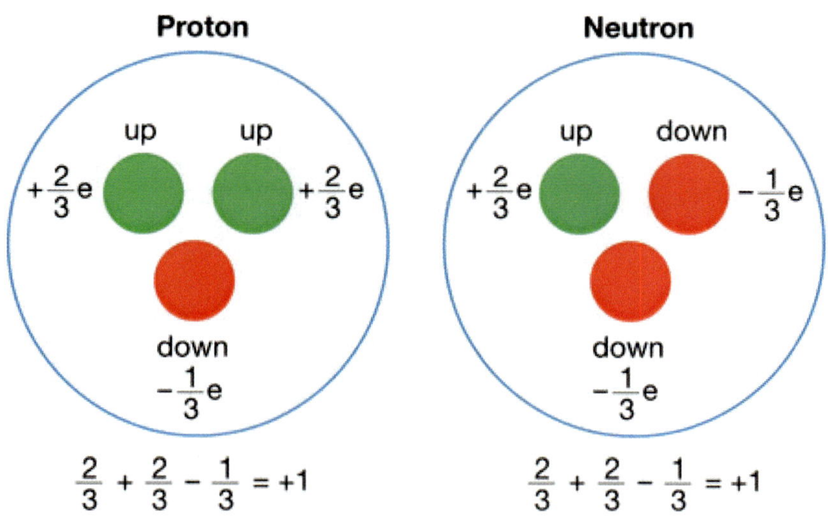

The up quark (u) is the lightest of all quarks, a type of elementary particle, and a major constituent of matter. It, along with the down quark, forms the proton and the neutron. It is part of the first generation of matter, has an electric charge of +2e/3 . Like all quarks, the up quark is an elementary fermion with spin S = 1⁄2. In 1964, Gell-Mann and Zweig (independently of each other) proposed the quark model, then consisting only of up, down, and strange quarks. However, while the quark model explained the Eightfold Way, no direct evidence of the existence of quarks was found until 1968 at the Stanford Linear Accelerator Center. Deep inelastic scattering experiments indicated that protons had substructure, and that protons made of more-fundamental particles explained the data (thus confirming the quark model).

The down quark (d) is the second-lightest of all quarks, a type of elementary particle, and a major constituent of matter. Together with the up quark, it forms the neutrons and protons of atomic nuclei. It is part of the first generation of matter, has an electric charge of -e/3 . Like all quarks, the down quark is an elementary fermion with spin S = 1/2. Also the peripheral velocity of spinning quarks is greater than the speed of light and under this condition the magnetic forces are stronger than the electric forces which contribute to the very strong QUARK-QUARK INTERACTION. Note that such electromagnetic interactions of natural laws invalidate the theory of the quantum chromodynamics (1973).

Especially in 1964 Gell-Mann after a taxonomy of particles suggested that both protons (p) and neutrons (n) consist of (uud) quarks and (dud) quarks respectively having fractional charges as

u = +2e/3 and d = -e/3. That is, uud = +e and dud = 0.

Of course such structures imply small charge distributions as

proton = (+Q = +4e/3, -q = -e/3) and neutron = (+Q = +2e/3, -Q = -2e/3)

which cannot lead to the nuclear structure. Actually, if we apply the fundamental charge-charge interaction of the well-established laws of electromagnetism on such small charge distributions, it would be impossible for us to get the simplest p-n structure of deuterium (D).

Meanwhile in 1933, Stern measured the magnetic moment of the proton to be 2.79 μN and in 1940 F. Bloch measured the neutron magnetic moment to be -1.91 μN. Such results deviate significantly from the predictions of Dirac's theory and invalidate both Yukawa's model and the simple quark model because a careful analysis of them provides considerable charge distributions due to a large number of quarks able to give the nuclear binding and structure by applying the well-established and fundamental laws of charge-charge interactions involving forces acting at a distance. In a paper of 2003 "Nuclear structure is governed by the fundamental laws of electromagnetism" (presented also at the 12th Symposium of the Hellenic Nuclear Physics Society) it is described that the charge distributions of protons and neutrons respectively by a careful analysis of the magnetic moments of nucleons and the deep inelastic scattering experiments. For example for the proton (p) the magnetic moment μ is given by

$$\mu/S = 2.793(e/M)$$

where S is the spin of proton, e the net charge of proton and M the mass of proton. Here we see that the above experimental relation cannot be consistent with the simple quark model with the scheme (uud) even in case in which the charge +Q = +4e/3 = 2u is along the periphery and the charge -q = -e/3 = 1d is in the center (deep inelastic scattering experiment). Clearly applying the electromagnetic laws for μ, and the laws of a rotating oblate spheroid (like the proton) we may write for μ and for the spin S (angular momentum) respectively as

$$\mu = i\pi R^2 = (Q\nu)\pi R^2 \, and \; S = t\, M\omega R^2 = tM(2\pi\nu)R^2$$

where t is a factor between a rotating sphere [$S = (0.4)M\omega R^2$] and a disc [$S = (0.5) M\omega R^2$]. That is $0.4 < t < 0.5$. Therefore $\mu / S = Q/2t = (2.793)e$. That is, for t = 0.47742 (oblate spheroid) we get for the proton along the periphery +Q = +8e/3 = 4u and in the center -q = -5e/3 = 5d. In the same way for the neutron we get - Q = -8e/3 = 8d along the periphery, and +Q = +8e/3 = 4u in the center. Surprisingly applications of electromagnetic laws on such experimental charge distributions which give for proton extra (4u,5d) charged quarks and for the neutron extra (8d,4u) charged quarks lead exactly to the simplest nuclear binding (-2.2246 MeV) of the deuterium. Moreover such extra charged quarks led to the discovery of 288 quarks in nucleons. As a result the proton has 93 (dud) neutral quark triads. Among them there are 4u charged quarks distributed along the periphery and 5d charged quarks limited in the center. Whereas the neutron has 92 (dud) neutral quark triads and among them are distributed 8d charged quarks along the periphery and 4u charged quarks limited in the center So, the new structure of protons and neutrons is given by

PROTON = [93(dud) + 4u +5d]. NEUTRON = [92(dud) + 8d + 4u]

After a careful analysis it was found that the masses of down quark (Md) and of up quark (Mu) as

Md = 3.69348645 MeV/c^2 = mass of 7.23 electrons and Mu = 2.40016645 MeV/c/^{2} = mass of 4.7 electrons

They give the total masses of nucleons as

Mass of neutron Mn = 939.565378 MeV/c^2 = mass of 1838.68 electrons

Mass of proton Mp = 938.272046 MeV/^{c2} = mass of 1836.15 electrons

Then the difference Mn - Mp = 1.293332 MeV/c^2 = mass of 2.53 electrons is exactly equal to Md-Mu. That is, Md - Mu = 7.23- 4.70 = mass of 2.53 electrons. For example n-p = d-u given by the following equation:

n - p = [92(dud) + 8d + 4u] - [93(dud) + 4u +5d] = 3d -(dud) = d-u = mass of 2.53 electrons.

Note that according to the law of energy and mass given by

$$\Delta w / \Delta m = \Delta E / \Delta M = c2$$

the increase of the electron energy ΔE = 1.293 MeV and the increase of the electron mass ΔM = mass of 2.53 electrons of the one emitting electron in the beta decay are due not to the relative motion of the emitting electron with respect to a falacious ethe or to an observer of special relativity but to the real absorption of the released energy Δw = 1.293 MeV and the mass defect Δm = mass of 2.53 electrons. (Experiments reject relativity).

Antineutrino Absorption and Emission

Here the antineutrino (ν^-) behaves like a neutron because it has negative charge along the periphery and positive one at the center. So it interacts electromagnetically with the up quark (u) having a positive charge (+2e/3) like a photon which as an electric dipole interacting with the charge (-e) of an electron in the photoelectric effect under a weak electromagnetic interaction. Therefore the antineutrino-quark interaction invalidates the so-called Electroweak Theory. In this case the absorption of the antineutrino by the proton gives a neutron and a positron under the two conservation laws of mass and energy as

$$\nu^- + p = n + e^+$$

$$or \quad \nu^- + \left[93(dud) + 4u + 5d\right] = [\ \left(92(dud) + 4u + 8d\ \right] + e^+$$

$$or \quad \nu^- + (dud) = 3d + e^+$$

$$or \quad \nu^- + u = d + e^+$$

According to the experiments the antineutrino in this case is an energetic particle with a mass equal to 1.8 MeV/c². Since the mass of the positron is equal to 0.51 MeV/c² and using my discovered values of the up and down quarks we can write the conservation law of mass in terms of MeV/c² as

$$1.8 + 2.4 = 3.69 + 0.51$$

In the same way we can write the conservation law of mass in the antineutrino emission as

$$n = p + e^- + \nu^-$$

$$or \ \left[\left(92(dud) + 4u + 8d\ \right] = \left[93(dud) + 4u + 5d\right] + e^- + \nu^-$$

$$or \ d = u + e^- + \nu^-$$

Since here the mass of the emitting energetic antineutrino is equal to 0.78 MeV/c² we can write the conservation law of mass in terms of MeV/c² as

$$3.69 = 2.4 + 0.51 + 0.78$$

Strange Quark

The strange quark or s quark (from its symbol, s) is the third lightest of all quarks, a type of elementary particle. Strange quarks are found in subatomic particles called hadrons. Example of hadrons containing strange quarks include kaons (K), strange D mesons (D_s), Sigma baryons (Σ), and other strange particles.

According to the IUPAP the symbol s is the official name, while strange is to be considered only as a mnemonic. The name sideways has also been used because the s quark has a I_3 value of 0 while the u ("up") and d ("down") quarks have values of $+\frac{1}{2}$ and $-\frac{1}{2}$ respectively.

Along with the charm quark, it is part of the second generation of matter, and has an electric charge of $\frac{1}{3}e$ and a bare mass of 95^{+5}_{-5} MeV/c². Like all quarks, the strange quark is an elementaryfermion with spin $\frac{1}{2}$, and experiences all four fundamental interactions: gravitation, electromagnetism, weak interactions, and strong interactions. The antiparticle of the strange quark is the strange antiquark (sometimes called *antistrange quark* or simply *antistrange*), which differs from it only in that some of its properties have equal magnitude but opposite sign.

The first strange particle (a particle containing a strange quark) was discovered in 1947 (kaons), but the existence of the strange quark itself (and that of the up and down quarks) was only postulated in 1964 by Murray Gell-Mann and George Zweig to explain the Eightfold Way classification scheme of hadrons. The first evidence for the existence of quarks came in 1968, in deep inelastic scattering experiments at the Stanford Linear Accelerator Center. These experiments confirmed the existence of up and down quarks, and by extension, strange quarks, as they were required to explain the Eightfold Way.

Charm Quark

The charm quark is one of six quarks that, along with leptons, form the basic building blocks of ordinary matter. It is hundreds of times more massive than the up and down quarks that make up protons and neutrons.

Theorists had predicted the existence of the charm quark to explain the absence of an expected particle interaction. When the charm quark did turn up, it was as a constituent of the J/psi particle, whose discovery in 1974 finally convinced physicists that quarks were real.

Particles containing a charm quark are known as either "charmed particles" or "charmonia." They have only a fleeting existence before decaying into more conventional particles. Many experiments have studied their properties. At facilities known as "charm factories," large numbers of charm-containing particles can be produced with little contamination from other types of particles. An electron-positron collider in Beijing, for example, is expected to ultimately produce 10 billion J/psi decays in one year's running time. This lcollevel of production allows scientists to observe subatomic processes with great precision, and may reveal subtle signs of new physics phenomena predicted by theorists.

Hadron

Hadron, any member of a class of subatomic particles that are built from quarks and thus react through the agency of the strong force. The hadrons embrace mesons, baryons (e.g., protons, neutrons, and sigma particles), and their many resonances. All observed subatomic particles are hadrons except for the gauge bosons of the fundamental interactions and the leptons. Except for protons and neutrons that are bound in atomic nuclei, all hadrons have short lives and are produced in the high-energy collisions of subatomic particles. The other three basic forces of nature also affect hadron behaviour: all hadrons are subject to gravitation; charged hadrons obey electromagnetic laws; and some hadrons break up by way of the weak force (as in radioactive decay), while others decay via the strong and the electromagnetic forces.

Properties

According to the quark model, the properties of hadrons are primarily determined by their so-called *valence quarks*. For example, a proton is composed of two up quarks (each with electric charge $+\frac{2}{3}$, for a total of $+\frac{4}{3}$ together) and one down quark (with electric charge $-\frac{1}{3}$). Adding these together yields the proton charge of +1. Although quarks also carry color charge, hadrons must have zero total color charge because of a phenomenon called color confinement. That is, hadrons must be "colorless" or "white". These are the simplest of the two ways: three quarks of different colors, or a quark of one color and an antiquarkcarrying the corresponding anticolor. Hadrons with the first arrangement are called baryons, and those with the second arrangement are mesons.

Hadrons, however, are not composed of just three or two quarks, because of the strength of the strong force. More accurately, strong force gluons have enough energy (E) to have resonances composed of massive (m) quarks ($E > mc^2$) . Thus, virtual quarks and antiquarks, in a 1:1 ratio, form the majority of massive particles inside a hadron. The two or three quarks that compose a hadron are the excess of quarks vs. antiquarks, and so too in the case of anti-hadrons (anti-particles). Because the virtual quarks are not stable wave packets (quanta), but an irregular and transient phenomenon, it is not meaningful to ask which quark is real and which virtual; only the excess is apparent from the outside in the form of a hadron. Massless virtual gluons compose the numerical majority of particles inside hadrons.

Like all subatomic particles, hadrons are assigned quantum numbers corresponding to the representations of the Poincaré group: $J^{PC}(m)$, where J is the spin quantum number, P the intrinsic parity (or P-parity), C the charge conjugation (or C-parity), and $m$the particle's mass. Note that the mass of a hadron has very little to do with the mass of its valence quarks; rather, due to mass–energy equivalence, most of the mass comes from the large amount of energy associated with the strong interaction. Hadrons may also carry flavor quantum numbers such as isospin (G parity), and strangeness. All quarks carry an additive, conserved quantum number called a baryon number (B), which is $+\frac{1}{3}$ for quarks and $-\frac{1}{3}$ for antiquarks. This means that baryons (groups of three quarks) have $B = 1$ whereas mesons have $B = 0$.

Hadrons have excited states known as resonances. Each ground state hadron may have several excited states; several hundreds of resonances have been observed in experiments. Resonances decay extremely quickly (within about 10^{-24} seconds) via the strong nuclear force.

In other phases of matter the hadrons may disappear. For example, at very high temperature and high pressure, unless there are sufficiently many flavors of quarks, the theory of quantum chromodynamics (QCD) predicts that quarks and gluons will no longer be confined within hadrons, "because the strength of the strong interaction diminishes with energy". This property, which is known as asymptotic freedom, has been experimentally confirmed in the energy range between 1 GeV (gigaelectronvolt) and 1 TeV (teraelectronvolt).

All free hadrons except the proton and antiproton are unstable.

Baryon

Baryon, any member of one of two classes of hadrons (particles built from quarks and thus experiencing the strong nuclear force). Baryons are heavy subatomic particles that are made up of three quarks. Both protons and neutrons, as well as other particles, are baryons. (The other class of hadronic particle is built from a quark and an antiquark and is called a meson.)

Baryons are characterized by a baryon number, *B*, of 1. Their antiparticles, called anti-baryons, have a baryon number of −1. An atom containing, for example, one proton and one neutron (each with a baryon number of 1) has a baryon number of 2. In addition to their differences in composition, baryons and mesons can be distinguished from one another by spin: the three quarks that make up a baryon can only produce half-integer values, while meson spins always add up to integer values.

Baryons are strongly interacting fermions; that is, they are acted on by the strong nuclear force and are described by Fermi–Dirac statistics, which apply to all particles obeying the Pauli exclusion principle. This is in contrast to the bosons, which do not obey the exclusion principle.

Baryons, along with mesons, are hadrons, particles composed of quarks. Quarks have baryon numbers of $B = \frac{1}{3}$ and antiquarks have baryon numbers of $B = -\frac{1}{3}$. The term "baryon" usually refers to *triquarks*—baryons made of three quarks ($B = \frac{1}{3} + \frac{1}{3} + \frac{1}{3} = 1$).

Other exotic baryons have been proposed, such as pentaquarks—baryons made of four quarks and one antiquark ($B = \frac{1}{3} + \frac{1}{3} + \frac{1}{3} + \frac{1}{3} - \frac{1}{3} = 1$), but their existence is not generally accepted. The particle physics community as a whole did not view their existence as likely in 2006, and in 2008, considered evidence to be overwhelmingly against the existence of the reported pentaquarks. However, in July 2015, the LHCb experiment observed two resonances consistent with pentaquark states in the $\Lambda_b^0 \rightarrow J/\psi K^- p$ decay, with a combined statistical significance of 15σ.

In theory, heptaquarks (5 quarks, 2 antiquarks), nonaquarks (6 quarks, 3 antiquarks), etc. could also exist.

Baryonic Matter

Nearly all matter that may be encountered or experienced in everyday life is baryonic matter, which includes atoms of any sort, and provides those with the property of mass. Non-baryonic matter, as implied by the name, is any sort of matter that is not composed primarily of baryons. This might include neutrinos and free electrons, dark matter, supersymmetric particles, axions, and black holes.

The very existence of baryons is also a significant issue in cosmology because it is assumed that the Big Bang produced a state with equal amounts of baryons and antibaryons. The process by which baryons came to outnumber their antiparticles is called baryogenesis.

Baryogenesis

Experiments are consistent with the number of quarks in the universe being a constant and, to be more specific, the number of baryons being a constant; in technical language, the total baryon number appears to be conserved. Within the prevailing Standard Model of particle physics, the number of baryons may change in multiples of three due to the action of sphalerons, although this is rare and has not been observed under experiment. Some grand unified theories of particle physics also predict that a single proton can decay, changing the baryon number by one; however, this has not yet been observed under experiment. The excess of baryons over antibaryons in the present universe is thought to be due to non-conservation of baryon number in the very early universe, though this is not well understood.

Properties

Isospin and Charge

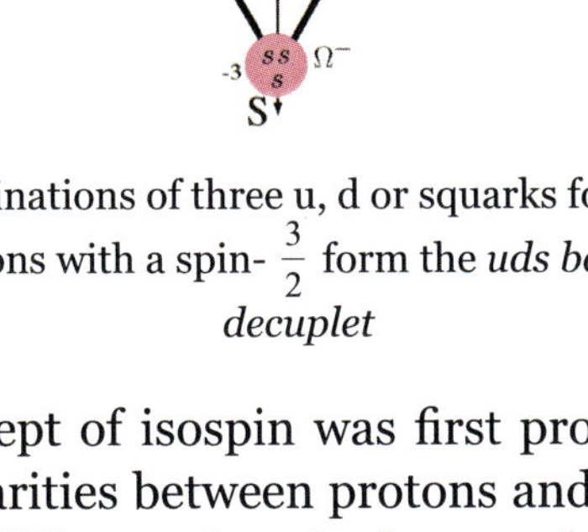

Combinations of three u, d or squarks forming baryons with a spin-$\frac{3}{2}$ form the *uds baryon decuplet*

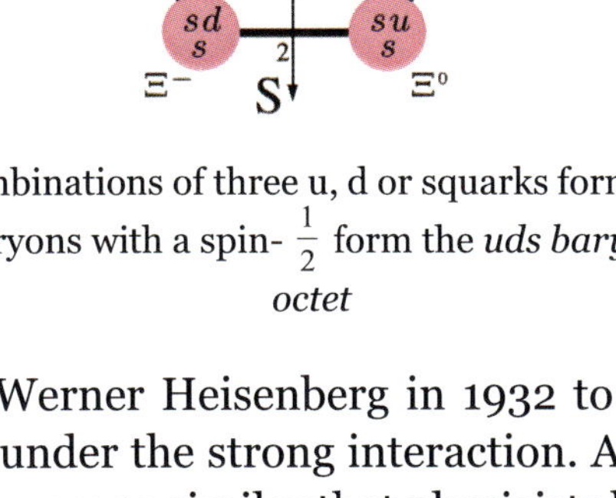

Combinations of three u, d or squarks forming baryons with a spin-$\frac{1}{2}$ form the *uds baryon octet*

The concept of isospin was first proposed by Werner Heisenberg in 1932 to explain the similarities between protons and neutrons under the strong interaction. Although they had different electric charges, their masses were so similar that physicists believed they were the same particle. The different electric charges were explained as being the result of some unknown excitation similar to spin. This unknown excitation was later dubbed *isospin* by Eugene Wigner in 1937.

This belief lasted until Murray Gell-Mann proposed the quark model in 1964 (containing originally only the u, d, and s quarks). The success of the isospin model is now understood to be the result of the similar masses of u and d quarks. Since u and d quarks have similar masses, particles made of the same number then also have similar masses. The exact specific u and d quark composition determines the charge, as u quarks carry charge $+\frac{2}{3}$ while d quarks carry charge $-\frac{1}{3}$. For example, the four Deltas all have different charges (Δ^{++}(uuu), Δ^{+} (uud), Δ^{o} (udd), Δ^{-} (ddd)), but have similar masses (~1,232 MeV/c²) as they are each made of a combination of three u and d quarks. Under the isospin model, they were considered to be a single particle in different charged states.

The mathematics of isospin was modeled after that of spin. Isospin projections varied in increments of 1 just like those of spin, and to each projection was associated a "charged state". Since the "Delta particle" had four "charged states", it was said to be of isospin $I = \frac{3}{2}$. Its "charged states" Δ^{++}, Δ^{+}, Δ^{o}, and Δ^{-}, corresponded to the isospin projections $I_3 = +\frac{3}{2}$, $I_3 = +\frac{1}{2}$, $I_3 = -\frac{1}{2}$, and $I_3 = -\frac{3}{2}$, respectively. Another example is the "nucleon particle". As there were two nucleon "charged states", it was said to be of isospin $\frac{1}{2}$. The positive nucleon N^{+} (proton) was identified with $I_3 = +1/2$ and the neutral nucleon N^{o} (neutron) with $I_3 = -\frac{1}{2}$. It was later noted that the isospin projections were related to the up and down quark content of particles by the relation:

$$I_3 = \frac{1}{2}[(n_{\mathrm{u}} - n_{\bar{\mathrm{u}}}) - (n_{\mathrm{d}} - n_{\bar{\mathrm{d}}})],$$

where the n's are the number of up and down quarks and antiquarks.

In the "isospin picture", the four Deltas and the two nucleons were thought to be the different states of two particles. However, in the quark model, Deltas are different states of nucleons (the N^{++} or N^{-} are forbidden by Pauli's exclusion principle). Isospin, although conveying an inaccurate picture of things, is still used to classify baryons, leading to unnatural and often confusing nomenclature.

Flavour Quantum Numbers

The strangeness flavour quantum number S was noticed to go up and down along with particle mass. The higher the mass, the lower the strangeness (the more s quarks). Particles could be described with isospin projections (related to charge) and strangeness (mass). As other quarks were discovered, new quantum numbers were made to have similar description of udc and udb octets and decuplets. Since only the u and d mass are similar, this description of particle mass and charge in terms of isospin and flavour quantum numbers works well only for octet and decuplet made of one u, one d, and one other quark, and breaks down for the other octets and decuplets (for example, ucb octet and decuplet). If the quarks all had the same mass, their behaviour would be called *symmetric*, as they would all behave in the same way to the strong interaction. Since quarks do not have the same mass, they do not interact in the same way (exactly like an electron placed in an electric field will accelerate more than a proton placed in the same field because of its lighter mass), and the symmetry is said to be broken.

It was noted that charge (Q) was related to the isospin projection (I_3), the baryon number (B) and flavour quantum numbers (S, C, B', T) by the Gell-Mann–Nishijima formula:

$$Q = I_3 + \frac{1}{2}(B + S + C + B' + T),$$

where S, C, B', and T represent the strangeness, charm, bottomness and topnessflavour quantum numbers, respectively. They are related to the number of strange, charm, bottom, and top quarks and antiquark according to the relations:

$$S = -\left(n_s - n_{\bar{s}}\right)$$

$$C = +\left(n_c - n_{\bar{c}}\right),$$

$$B' = -(n_b - n_{\bar{b}}),$$

$$T = +(n_t - n_{\bar{t}}),$$

meaning that the Gell-Mann–Nishijima formula is equivalent to the expression of charge in terms of quark content:

$$Q\frac{2}{3}\left[\left(n_u - n_{\bar{u}}\right) + \left(n_c - n_{\bar{c}}\right) + \left(n_t - n_{\bar{t}}\right)\right] - \frac{1}{3}\left[\left(n_d - n_{\bar{d}}\right) + \left(n_s - n_{\bar{s}}\right) + \left(n_b - n_{\bar{b}}\right)\right].$$

Spin, Orbital Angular Momentum and Total Angular Momentum

Spin (quantum number S) is a vector quantity that represents the "intrinsic" angular momentum of a particle. It comes in increments of $\frac{1}{2}$ ħ (pronounced "h-bar"). The ħ is often dropped because it is the "fundamental" unit of spin, and it is implied that "spin 1" means "spin 1 ħ". In some systems of natural units, ħ is chosen to be 1, and therefore does not appear anywhere.

Quarks are fermionic particles of spin $\frac{1}{2}$ ($S = \frac{1}{2}$). Because spin projections vary in increments of 1 (that is 1 ħ), a single quark has a spin vector of length $\frac{1}{2}$, and has two spin projections ($S_z = +\frac{1}{2}$ and $S_z = -\frac{1}{2}$). Two quarks can have their spins aligned, in which case the two spin vectors add to make a vector of length $S = 1$ and three spin projections

($S_z = +1$, $S_z = 0$, and $S_z = -1$). If two quarks have unaligned spins, the spin vectors add up to make a vector of length $S = 0$ and has only one spin projection ($S_z = 0$), etc. Since baryons are made of three quarks, their spin vectors can add to make a vector of length $S = \frac{3}{2}$, which has four spin projections ($S_z = +\frac{3}{2}$, $S_z = +\frac{1}{2}$, $S_z = -\frac{1}{2}$, and $S_z = -\frac{3}{2}$), or a vector of length $S = \frac{1}{2}$ with two spin projections ($S_z = +\frac{1}{2}$, and $S_z = -\frac{1}{2}$).

There is another quantity of angular momentum, called the orbital angular momentum(azimuthal quantum number L), that comes in increments of 1 ħ, which represent the angular moment due to quarks orbiting around each other. The total angular momentum(total angular momentum quantum number J) of a particle is therefore the combination of intrinsic angular momentum (spin) and orbital angular momentum. It can take any value from $J = |L - S|$ to $J = |L + S|$, in increments of 1.

Baryon angular momentum quantum numbers for L = 0, 1, 2, 3				
Spin (S)	**Orbital angular momentum (L)**	**Total angular momentum (J)**	**Parity (P)**	**Condensed notation (J^P)**
$\frac{1}{2}$	0	$\frac{1}{2}$	+	$\frac{1}{2}^+$
	1	$\frac{3}{2}, \frac{1}{2}$	–	$\frac{3}{2}^-, \frac{1}{2}^-$
	2	$\frac{5}{2}, \frac{3}{2}$	+	$\frac{5}{2}^+, \frac{3}{2}^+$
	3	$\frac{7}{2}, \frac{5}{2}$	–	$\frac{7}{2}^-, \frac{5}{2}^-$
$\frac{3}{2}$	0	$\frac{3}{2}$	+	$\frac{3}{2}^+$
	1	$\frac{5}{2}, \frac{3}{2}, \frac{1}{2}$	–	$\frac{5}{2}^-, \frac{3}{2}^-, \frac{1}{2}^-$
	2	$\frac{7}{2}, \frac{5}{2}, \frac{3}{2}, \frac{1}{2}$	+	$\frac{7}{2}^+, \frac{5}{2}^+, \frac{3}{2}^+, \frac{1}{2}^+$
	3	$\frac{9}{2}, \frac{7}{2}, -, \frac{3}{2}$	–	$\frac{9}{2}^-, \frac{7}{2}^-, \frac{5}{2}^-, \frac{3}{2}^-$

Particle physicists are most interested in baryons with no orbital angular momentum ($L = 0$), as they correspond to ground states—states of minimal energy. Therefore, the two groups of baryons most studied are the $S = \frac{1}{2}$; $L = 0$ and $S = \frac{3}{2}$; $L = 0$, which corresponds to $J = \frac{1}{2}^{+}$ and $J = \frac{3}{2}^{+}$, respectively, although they are not the only ones. It is also possible to obtain $J = \frac{3}{2}^{+}$ particles from $S = \frac{1}{2}$ and $L = 2$, as well as $S = \frac{3}{2}$ and $L = 2$. This phenomenon of having multiple particles in the same total angular momentum configuration is called *degeneracy*. How to distinguish between these degenerate baryons is an active area of research in baryon spectroscopy.

Parity

If the universe were reflected in a mirror, most of the laws of physics would be identical—things would behave the same way regardless of what we call "left" and what we call "right". This concept of mirror reflection is called intrinsic parity" or simply "parity" (*P*). Gravity, the electromagnetic force, and the strong interaction all behave in the same way regardless of whether or not the universe is reflected in a mirror, and thus are said to conserve parity (P-symmetry). However, the weak interaction does distinguish "left" from "right", a phenomenon called parity violation (P-violation).

Based on this, if the wavefunction for each particle (in more precise terms, the quantum field for each particle type) were simultaneously mirror-reversed, then the new set of wavefunctions would perfectly satisfy the laws of physics (apart from the weak interaction). It turns out that this is not quite true: for the equations to be satisfied, the wavefunctions of certain types of particles have to be multiplied by −1, in addition to being mirror-reversed. Such particle types are said to have negative or odd parity ($P = -1$, or alternatively $P = -$), while the other particles are said to have positive or even parity ($P = +1$, or alternatively $P = +$).

For baryons, the parity is related to the orbital angular momentum by the relation:

$$P = (-1)^{L}.$$

As a consequence, baryons with no orbital angular momentum ($L = 0$) all have even parity ($P = +$).

Nomenclature

Baryons are classified into groups according to their isospin (*I*) values and quark (*q*) content. There are six groups of baryons—nucleon (N), Delta (Δ), Lambda (Λ), Sigma (Σ), Xi (Ξ), and Omega (Ω). The rules for classification are defined by the Particle Data Group. These rules consider the up (u), down (d) and strange (s) quarks to be *light* and the charm (c), bottom (b), and top (t) quarks to be *heavy*. The rules cover all the particles that can be made from three of each of the six quarks, even though baryons made of top quarks are not expected to exist because of the top quark's short lifetime. The rules do not cover pentaquarks.

- Baryons with three $u\, and\, /\, or\, d\, quarks\ are\, N's\ (I=\frac{1}{2})\ or\, \Delta's\ (I=\frac{3}{2})$.
- Baryons with two u and/or d quarks are $\Lambda's\ (I=0)\ or\, \Sigma's\ (I=1)$. If the third quark is heavy, its identity is given by a subscript.
- Baryons with one u or d quark are $\Xi's\ (I=\frac{1}{2})$. One or two subscripts are used if one or both of the remaining quarks are heavy.
- Baryons with no u or d quarks are $\Omega's\ (I=0)$, and subscripts indicate any heavy quark content.
- Baryons that decay strongly have their masses as part of their names. For example, Σ^o does not decay strongly, but $\Delta^{++}(1232)$ does.

It is also a widespread (but not universal) practice to follow some additional rules when distinguishing between some states that would otherwise have the same symbol.

- Baryons in total angular momentum $J = \frac{3}{2}$ configuration that have the same symbols as their $J = \frac{1}{2}$ counterparts are denoted by an asterisk (*).
- Two baryons can be made of three different quarks in $J = \frac{1}{2}$ configuration. In this case, a prime (′) is used to distinguish between them.
- *Exception*: When two of the three quarks are one up and one down quark, one baryon is dubbed Λ while the other is dubbed Σ.

Quarks carry a charge, so knowing the charge of a particle indirectly gives the quark content. For example, the rules above say that a Λ^+_c contains a c quark and some combination of two u and/or d quarks. The c quark has a charge of ($Q = +\frac{2}{3}$), therefore the other two must be a u quark ($Q = +\frac{2}{3}$), and a d quark ($Q = -\frac{1}{3}$) to have the correct total charge ($Q = +1$).

Table of Baryons

Particle	Symbol	Makeup	Rest mass MeV/c²	Spin	B	S	Lifetime (seconds>	Decay Modes
Proton	p	uud	938.3	1/2	+1	0	Stable	...
Neutron	n	ddu	939.6	1/2	+1	0	920	$pe^-\nu_e$
Lambda	Λ^o	uds	1115.6	1/2	+1	-1	2.6 x10^{-10}	$p\pi^-$, $n\pi^o$
Sigma	Σ^+	uus	1189.4	1/2	+1	-1	0.8 x10^{-10}	$p\pi^o$, $n\pi^+$
Sigma	Σ^o	uds	1192.5	1/2	+1	-1	6x10^{-20}	$\Lambda^o\gamma$
Sigma	Σ^-	dds	1197.3	1/2	+1	-1	1.5 x10^{-10}	$n\pi^-$

Delta	Δ^{++}	uuu	1232	3/2	+1	0	0.6 $\times10^{-23}$	$p\pi^+$
Delta	Δ^{+}	uud	1232	3/2	+1	0	0.6 $\times10^{-23}$	$p\pi^o$
Delta	Δ^{o}	udd	1232	3/2	+1	0	0.6 $\times10^{-23}$	$n\pi^o$
Delta	Δ^{-}	ddd	1232	3/2	+1	0	0.6 $\times10^{-23}$	$n\pi^-$
Xi Cascade	Ξ^{o}	uss	1315	1/2	+1	-2	2.9 $\times10^{-10}$	$\Lambda^o\pi^o$
Xi Cascade	Ξ^{-}	dss	1321	1/2	+1	-2	1.64 $\times10^{-10}$	$\Lambda^o\pi^-$
Omega	Ω^{-}	sss	1672	3/2	+1	-3	0.82 $\times10^{-10}$	$\Xi^o\pi^-$, Λ^oK^-
Lambda	Λ^{+}_{c}	udc	2281	1/2	+1	0	2×10^{-13}	...

As an example of the notation in the column of decay modes, the sigma + decay mode is written

$$p\pi^0,\ n\pi^+$$

which implies there are two common decay modes,

$$\Sigma^+ \rightarrow p + \pi^0 \quad and \quad \Sigma^+ \rightarrow n + \pi^+$$

Particles which decay by the strong interaction typically do so on a time scale of about 10-23 seconds, and some of the baryons listed in the table above have lifetimes on the order of 10-10 seconds. The reason for those longer lifetimes is that they are forbidden by some kind of conservation law from decaying by the strong interaction, and must decay by the weak interaction. These decays involve quark transformations which can be represented by Feynman diagrams.There are dozens of other baryons (i.e., combinations of three quarks), but most of them decay by the strong interaction and are so short lived that they don't leave a measurable track in a bubble chamber or other detection device. These baryons have been detected as "resonances" or peaks in the probability for particle interactions.

Nucleon

The atom is electrically neutral due to equal balance of negative charged particles electrons and positively charged particles. And the neutrons are neutral particles. The negative particles are revolved around a nucleus and nucleus contains positive particles protons and neutral neutrons. The total number of neutrons and protons is called Nucleons and this is the mass number while the number of only protons is called atomic number of elements.

Nucleons are held up together inside the nucleus due to nuclear force. Nucleons are further made up of three quarks. These quarks are help up together due to strong interactions.

Proton and neutron can also be found alone, not a part of another nucleus. A proton is the nucleus of H – 1 atom or hydrogen atom. Neutrons can be found in nuclear reactions. A neutron is unstable in nature. As said above nucleon is made of three quarks. They can be up or down quark. A proton consists of one down quark and two up quarks. A neutron consists of one up quark and two down quarks. The strong interaction between the quarks is carried by a particle called gluon.

Charge of up quark is $\frac{+2e}{3}$ and that of down quark is $\frac{-e}{3}$. Hence the charge of proton is +e and that of neutron is 0.. Therefore neutron is neutral in nature. Mass of neutron and proton is almost equal. Mass of neutron is 1.6749×10^{-27} kg and that of proton is 1.6726×10^{-27} kg.

Proton and neutron together forms an isospin doublet. This is why their masses is approximately equal. Neutron is 0.1 percent heavier than proton. Proton and neutron both have same spin i.e. $\frac{1}{2}$. Therefore they lie in the category of fermions and hence Pauli's exclusion principle is applicable on them. According to this they cannot lie in the same quantum state. Therefore they spread out in the nuclear shells similar to electron shells.

A neutron is unstable and hence undergoes beta ($\beta-$) decay and converts into proton releasing an electron and electron antineutrino. While a proton is stable in nature and has larger half life. Inside a nucleus proton and neutron may or may not undergo decay process depending on the nuclide.

Both protons and neutrons have corresponding antiparticles called as antiproton and the antineutron respectively. These anti particles have the same mass but opposite charge as the proton and neutron respectively. They interact in the same way as the nucleons. In particular, anti nucleons can bind to form an "anti nucleus". Till now, scientists have created anti deuterium and anti helium nuclei.

Nucleon Number

Nucleon number is also known as mass number. Nucleon number of an atom is given by the sum of total number of protons and neutrons or total number of nucleons present in the nucleus of that atom.

For each different isotope of an element, mass number is different. Mass number is different from the atomic number. Atomic number is the total number of protons present in the nucleus of an atom. Therefore total number of neutrons can be found out by the difference of mass number and atomic number.

For an atom:

$$\text{Mass number}\,(A) = \text{atomic number}\,(Z) + \text{number of neutrons}\,(N)$$

In a radioactive decay, mass number of an atom changes. Different types of radioactive decays are characterized by the change in atomic number and nucleon number. This characterization is done according to *Fajans and Soddy's law of radioactive displacement.*

For e.g. :

- U-238 undergo radioactive decay by alpha decay, where the number of protons and neutrons both decreases by two in the form of an alpha particle. Therefore the number of neutrons and atomic number decreases by 2 from 146 to 144 and from 92 to 90 respectively, hence the nucleon number decreases by 4 from 238 to 234. This results in an atom of thorium-234 and an alpha particle .
- On the other side, carbon-14 undergoes decays naturally by beta decay. Here one neutron is transformed into a proton by emitting an electron and anti particle of neutron (anti - neutrino). Therefore the atomic number increases by 1 from 6 to 7 and the nucleon or mass number remains the same i.e. 14, whereas the number of neutrons decreases by 1 from 8 to 7. This results in the formation of nitrogen-14 atom, with 7 protons and neutrons.

The nucleon number gives helps in estimation of the isotopic mass measured in atomic mass unit (u). For C – 12, the isotopic mass is exactly 12 because the atomic mass unit is 1/12 of the mass of C- 12 . The isotopic mass for other isotopes is normally within 0.1 u of the mass number. For example, nucleon number of Cl – 35 is 35 and its isotopic mass is equal to 34.96885. The difference between mass number and isotopic mass which is known as the Mass Defect is mainly due to two reasons:

- The neutron inside atoms is little heavier than the proton which increases the mass of a nucleus with more number of neutrons than protons relative to the atomic mass unit scale based on C - 12 with equal numbers of protons and neutrons.
- Second reason is that the nuclear binding energy varies from nucleus to nucleus. A nucleus with higher binding energy has a lesser total energy resulting in a lower mass according to Einstein's mass-energy relation $E = mc^2$. For Cl – 35, the isotopic mass is less than 35, therefore it must be the dominant factor.

Mass of Nucleon

A nucleon is of two types: proton and neutron. Protons and neutrons have approximately the same mass, mass of neutron is 1.6749×10^{-27} kg and that of proton is 1.6726×10^{-27} kg. Proton and neutron together forms an isospin doublet. This is why their masses is approximately equal. Neutron is 0.1 percent heavier than proton. So we won't make any difference between proton and neutron. We will refer to them by the composite term "nucleon", because they are found in the nuclei of the atoms. So, if matter is made of small atoms, a question arises that where is the mass located in atoms?

The electrons are very light as compared to nucleons. 1 nucleon weighs as much as two thousand electrons. For all practical purposes, the mass of atoms is considered as the mass of nucleons. It is pondered upon that nucleons consist of 3 visible quarks, which can be considered right. It can be regarded as logical to visualize that mass of each quark is $\frac{1}{3}$ times mass of an individual nucleon, but that›s not right actually. The mass of the 3 quarks together in the nucleons weighs only about 1 - 2 % of the mass of the nucleons, and there is a basic query that where did the rest 89 % come from ?

First, we have to think that nucleons are not static objects with 3 ingredients i.e. quarks. A nucleon is made of 3 very light quarks which are held together by the aid of interaction called strong interaction is is binding. These three quarks move at very high velocities inside the nucleon. To understand this, imagine 3 ping balls in a jackpot machine. These balls are not the most important thing to consider; rather, one should focus on what's forcing them to move. Similarly, think of nucleons as three quark flecks which are tossed rigorously inside a small subatomic tornado. Here tornado is far more important to consider than the tiny flecks.

This can be thought of Einstein's energy mass equation, $E = mc^2$. This equation predicts the mass and energy are equal in nature to each other. talking about the mass of nucleons, we will notice that around 98 percent of the mass of the universe does not exist as the mass in the usual way we think about it but rather the mass exists in the form of energy. The mass of the quarks themselves comes from a different source. The mass of the quarks and leptons was given by the scientist Higgs Boson. However, it's not the dominant source of mass in the universe. The mass we do have is due to the well understood nuclear physics (due to strong nuclear force or strong interaction).

Energy per Nucleon

Conventionally mass of a nucleus is sum of mass all nucleons i.e sum of mass of protons and neutrons present in the nucleus. The mass of a nucleus is less than the summation of mass of the individual nucleons that make up that nucleus. Here comes the concept of nuclear binding energy.

Nuclear binding energy is the energy needed to break the nucleus into its individual nucleons. This can be also expressed as the amount of energy released when the nucleus is formed by combining separate nucleons.

Binding energy is equal to the potential nuclear energy decrement of the nucleons when they combine to form nucleus. This is equivalent to the work done on the nucleons by the strong interaction or strong nuclear force.

Binding energy is the energy that makes the nucleons to be hold together and is associated with the strong interaction.

The mass of a nucleus is less than the summation of mass of the individual nucleons that form that nucleus. The mass difference (ΔM) between the mass of nucleus and total mass of individual nucleons is equivalent to the binding energy of the nucleus.

Einstein has given an equation which shows the relationship between mass difference and binding energy as follows :

$$E_{binding} = \Delta M c^2$$

(There exist is a factor for conversion of mass (u) to energy in MeV on the AQA board data sheet which saves us converting the mass into kg and then using $E = mc^2$ to find out the energy in joules and then convert the energy in J to MeV).

ΔM is difference between mass of nucleus and sum of mass of individual nucleons

$$\Delta M = Z \times M_p + (A - Z) \times M_n - M_{nucleus}$$

Where:

ΔM is mass difference

M_p is the mass of a proton

M_n is the mass of a neutron

$M_{nucleus}$ is mass of the formed nucleus

Z is atomic number (number of proton present in the nucleus)

A is nucleon number or mass number (total number of protons and neutrons present in the nucleus)

Binding Energy / Nucleon: We can work out the binding energy per nucleon and the number of nucleons if we know the binding energy in a nucleus. Binding energy per nucleon is the average energy needed to remove each nucleon.

Stability of a nucleus depends on the binding energy per nucleon. Stability of nucleus is directly proportional to the binding energy per nucleon. Stability increases with the increase in binding energy per nucleon.

We can draw a graph of binding energy per nucleon versus nucleon number which gives a smooth curve - with one remarkable anomaly - helium (^4He) the binding energy per nucleon is (28.38 / 4) M eV = 7.1 M eV

This is not what we would expect on the general curve.

Binding Energy Curve

The peak in the binding energy curve in the region of stability near iron is because of the fact that either the breaking up of heavier nuclei by fission or combining of lighter nuclei by fusion will give nuclei which are more tightly bound i.e. less mass per nucleon.

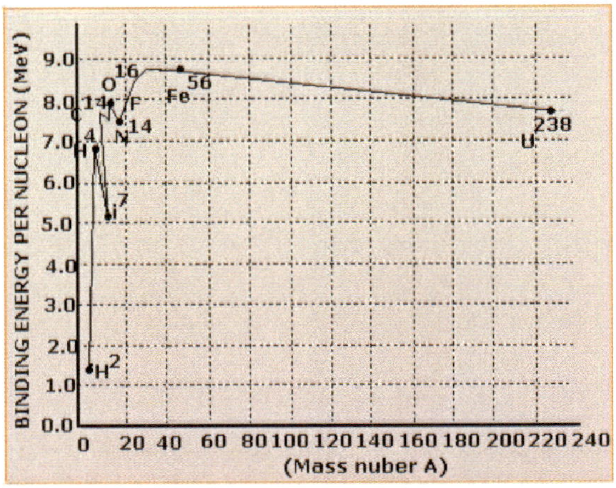

Properties

Quark Composition of a Nucleon

Proton (p): uud Neutron (n): udd Antiproton (p): uud Antineutron (n): udd

A proton (p) is composed of two up quarks (u) and one down quark (d): uud. A neutron (n) has one up quark (u) and two down quarks (d): udd. An antiproton (p) has two up antiquarks (u) and one down antiquark (d): uud. An antineutron (n) has one up antiquark (u) and two down antiquarks (d): udd. The color charge (color assignment) of individual quarks is arbitrary, but all three colors (red, green, blue) must be present.

Protons and neutrons are best known in their role as nucleons, i.e., as the components of atomic nuclei, but they also exist as free particles. Free neutrons are unstable, with a half-life of around 13 minutes, but they are common in nature and have important applications. Singular protons, not bound to other nucleons, are usually regarded as the nuclei of hydrogen atoms or ions, but in some extreme cases (cosmic rays, proton beams), they may be regarded as free protons.

Neither the proton nor neutron is an elementary particle, meaning each is composed of smaller parts, namely three quarks each. A proton is composed of two up quarks and one down quark, while the neutron has one up quark and two down quarks. Quarks are held together by the strong force, or equivalently, by gluons, which mediate the strong force.

An up quark has electric charge $^{+2}/_3\, e$, and a down quark has charge $^{-1}/_3\, e$, so the summed electric charges of proton and neutron are $+e$ and 0, respectively. Thus, the neutron has a charge of 0 (zero), and therefore is eletrically neutral; indeed, the term "neutron" comes from the fact that a neutron is electrically neutral.

The mass of the proton and neutron is quite similar: The proton is 1.6726×10^{-27} kg or 938.27 MeV/c^2, while the neutron is 1.6749×10^{-27} kg or 939.57 MeV/c^2. The neutron is roughly 0.13% heavier. The similarity in mass can be explained roughly by the slight difference in masses of up quarks and down quarks composing the nucleons. However, a detailed explanation remains an unsolved problem in particle physics.

The spin of both protons and neutrons is $^1/_2$, which means they are fermions and, like electrons (and unlike bosons), are subject to the Pauli exclusion principle, a very important phenomenon in nuclear physics: protons and neutrons in an atomic nucleus cannot all be in the same quantum state; instead they spread out into nuclear shells analogous to electron shells in chemistry. Also important, this spin (of proton and neutron) is the source of nuclear spinin larger nuclei. Nuclear spin is best known for its crucial role in the NMR/MRI technique for chemical and biochemical analyses.

The magnetic moment of a proton, denoted μ_p, is 2.79 nuclear magnetons (μ_N), while the magnetic moment of a neutron is $\mu_n = -1.91\ \mu_N$. These parameters are also important in NMR/MRI.

Stability

A neutron in free state is an unstable particle, with a half-life around ten minutes. It undergoes β^-decay (a type of radioactive decay) by turning into a proton while emitting an electron and an electron antineutrino. A proton by itself is thought to be stable, or at least its lifetime is too long to measure. This is an important discussion in particle physics.

Inside a nucleus, on the other hand, combined protons and neutrons (nucleons) can be stable or unstable depending on the nuclide, or nuclear species. Inside some nuclides, a neutron can turn into a proton (producing other particles) as described above; the reverse can happen inside other nuclides, where a proton turns into a neutron (producing other particles) through β^+ decay, or electron capture. And inside still other nuclides, both protons and neutrons are stable and do not change form.

Antinucleons

Both nucleons have corresponding antiparticles: the antiproton and the antineutron, which have the same mass and opposite charge as the proton and neutron respectively, and they interact in the same way. (This is generally believed to be *exactly* true, due to CPT symmetry. If there is a difference, it is too small to measure in all experiments to date.) In particular, antinucleons can bind into an "antinucleus". So far, scientists have created antideuterium and antihelium-3 nuclei.

Meson

Meson, any member of a family of subatomic particles composed of a quark and an antiquark. Mesons are sensitive to the strong force, the fundamental interaction that binds the components of the nucleus by governing the behaviour of their constituent quarks. Predicted theoretically in 1935 by the Japanese physicist Yukawa Hideki, the existence of mesons was confirmed in 1947 by a team led by the English physicist Cecil Frank Powell with the discovery of the pi-meson (pion) in cosmic-rayparticle interactions. More than 200 mesons have been produced and characterized in the intervening years, most in high-energy particle-accelerator experiments. All mesons are unstable, with lifetimes ranging from 10^{-8} second to less than 10^{-22} second. They also vary widely in mass, from 140 megaelectron volts (MeV; 10^6 eV) to nearly 10 gigaelectron volts (GeV; 10^9 eV). Mesons serve as a useful tool for studying the properties and interactions of quarks.

Despite their instability, many mesons last long enough (a few billionths of a second) to be observed with particle detectors, making it possible for researchers to reconstruct the motions of quarks. Any model attempting to explain quarks must correctly elucidate the behaviour of mesons. One of the early successes of the Eightfold Way—a forerunner of modern quark models devised by the physicists Murray Gell-Mann and Yuval Ne'eman—was the prediction and subsequent discovery of the eta-meson (1962). Some years later the decay rate of the pi-meson into two photons was used to support the hypothesis that quarks can take on one of three "colours." Studies of the competing decay modes of K-mesons, which occur via the weak force, have led to a better understanding of parity (the property of an elementary particle or physical system that indicates whether its mirror image occurs in nature) and its nonconservation in the weak interaction. CP violation(the violation of the combined conservation lawsassociated with charge and parity) was discovered first in the K-meson system and is under investigation in B-mesons (which contain bottom quarks).

Mesons also provide a means of identifying new quarks. The J/psi particle, discovered independently by teams led by the American physicists Samuel C.C. Ting and Burton Richter in 1974, proved to be a meson made up of a charm quark and its antiquark. (Up to this time, three quark types had been postulated—up, down, and strange.) It was the first manifestation of charm, a new quantum number the existence of which implies that quarks are related in pairs. The subsequent discovery of another heavy meson, called upsilon, revealed the existence of the bottom quarkand its accompanying antiquark and gave rise to speculation about the existence of a companion particle, the top quark. This sixth quark type, or "flavour," was discovered in 1995. Conclusive proof of its existence culminated the search for one of the last missing pieces in the Standard Model of particle physics, which describes the fundamental particles and their interactions.

Spin, Orbital Angular Momentum and Total Angular Momentum

Spin (quantum number S) is a vector quantity that represents the "intrinsic" angular momentum of a particle. It comes in increments of $\frac{1}{2}\hbar$. The $\hbar$ is often dropped because it is the "fundamental" unit of spin, and it is implied that "spin 1" means "spin 1 $\hbar$". (In some systems of natural units, $\hbar$ is chosen to be 1, and therefore does not appear in equations.)

Quarks are fermions—specifically in this case, particles having spin $\frac{1}{2}$ ($S = \frac{1}{2}$). Because spin projections vary in increments of 1 (that is 1 $\hbar$), a single quark has a spin vector of length $\frac{1}{2}$, and has two spin projections ($S_z = +\frac{1}{2}$ and $S_z = -\frac{1}{2}$). Two quarks can have their spins aligned, in which case the two spin vectors add to make a vector of length $S = 1$ and three spin projections ($S_z = +1$, $S_z = 0$, and $S_z = -1$), called the spin-1triplet. If two quarks have unaligned spins, the spin vectors add up to make a vector of length S = 0 and only one spin projection ($S_z = 0$), called the spin-0 singlet. Because mesons are made of one quark and one antiquark, they can be found in triplet and singlet spin states.

There is another quantity of quantized angular momentum, called the orbital angular momentum (quantum number L), that comes in increments of 1 $\hbar$, which represent the angular momentum due to quarks orbiting around each other. The total angular momentum (quantum number J) of a particle is therefore the combination of intrinsic angular momentum (spin) and orbital angular momentum. It can take any value from $J = |L - S|$ to $J = |L + S|$, in increments of 1.

Meson angular momentum quantum numbers for L = 0, 1, 2, 3				
S	L	J	P	J^P
0	0	0	−	0^-
	1	1	+	1^+
	2	2	−	2^-
	3	3	+	3^+
1	0	1	−	1^-
	1	2, 1, 0	+	$2^+, 1^+, 0^+$
	2	3, 2, 1	−	$3^-, 2^-, 1^-$
	3	4, 3, 2	+	$4^+, 3^+, 2^+$

Particle physicists are most interested in mesons with no orbital angular momentum ($L = 0$), therefore the two groups of mesons most studied are the $S = 1$; $L = 0$ and $S = 0$; $L = 0$, which corresponds to $J = 1$ and $J = 0$, although they are not the only ones. It is also possible to obtain $J = 1$ particles from $S = 0$ and $L = 1$. How to distinguish between the $S = 1$, $L = 0$ and $S = 0$, $L = 1$ mesons is an active area of research in meson spectroscopy.

Parity

If the universe were reflected in a mirror, most of the laws of physics would be identical—things would behave the same way regardless of what we call "left" and what we call "right". This concept of mirror reflection is called parity (*P*). Gravity, the electromagnetic force, and the strong interaction all behave in the same way regardless of whether or not the universe is reflected in a mirror, and thus are said to conserve parity (P-symmetry). However, the weak interaction does distinguish "left" from "right", a phenomenon called parity violation (P-violation).

Based on this, one might think that, if the wavefunction for each particle (more precisely, the quantum field for each particle type) were simultaneously mirror-reversed, then the new set of wavefunctions would perfectly satisfy the laws of physics (apart from the weak interaction). It turns out that this is not quite true: In order for the equations to be satisfied, the wavefunctions of certain types of particles have to be multiplied by −1, in addition to being mirror-reversed. Such particle types are said to have *negative* or *odd* parity ($P = -1$, or alternatively $P = -$), whereas the other particles are said to have *positive* or *even* parity ($P = +1$, or alternatively $P = +$).

For mesons, the parity is related to the orbital angular momentum by the relation:

$$P = (-1)^{L+1}$$

where the L is a result of the parity of the corresponding spherical harmonic of the wavefunction. The "+ 1" comes from the fact that, according to the Dirac equation, a quark and an antiquark have opposite intrinsic parities. Therefore, the intrinsic parity of a meson is the product of the intrinsic parities of the quark (+1) and antiquark (−1). As these are different, their product is −1, and so it contributes the "+ 1" that appears in the exponent.

As a consequence, all mesons with no orbital angular momentum ($L = 0$) have odd parity ($P = -1$).

C-parity

C-parity is only defined for mesons that are their own antiparticle (i.e. neutral mesons). It represents whether or not the wavefunction of the meson remains the same under the interchange of their quark with their antiquark. If

$$|q\bar{q}\rangle = |\bar{q}q\rangle$$

then, the meson is "C even" (C = +1). On the other hand, if

$$|q\bar{q}\rangle = -|\bar{q}q\rangle$$

then the meson is "C odd" (C = −1).

C-parity rarely is studied on its own, but more commonly in combination with P-parity into CP-parity. CP-parity was thought to be conserved, but was later found to be violated in weak interactions.

G-parity

G parity is a generalization of the C-parity. Instead of simply comparing the wavefunction after exchanging quarks and antiquarks, it compares the wavefunction after exchanging the meson for the corresponding antimeson, regardless of quark content.

If

$$|q_1\bar{q}_2\rangle = |\bar{q}_1 q_2\rangle$$

then, the meson is "G even" (G = +1). On the other hand, if

$$|q_1\bar{q}_2\rangle = -|\bar{q}_1 q_2\rangle$$

then the meson is "G odd" (G = −1).

Isospin and Charge

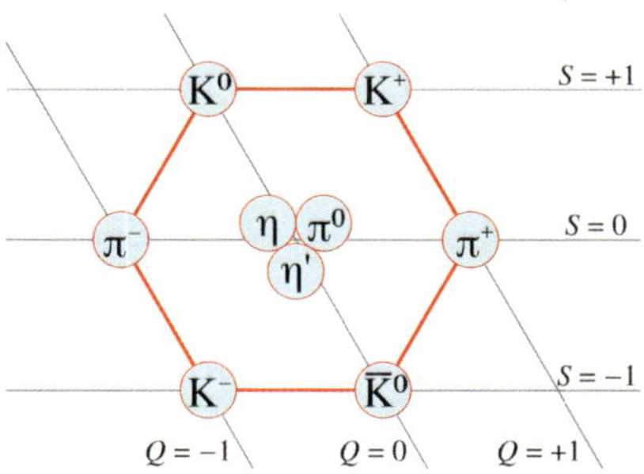

Combinations of one u, d or s quarks and one u, d, or s antiquark in $J^P = 0^-$ configuration form a nonet.

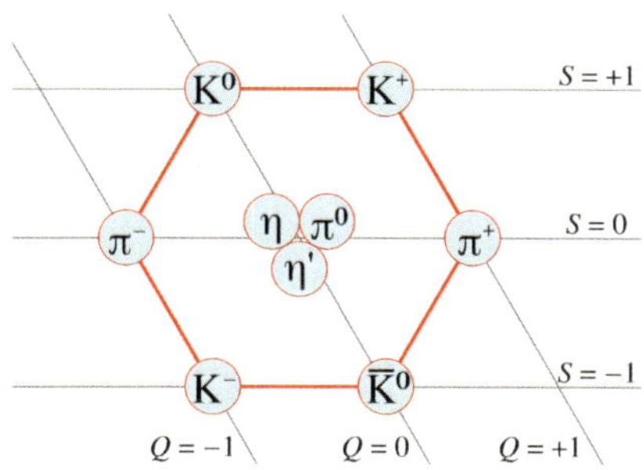

Combinations of one u, d or s quarks and one u, d, or s antiquark in $J^P = 1^-$ configuration also form a nonet.

The concept of isospin was first proposed by Werner Heisenberg in 1932 to explain the similarities between protons and neutrons under the strong interaction. Although they had different electric charges, their masses were so similar that physicists believed that they were actually the same particle. The different electric charges were explained as being the result of some unknown excitation similar to spin. This unknown excitation was later dubbed *isospin* by Eugene Wigner in 1937. When the first mesons were discovered, they too were seen through the eyes of isospin and so the three pions were believed to be the same particle, but in different isospin states.

This belief lasted until Murray Gell-Mann proposed the quark model in 1964 (containing originally only the u, d, and s quarks). The success of the isospin model is now understood to be the result of the similar masses of the u and d quarks. Because the u and d quarks have similar masses, particles made of the same number of them also have similar masses. The exact specific u and d quark composition determines the charge, because u quarks carry charge $+\frac{2}{3}$ whereas d quarks carry charge $-\frac{1}{3}$. For example, the three pions all have different charges (π^+ ($u\bar{d}$), π^0 (a quantum superposition of $u\bar{u}$ and

$d\bar{d}$ states), π^- ($d\bar{u}$)), but have similar masses (c. 140 MeV/c^2) as they are each made of a same number of total of up and down quarks and antiquarks. Under the isospin model, they were considered to be a single particle in different charged states.

The mathematics of isospin was modeled after that of spin. Isospin projections varied in increments of 1 just like those of spin, and to each projection was associated a "charged state". Because the "pion particle" had three "charged states", it was said to be of isospin $I = 1$. Its "charged states" π^+, π^0, and π^-, corresponded to the isospin projections $I_3 = +1$, $I_3 = 0$, and $I_3 = -1$ respectively. Another example is the "rho particle", also with three charged states. Its "charged states" ρ^+, ρ^0, and ρ^-, corresponded to the isospin projections $I_3 = +1$, $I_3 = 0$, and $I_3 = -1$ respectively. It was later noted that the isospin projections were related to the up and down quark content of particles by the relation

$$I_3 = \frac{1}{2}[(n_u - n_{\bar{u}}) - (n_d - n_{\bar{d}})],$$

where the n's are the number of up and down quarks and antiquarks.

In the "isospin picture", the three pions and three rhos were thought to be the different states of two particles. However, in the quark model, the rhos are excited states of pions. Isospin, although conveying an inaccurate picture of things, is still used to classify hadrons, leading to unnatural and often confusing nomenclature. Because mesons are hadrons, the isospin classification is also used, with $I_3 = +\frac{1}{2}$ for up quarks and down antiquarks, and $I_3 = -\frac{1}{2}$ for up antiquarks and down quarks.

Flavour Quantum Numbers

The strangeness quantum number S was noticed to go up and down along with particle mass. The higher the mass, the lower the strangeness (the more s quarks). Particles could be described with isospin projections (related to charge) and strangeness (mass). As other quarks were discovered, new quantum numbers were made to have similar description of udc and udb nonets. Because only the u and d mass are similar, this description of particle mass and charge in terms of isospin and flavour quantum numbers only works well for the nonets made of one u, one d and one other quark and breaks down for the other nonets (for example ucb nonet). If the quarks all had the same mass, their behaviour would be called *symmetric*, because they would all behave in exactly the same way with respect to the strong interaction. However, as quarks do not have the same mass, they do not interact in the same way (exactly like an electron placed in an electric field will accelerate more than a proton placed in the same field because of its lighter mass), and the symmetry is said to be broken.

It was noted that charge (Q) was related to the isospin projection (I_3), the baryon number (B) and flavour quantum numbers (S, C, B', T) by the Gell-Mann–Nishijima formula:

$$Q = I_3 + \frac{1}{2}(B + S + C + B' + T),$$

where S, C, B', and T represent the strangeness, charm, bottomness and topnessflavour quantum numbers respectively. They are related to the number of strange, charm, bottom, and top quarks and antiquark according to the relations:

$$S = -(n_s - n_{\bar{s}})$$

$$C = +(n_c - n_{\bar{c}})$$

$$B' = -(n_b - n_{\bar{b}})$$

$$T = +(n_t - n_{\bar{t}}),$$

meaning that the Gell-Mann–Nishijima formula is equivalent to the expression of charge in terms of quark content:

$$Q = \frac{2}{3}[(n_u - n_{\bar{u}}) + (n_c - n_{\bar{c}}) + (n_t - n_{\bar{t}})] - \frac{1}{3}[(n_d - n_{\bar{d}}) + (n_s - n_{\bar{s}}) + (n_b - n_{\bar{b}})].$$

Classification

Mesons are classified into groups according to their isospin (I), total angular momentum(J), parity (P), G-parity (G) or C-parity (C) when applicable, and quark (q) content. The rules for classification are defined by the Particle Data Group, and are rather convoluted.

Types of Meson

Mesons are classified into types according to their spin configurations. Some specific configurations are given special names based on the mathematical properties of their spin configuration.

Types of mesons					
Type	***S***	***L***	***P***	***J***	J^P
Pseudoscalar meson	0	0	−	0	0^-
Pseudovector meson	0, 1	1	+	1	1^+
Vector meson	1	0, 2	−	1	1^-
Scalar meson	1	1	+	0	0^+
Tensor meson	1	1, 3	+	2	2^+

Nomenclature

Flavourless Mesons

Flavourless mesons are mesons made of pair of quark and antiquarks of the same flavour (all their flavour quantum numbers are zero: $S = 0$, $C = 0$, $B' = 0$, $T = 0$). The rules for flavourless mesons are:

Nomenclature of flavourless mesons

$q\bar{q}$ content	$J^{PC\dagger}$ → I ↓	$0^{-+}, 2^{-+}, 4^{-+}, \ldots$	$1^{+-}, 3^{+-}, 5^{+-}, \ldots$	$1^{--}, 2^{--}, 3^{--}, \ldots$	$0^{++}, 1^{++}, 2^{++}, \ldots$
$u\bar{d}$ $\frac{u\bar{u}-d\bar{d}}{\sqrt{2}}$ $d\bar{u}$	1	π^+ π^0 π^-	b^+ b^0 b^-	ρ^+ ρ^0 ρ^-	a^+ a^0 a^-
Mix of $u\bar{u}$, $d\bar{d}$, $s\bar{s}$	0	η η'	h h′	ω ϕ	f f′
$c\bar{c}$	0	η_c	h_c	$\psi^{\dagger\dagger}$	χ_c
$b\bar{b}$	0	η_b	h_b	Υ	χ_b
$t\bar{t}$	0	η_t	h_t	θ	χ_t

† ^ The C parity is only relevant to neutral mesons.

†† ^ For $J^{PC}=1^{--}$, the ψ is called the J/ψ

In addition:

- When the spectroscopic state of the meson is known, it is added in parentheses.
- When the spectroscopic state is unknown, mass (in MeV/c^2) is added in parentheses.
- When the meson is in its ground state, nothing is added in parentheses.

Flavoured Mesons

Flavoured mesons are mesons made of pair of quark and antiquarks of different flavours. The rules are simpler in this case: the main symbol depends on the heavier quark, the superscript depends on the charge, and the subscript (if any) depends on the lighter quark. In table form, they are:

Nomenclature of flavoured mesons

antiquark → quark ↓	up	down	charm	strange	top	bottom
up	—	[21]	$\bar{D}^0$	K^+	$\bar{T}^0$	B^+
down	[21]	—	D^-	K^0	T^-	B^0
charm	D^0	D^+	—	D_s^+	$\bar{T}_c^0$	B_c^+
strange	K^-	$\bar{K}^0$	D_s^-	—	T_s^-	B_s^0
top	T^0	T^+	T_c^0	T_s^+	—	T_b^+
bottom	B^-	$\bar{B}^0$	B_c^-	$\bar{B}_s^0$	T_b^-	—

In addition:

- If J^P is in the "normal series" (i.e., $J^P = 0^+, 1^-, 2^+, 3^-, \ldots$), a superscript $*$ is added.
- If the meson is not pseudoscalar ($J^P = 0^-$) or vector ($J^P = 1^-$), J is added as a subscript.
- When the spectroscopic state of the meson is known, it is added in parentheses.
- When the spectroscopic state is unknown, mass (in MeV/c^2) is added in parentheses.
- When the meson is in its ground state, nothing is added in parentheses.

Exotic Mesons

There is experimental evidence for particles that are hadrons (i.e., are composed of quarks) and are color-neutral with zero baryon number, and thus by conventional definition are mesons. Yet, these particles do not consist of a single quark/antiquark pair, as all the other conventional mesons discussed above do. A tentative category for these particles is exotic mesons.

There are at least five exotic meson resonances that have been experimentally confirmed to exist by two or more independent experiments. The most statistically significant of these is the Z(4430), discovered by the Belle experiment in 2007 and confirmed by LHCb in 2014. It is a candidate for being a tetraquark: a particle composed of two quarks and two antiquarks.

Mesons

Particle	Symbol	Anti-particle	Makeup	Rest mass MeV/c²	S	C	B	Lifetime	Decay Modes
Pion	π^+	π^-	ud	139.6	0	0	0	2.60 x10^{-8}	$\mu^+\nu_\mu$
Pion	π^0	Self	$\frac{u\bar{u} - d\bar{d}}{\sqrt{2}}$	135.0	0	0	0	0.83 x10^{-16}	2γ
Kaon	K^+	K^-	us	493.7	+1	0	0	1.24 x10^{-8}	$\mu^+\nu_\mu$, $\pi^+\pi^0$
Kaon	K^0_S	K^0_S	1*	497.7	+1	0	0	0.89 x10^{-10}	$\pi^+\pi^-$,$2\pi^0$
Kaon	K^0_L	K^0_L	1*	497.7	+1	0	0	5.2 x10^{-8}	$\pi^+e^-\nu_e$
Eta	η^0	Self	2*	548.8	0	0	0	<10^{-18}	2γ, 3μ
Eta prime	$\eta^{0'}$	Self	2*	958	0	0	0	...	$\pi^+\pi^-\eta$
Rho	ρ^+	ρ^-	ud	770	0	0	0	0.4 x10^{-23}	$\pi^+\pi^0$

Rho	ρ^0	Self	uu, dd	770	0	0	0	0.4 $\times 10^{-23}$	$\pi^+\pi^-$
Omega	ω^0	Self	uu, dd	782	0	0	0	0.8 $\times 10^{-22}$	$\pi^+\pi^-\pi^0$
Phi	ϕ	Self	ss	1020	0	0	0	20 $\times 10^{-23}$	K^+K^-, K^0K^0
D	D^+	D^-	cd	1869.4	0	+1	0	10.6 $\times 10^{-13}$	K + _, e + _
D	D^0	D^0	cu	1864.6	0	+1	0	4.2 $\times 10^{-13}$	[K,μ,e] + _
D	D^+_s	D^-_s	cs	1969	+1	+1	0	4.7 $\times 10^{-13}$	K + _
J/Psi	J/ψ	Self	cc	3096.9	0	0	0	0.8 $\times 10^{-20}$	e^+e^-, $\mu^+\mu^-$...
B	B^-	B^+	bu	5279	0	0	-1	1.5 $\times 10^{-12}$	D^0 + _
B	B^0	B^0	db	5279	0	0	-1	1.5 $\times 10^{-12}$	D^0 + _
B_s	B_s^0	B_s^0	sb	5370	-1	0	-1	...	B^-_s + _
Upsilon	ϒ	Self	bb	9460.4	0	0	0	1.3 $\times 10^{-20}$	e^+e^-, $\mu^+\mu^-$..

Note that an underline indicates an antiparticle

Color Charge

Color charge is the 3-valued hidden quantum number carried by quarks, antiquarks and gluons. Color charge has a 3 valuedness that we associate with the group $SU(3)_{color}$. Color charge is hidden in the sense that only singlets of $SU(3)_{color}$ that are *neutral* occur in nature (at least macroscopically and at low temperatures). The strongly interacting color-neutral particles composed of quarks, antiquarks and gluons that occur in nature are called *hadrons*. (The word *color* in this context is purely colloquial and has no relation to the color that we see with our eyes in everyday life.)

Color charge has two aspects: (a) as a quantum number that labels states of quarks, antiquarks and gluons: hadrons are in the singlet of $SU(3)_{color}$ as a global symmetry group and (b) as the source of the strong color force acting between quarks associated with $SU(3)_{color}$ as a local gauge group. Each of these is analogous to aspects of electric charge: (a) as a quantum number that counts the amount of electric charge in a state: neutral atoms have zero electric charge under U(1) as a global symmetry group, (b) as the source of electromagnetic forces associated with U(1) as a local gauge group acting between electrically charged particles.

O.W. Greenberg introduced the aspect of color charge as a quantum number in 1964 (Greenberg 1964). Y. Nambu, (Nambu 1966) and M.-Y. Han and Y. Nambu (Han and Nambu 1965) introduced the aspect of color charge as the source of the force between quarks in 1965 associated with the local gauge group $SU(3)_{color}$.

Red, Green and Blue

In quantum chromodynamics, a quark's color can take one of three values or charges: red, green, and blue. An antiquark can take one of three anticolors: called antired, antigreen, and antiblue (represented as cyan, magenta and yellow, respectively). Gluons are mixtures of two colors, such as red and antigreen, which constitutes their color charge. QCD considers eight gluons of the possible nine color–anticolor combinations to be unique.

The following illustrates the coupling constants for color-charged particles:

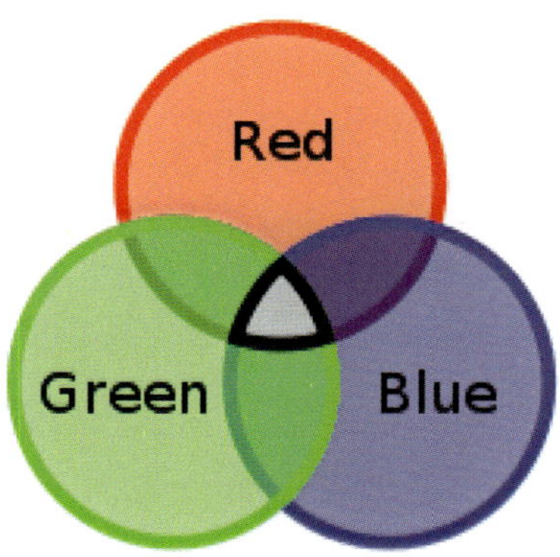

The quark colors (red, green, blue) combine to be colorless

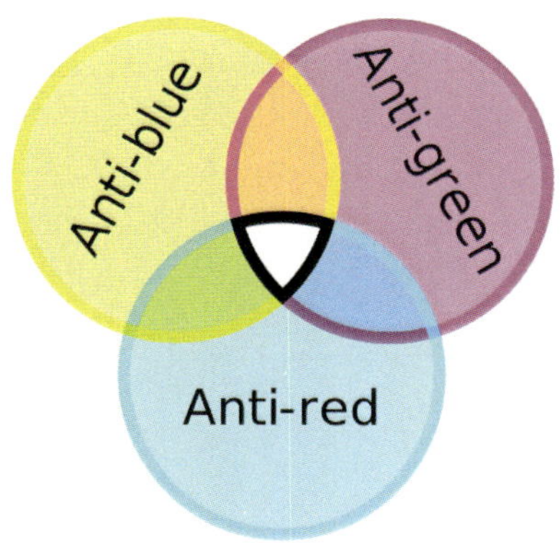

The quark anticolors (antired, antigreen, antiblue) also combine to be colorless

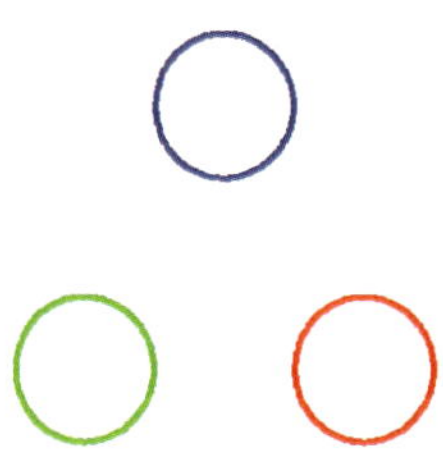
A hadron with 3 quarks (red, green, blue) before a color change

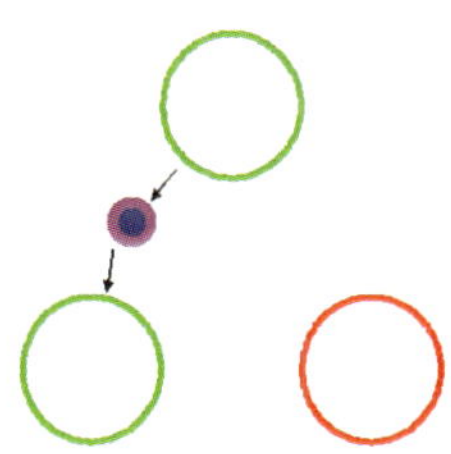
Blue quark emits a blue-antigreen gluon

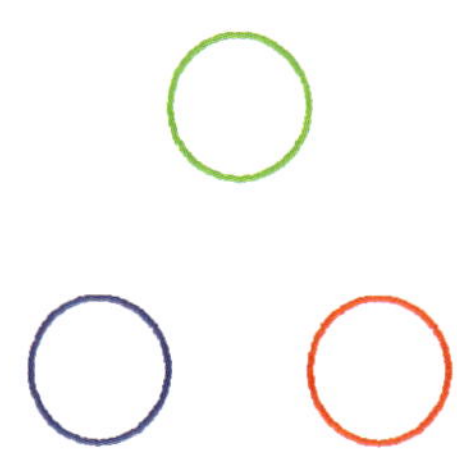
Green quark has absorbed the blue-antigreen gluon and is now blue; color remains conserved

Field Lines from Color Charges

Analogous to an electric field and electric charges, the strong force acting between color charges can be depicted using field lines. However, the color field lines do not arc outwards from one charge to another as much, because they are pulled together tightly by gluons (within 1 fm). This effect confines quarks within hadrons.

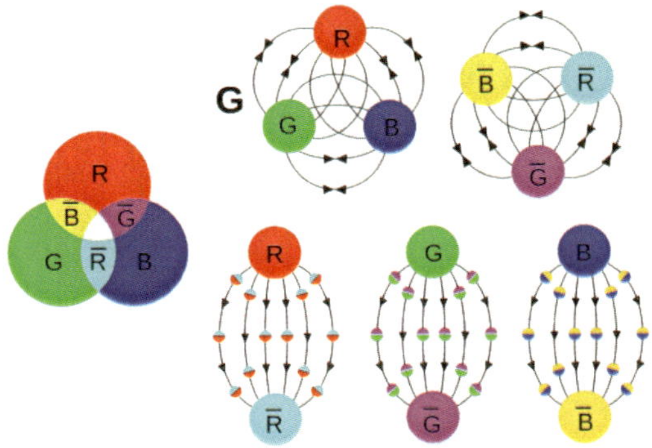

Fields due to color charges, as in quarks (G is the gluon field strength tensor). These are "colorless" combinations. Top: Color charge has "ternary neutral states" as well as binary neutrality (analogous to electric charge). Bottom: Quark/antiquark combinations.

Coupling Constant and Charge

In a quantum field theory, a coupling constant and a charge are different but related notions. The coupling constant sets the magnitude of the force of interaction; for example, in quantum electrodynamics, the fine-structure constant is a coupling constant. The charge in a gauge theory has to do with the way a particle transforms under the gauge symmetry; i.e., its representation under the gauge group. For example, the electron has charge −1 and the positron has charge +1, implying that the gauge transformation has opposite effects on them in some sense. Specifically, if a local gauge transformation $\phi(x)$ is applied in electrodynamics, then one finds:

$$A_\mu \to A_\mu + \partial_\mu \phi(x)\,,\quad \psi \to \exp[+iQ\phi(x)]\psi \;\text{ and }\; \bar{\psi} \to \exp[-iQ\phi(x)]\bar{\psi}$$

where A_μ is the photon field, and ψ is the electron field with $Q = -1$ (a bar over ψ denotes its antiparticle — the positron). Since QCD is a non-abelian theory, the representations, and hence the color charges, are more complicated.

Color as a Quantum Number that Labels States

The role of color charge as a quantum number that is neutral in hadrons is analogous to the role of electric charge as a quantum number that is neutral in un-ionized atoms. Each flavor of quark and antiquark carries the 3-valued color charge. For quarks the color charge transforms as a 3_α under the global $SU(3)_{color}$; for antiquarks the color charge transforms as a $3^{*\beta}$. Gluons transform as the traceless part of an 8^β_α. States composed of products of quarks, antiquarks and gluons can be reduced into irreducible representations of $SU(3)_{color}$.

The only states that occur macroscopically (and at low temperature) are the singlets of $SU(3)_{color}$. These are hadrons, the color neutral states. For example, mesons have a term $q_\alpha \bar{q}^\alpha$ as the leading constituent, as well as higher components with additional gluons and quark-antiquark pairs such as $q_\alpha G^\alpha_\beta \bar{q}^\beta, q_\alpha \bar{q}^\beta q_\beta \bar{q}^\alpha$ and $q_\alpha \bar{q}^\beta G^\gamma_\beta q_\gamma \bar{q}^\delta G^\alpha_\delta$ constructed so that the state is a color singlet. Baryons have a leading term $\epsilon^{\alpha\beta\gamma}q_\alpha q_\beta q_\gamma$ as well as terms with additional gluons, quarks and antiquarks such as $\epsilon^{\beta\delta\sigma}q_\alpha G^\alpha_\beta q_\gamma G^\gamma{}_\delta q_\sigma$.

Color as the Source of the Strong Color Force

Color charge plays a second role in connection with the strong force between color-charge carrying particles, just as electric charge plays a second role in connection with the electromagnetic force between electrically charged particles. The electric force is mediated by photons, which are the quanta of the electromagnetic field associated with the local gauge group U(1). Photons do not carry electric charge. In contrast, the strong or color force is associated with the nonabelian local gauge group $SU(3)_{color}$. Because the color charges do not commute with each other, i.e. are nonabelian, the gluons carry color charges, roughly equivalent to the combined charges of a quark and an antiquark.

Quarks carry the the fundamental 3 representation of $SU(3)_{color}$. Antiquarks carry the complex conjugate $3^\star$ representation of the group. Gluons, the mediators of the color force, carry the 8-dimensional adjoint representation of the color group. This adjoint representation is analogous to the traceless product of the $SU(3)_{color}$ representations of a quark and an antiquark. The strong force is mediated by the gluons which are quanta of the color or quantum chromodynamic field associated with the local gauge group $SU(3)_{color}$ in analogy to photons as mediators of the electromagnetic force associated with the U(1) gauge theory. However, the gluons interact directly with each other as well as with quarks and antiquarks, in contrast to photons, which interact directly only with electrically charged particles.

Color Force as a Paradigm Shift from Meson Forces as the Source of Strong Interactions

The discovery of quark color in 1964 and the gauge theory, $SU(3)_{color}$, in 1965 changed our understanding of the strong interaction in a qualitative way. Before the discovery of the color carried by quarks, the strong interaction was thought to be mediated by the exchange of mesons, such as the $\pi^{+,0,-}$, and the ω^0, $\rho^{+,0,-}$, ϕ 0 mesons. With discoveries of quarks and color we now understand the strong interaction to be connected with the $SU(3)_{color}$ gauge theory. We do not believe, however, that single exchange of gluons, which might be expected from perturbation theory, is the main mechanism. Indeed, meson exchange may also play a role in the strong interaction.

Flavor Independence of Color

Hadrons are singlets (i.e., neutral) under the color group, so their color charges are hidden. The coupling of quarks to gluons is independent of the flavor of the quark. Further, the coupling of quarks to photons is also independent of the electric charge of the quark. This latter property is required for the exact conservation of both color charge and electric charge. As stated above, quarks carry fractional values of the electric charge, e, of the proton; however hadrons, which only occur in color singlets, carry integer values of electric charge.

Consequences of the Nonabelian Color Force

The non-abelian character of the color field has profound consequences for the color force. The color force becomes weak (as the reciprocal of the logarithm of the energy) at high energy or short distance, and it becomes strong at low energy or long distance. The weakness at high energy, called asymptotic freedom, provides a justification for the quasi-free behavior of quarks and gluons in the parton model of hadrons, which is useful in describing high-energy scattering. The strength of the color force at low energy leads to the confinement of quarks and gluons, discussed above. This provides an explanation for the observed absence of particles with fractional values of electric charge, despite the fractional values of electric charge carried by quarks.

Road from Color Charge to Quantum Chromodynamics (QCD)

Crucial steps along this road include the demonstration of renormalizability for non-abelian gauge theories, the discovery that these theories are asymptotically free, and the importance of effects outside the realm of perturbation theory.

Empirical Evidence for Color Charge

Ground State Baryons

The first observational evidence for color charge was the permutation symmetry of the quark quantum numbers in the ground state baryons. The SU(6) model of F. Gürsey and L.A. Radicati (Gürsey and Radicati 1964) discussed below placed the ground state baryons in the 56 representation of SU(6). This representation is symmetric under permutations of the quark quantum numbers. Since the quarks have spin 1/2 they should be fermions according to the spin-statistics theorem and should be antisymmetric under permutations of their quantum numbers. This contradiction is the spin-statistics paradox. The introduction of the three-valued hidden color charge by Greenberg in 1964 resolved this contradiction by making the state of the 3-valued color charge carried by quarks antisymmetric. It justified the symmetry of the baryons under permutations of their visible degrees of freedom, which are the space, spin and flavor degrees of freedom.

The Symmetric Quark Model for Baryons

Greenberg constructed the *symmetric quark model* for baryons using the antisymmetry of the color degree of freedom to justify symmetry of the baryon wave functions in the space, spin and flavor degree of freedom. He gave a table of baryon states for quarks in the s and p states. With M. Resnikoff (Greenberg and Resnikoff 1967) he gave a detailed analysis of baryons in the $(56,1^+)$ and $(70,1^-)$ supermultiplets, where the notation is (dimSU(6),L^P) . This model is the starting point for the study of baryon spectroscopy. For mesons, since there is only one quark or antiquark in the constituent model, the statistics of the quarks is irrelevant.

Neutral Pion Decay to Photons

Further observational evidence for color charge came from the decay rate for $\pi^0 \rightarrow \gamma\gamma$ via the axial anomaly. The 3-valued color charge enters as the square of the number of colors and provides a factor of 9 that brings the theoretical calculation of the decay rate into agreement with experiment.

Electron-positron Annihilation to Hadrons

Another experimental evidence came from the ratio $\sigma\left(e^+ e^- \rightarrow hadrons\right) / \sigma\left(e^+ e^- \rightarrow \mu^+ \mu^-\right)$. Here the 3-valued color charge provides a factor of 3 that brings theory in agreement with experiment. These two tests of color, as well as the permutation symmetry of the quarks in baryons, test the color quantum numbers of the quarks, but do not (in lowest order) test the nonabelian nature of the color force.

Empirical Tests of the Nonabelian Color Force

Among the tests of the nonabelian nature of the color force are asymptotic freedom, permanent confinement of quarks, jets in high-energy scattering, and many precision tests of the standard model. Asymptotic freedom, the decrease of the color force at short distance or high energy, results from the nonabelian interaction of gluons with quarks. Virtual gluons antishield the color charge of a quark (or antiquark) at short distance and thus reduce the color force at short distances. At long distance the growing potential interaction energy of quarks and antiquarks prevents the isolation of individual quarks or antiquarks. Jets in high-energy scattering reflect the transient production of quarks which hadronize to produce strongly interacting particles. Precision tests of quantum chromodynamics, such as the running of coupling constants, also depend on the nonabelian nature of the color interaction.

Historical Developments that Led to Color

Precursors of Quarks

The idea of constructing known particles out of other, more basic particles, has a long history, going back to the effort of M. Born and N. Nagendra Nath (Born and Nagendra Nath 1936) to construct the photon from neutrinos in the 1930's and the effort of E. Fermi and C.N. Yang (Fermi and Yang 1949) to construct pions from protons and neutrons in the 1040's. S. Sakata (Sakata 1956) proposed a model in 1956 with the proton, neutron and lambda baryon were taken as the basic particles. The Sakata model failed to account for the octet of baryons $(p, n, \Lambda, ^{\Sigma +,0,-}, \Xi^{0,-})$.

Quarks

Current Quarks

M. Gell-Mann (Gell-Mann 1964) suggested a model with three *quarks,* $u^{\frac{2}{3}}, d^{-\frac{1}{3}}, s^{-\frac{1}{3}}$ as the fundamental objects in 1964. Gell-Mann's model has the radical departure that the quark electric charges, shown as superscripts above, are fractions of the proton charge e. In addition the quarks carry baryon numbers 1/3 of the baryon number of the nucleon. Particles with either of these fractional values have never been observed. Gell-Mann gave expressions for the vector electromagnetic current and the vector and axial weak currents in terms of quark fields. He showed that quark currents could give the Cabibbo model of weak interactions and also build up the SU(3)×SU(3) current algebra. Gell-Mann's original model used the three *flavors,* u, d, s that were known in 1964. Three additional flavors, c, t, b have since been discovered.

Constituent Quarks

G. Zweig (Zweig 1964) independently introduced a model with fractional electric and baryon charges with emphasis on the constituent point of view. Zweig called the fundamental objects in his model *aces,* a name that was superseded by quarks. He gave an extensive analysis of the space-time and group theory structure of baryons and mesons on the basis of his model. He gave the distinctions between the predictions of the earlier eightfold way and his model. He also gave a reason for the suppression of the decay $\phi \rightarrow \rho\pi$ which is an example of the Okubo, Zweig and Iizuka rule (Okubo 1963, Zweig 1964, Iizuka 1966) that can be described in terms of quark line diagrams. Zweig adopted the point of view that his aces are real physical particles. Both Gell-Mann and Zweig assumed the quarks have spin 1/2 in order that mesons can be constructed from $(q\bar{q})$ and baryons can be constructed from (qqq) ; however spin did not enter their models in a deeper way.

Incorporation of Spin in the Quark Model

In the same year, 1964, that Gell-Mann and Zweig independently suggested quarks and aces, F. Gürsey and L.A. Radicati (Gürsey and Radicati 1964) incorporated spin in the model in a more intrinsic way. Gürsey and Radicati considered the quarks to be a 6 of an $SU(6)$ group that included the $SU(3)_{flavor}$ of the original quark model with the 3 flavors u, d, s and the $SU(2)_{spin}$ of the quark spin. Because baryons are composed of 3 quarks, Gürsey and Radicati considered the reduction of $6 \otimes 6 \otimes 6 \rightarrow 56 + 70 + 70 + 20$ into irreducibles of $SU(6)$. Gürsey and Radicati placed the ground state baryons in the 56 representation of $SU(6)$ which includes the known nucleon octet and delta decuplet

$$56 \rightarrow (8, 1/2) + (10, 3/2) \, under \, SU(6)_{fS} \rightarrow SU(3)_f \times SU(2)_s,$$

where the nucleon octet is $\left(p^+, n^0, \Lambda^0, \Sigma^+, \Sigma^0, \Sigma^-, \Xi^0, \Xi^-\right)$ and the delta decuplet is $(\Delta^{++}, \Delta^+, \Delta^0, \Delta^-, Y_1^{*+}, Y_1^{*0}, Y_1^{*-}, \Xi^{*0}, \Xi^{*-}, \Omega^-)$. The 56 representation has the three quarks in a symmetric state under permutations in contradiction with what one would expect. The spin-statistics theorem requires that spin 1/2 quarks obey the Pauli exclusion principle and occur in the 20 representation which is antisymmetric under permutations if there are no additional degrees of freedom carried by quarks.

The spin-statistics Paradox

Gürsey and Radicati noted the symmetry of the 56 representation for the ground state baryons and suggested that this would imply that the forces between the quarks are repulsive. R.H. Dalitz (Dalitz 1966) supported the idea that an antisymmetric space wavefunction for the ground state baryons would allow the $SU(3)$ and spin state to be symmetric without violating the spin-statistics connection. Many physicists were skeptical of the reality of quarks. If quarks were merely a mathematical device then perhaps their statistics could be ignored. Greenberg (Greenberg 1964) proposed the solution to the spin-statistics paradox that has been confirmed by experiment and was ultimately accepted by the physics community. He suggested that each flavor of quark comes in three varieties, colloquially called *colors*. The quark color degree of freedom, can be taken treated as the fundamental 3 of a new $SU(3)_{color}$ symmetry. Then in a neutral state, i.e. in the singlet of $SU(3)_{color}$, the color degree of freedom is antisymmetric. The quarks as fermions are then antisymmetric under permutations of all their degrees of freedom. The total antisymmetry comes about from the product of the symmetry of the space, spin and flavor degrees of freedom and the antisymmetry of the color degree of freedom. (Greenberg introduced color using parafermi statistics of order 3 and showed that the generalized spin-statistics theorem for parastatistics is obeyed.)

Skepticism about Quarks and Color

The idea of quarks was received with skepticism because particles with fractional electric charges and baryon number had never been observed. The suggestion that, in addition, quarks carry a new hidden three-valued charge was received with even greater skepticism. Because of this quarks and color were not accepted generally by the physics community until the discovery of the J/Ψ and other hadrons that carry charm quarks in 1974.

For the mesons composed of a quark and an antiquark the quark statistics is irrelevant.

The Relation between Parafermi Statistics and Explicit Color

The relation between the order 3 of parafermi statistics for quarks and the order 3 for the explicit color first proposed by Nambu and by Han and Nambu was clarified by Greenberg and D. Zwanziger in 1966 (Greenberg and Zwanziger 1966). They showed that the states that are bosons or fermions in the parastatistics model are in 1-to1 correspondence with the states that are color singlets in the $SU(3)_{color}$ model. Further analysis of this question was given by Drühl, Haag and Roberts (Drühl, Haag and Roberts 1970).

References

- M. Gell-Mann (1995). The Quark and the Jaguar: Adventures in the Simple and the Complex. Henry Holt and Co. p. 180. ISBN 978-0-8050-7253-2
- "Belle Discovers a New Type of Meson" (Press release). KEK. 2007. Archived from the original on 2009-01-22. Retrieved 2009-06-20
- E. D. Bloom; et al. (1969). "High-Energy Inelastic e–pScattering at 6° and 10°". Physical Review Letters. 23 (16): 930–934. Bibcode:1969PhRvL..23..930B. doi:10.1103/PhysRevLett.23.930
- W.-M. Yao; et al. (Particle Data Group) (2006). "Review of Particle Physics: Pentaquark Update" (PDF). Journal of Physics G. 33 (1): 1–1232. arXiv:astro-ph/0601168. Bibcode:-2006JPhG...33....1Y. doi:10.1088/0954-3899/33/1/001
- J. D. Barrow (1997) [1994]. "The Singularity and Other Problems". The Origin of the Universe (Reprint ed.). Basic Books. ISBN 978-0-465-05314-8
- J. I. Friedman. "The Road to the Nobel Prize". Huế University. Archived from the original on 2008-12-25. Retrieved 2008-09-29
- B. J. Bjorken; S. L. Glashow (1964). "Elementary Particles and SU(4)". Physics Letters. 11 (3): 255–257. Bibcode:1964PhL....11..255B. doi:10.1016/0031-9163(64)90433-0
- K. Nakamura; et al. (Particle Data Group) (2010). "Review of Particles Physics: The CKM Quark-Mixing Matrix" (PDF). Journal of Physics G. 37: 075021. Bibcode:-2010JPhG...37g5021N. doi:10.1088/0954-3899/37/7A/075021
- S.S.M. Wong (1998a). "Chapter 2—Nucleon Structure". Introductory Nuclear Physics (2nd ed.). New York (NY): John Wiley & Sons. pp. 21–56. ISBN 0-471-23973-9
- D.M. Manley (2005). "Status of baryon spectroscopy". Journal of Physics: Conference Series. 5: 230–237. Bibcode:2005JPhCS...9..230M. doi:10.1088/1742-6596/9/1/043
- H. Garcilazo; J. Vijande & A. Valcarce (2007). "Faddeev study of heavy-baryon spectroscopy". Journal of Physics G. 34 (5): 961–976. arXiv:hep-ph/0703257. doi:10.1088/0954-3899/34/5/014
- "The Nobel Prize in Physics 1999, Press Release" (Press release). The Nobel Foundation. 12 October 1999. Retrieved 2009-09-10

4

Lepton: An Elementary Particle

Leptons are elementary particles, with half-integer spin. Its two main classes are charged leptons and neutral leptons. The topics elaborated in this chapter will help in providing a better perspective about the different kinds of neutrinos, such as electron neutrino, muon neutrino, and tau neutrino. It also sheds light on the concept of lepton number and explains Koide formula.

Lepton, any member of a class of subatomic particles that respond only to the electromagnetic force, weak force, and gravitational force and are not affected by the strong force. Leptons are said to be elementary particles; that is, they do not appear to be made up of smaller units of matter. Leptons can either carry one unit of electric charge or be neutral. The charged leptons are the electrons, muons, and taus. Each of these types has a negative charge and a distinct mass. Electrons, the lightest leptons, have a mass only $^{1}/_{1,840}$ that of a proton. Muons are heavier, having more than 200 times as much mass as electrons. Taus, in turn, are approximately 3,700 times more massive than electrons. Each charged lepton has an associated neutral partner, or neutrino (i.e., electron-, muon-, and tau-neutrino), that has no electric charge and no significant mass. Moreover, all leptons, including the neutrinos, have antiparticles c alled antileptons. The mass of the antileptons is identical to that of the leptons, but all of the other properties are reversed.

A third characteristic feature of leptons, in addition to their charge and mass properties, is their intrinsic angular momentum, or spin. Leptons are classified within a larger group of subatomic particles, the fermions, which are characterized by half-integer values of their spin. The total number of leptons appears to remain the same in every particle reaction. Mathematically, total lepton number L (the number of leptons minus the number of antileptons) is constant. In addition, a conservation law for leptons of each type seems to hold; the number of electrons and electron-neutrinos, for example, is conserved separately from the number of muons and muon-neutrinos. The current limit of violation of this conservation law is one part per million.

The word lepton comes from the Greek *leptos*, which means “small”, “fine”, or “thin”. The first recorded use of the word was by physicist Leon Rosenfeld in his book *Nuclear*

Forces (1948). In the book, he attributed the use of the word to a suggestion made by Danish chemist and physicist Prof. Christian Moller.

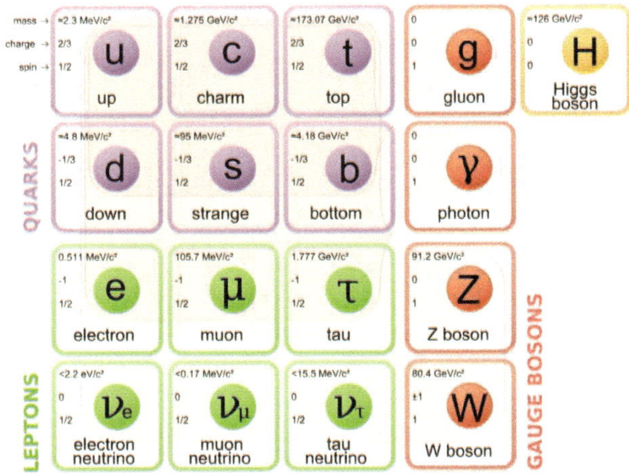

The Standard Model of Particle Physics, showing all known elementary particles.

The term was chosen to refer to particles of small mass, since the only known leptons in Rosenfeld's time were muons. These elementary particles are over 200 times more massive than electrons, but have only about one-ninth the the mass of a proton. Along with quarks, leptons are the basic building blocks of matter, and are therefore seen as "elementary particles".

The Types of Leptons

According to the Standard Model, there are six different types of leptons. These include the Electron, the Muon, and Tau particles, as well as their associated neutrinos (i.e. electron neutrino, muon neutrino, and tau neutrino). Leptons have negative charge and a distinct mass, whereas their neutrinos have a neutral charge.

Electrons are the lightest, with a mass of 0.000511 gigaelectronvolts (GeV), while Muons have a mass of 0.1066 Gev and Tau particles (the heaviest) have a mass of 1.777 Gev. The different varieties of the elementary particles are commonly called "flavors". While each of the three lepton flavors are different and distinct (in terms of their interactions with other particles), they are not immutable.

A neutrino can change its flavor, a process which is known as "neutrino flavor oscillation". This can take a number of forms, which include solar neutrino, atmospheric neutrino, nuclear reactor, or beam oscillations. In all observed cases, the oscillations were confirmed by what appeared to be a deficit in the number of neutrinos being created.

One observed cause has to do with "muon decay" a process where muons change their flavor to become electron neutrinos or tau neutrinos – depending on the circumstances. In addition, all three leptons and their neutrinos have an associated antiparticle (antilepton).

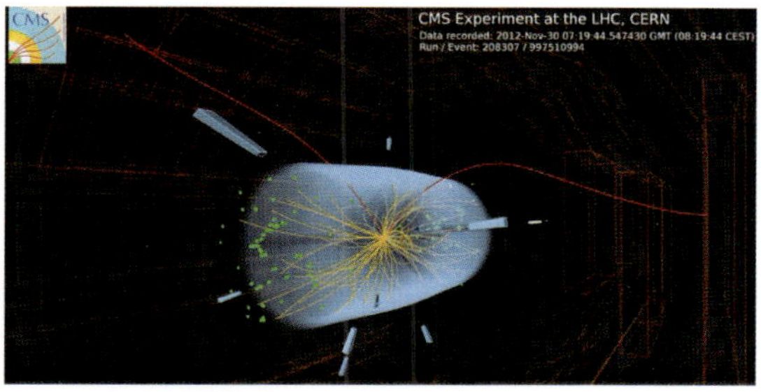

Muons, a type of lepton, shown being produced by the Large Hadron Collider.

For each, the antileptons have an identical mass, but all of the other properties are reversed. These pairings consist of the electron/positron, muon/antimuon, tau/antitau, electron neutrino/electron antineutrino, muon neutrino/muan antinuetrino, and tau neutrino/tau antineutrino.

The present Standard Model assumes that there are no more than three types (aka. "generations") of leptons with their associated neutrinos in existence. This accords with experimental evidence that attempts to model the process of nucleosynthesis after the Big Bang, where the existence of more than three leptons would have affected the abundance of helium in the early Universe.

Properties

Spin and Chirality

Left-handed and right-handed helicities

Leptons are spin-$\frac{1}{2}$ particles. The spin-statistics theorem thus implies that they are fermions and thus that they are subject to the Pauli exclusion principle: No two leptons of the same species can be in exactly the same state at the same time. Furthermore, it means that a lepton can have only two possible spin states, namely up or down.

A closely related property is chirality, which in turn is closely related to a more easily visualized property called helicity. The helicity of a particle is the direction of its spin relative to its momentum; particles with spin in the same direction as their momentum are called *right-handed* and otherwise they are called *left-handed*. When a particle is massless, the direction of its momentum relative to its spin is frame independent, while for massive particles it is possible to 'overtake' the particle by a Lorentz transformation flipping the helicity. Chirality is a technical property (defined through the transformation behaviour under the Poincaré group) that agrees with helicity for (approximately) massless particles and is still well defined for massive particles.

In many quantum field theories, such as quantum electrodynamics and quantum chromodynamics, left- and right-handed fermions are identical. However, the Standard Model's Weak interaction, treats left-handed and right-handed fermions are asymmetrically: Only left-handed fermions (and right-handed anti-fermions) participate in the weak interaction. This is an example of parity violation explicitly written into the model. In the literature, left-handed fields are often denoted by a capital *L* subscript (e.g. the normal electron: e_L^-) and right-handed fields are denoted by a capital *R* subscript (e.g. a positron e_R^+).

Right-handed neutrinos and left-handed anti-neutrinos have no possible interaction with other particles and so are not a functional part of the Standard Model, although their exclusion is not a strict requirement; they are sometimes listed in particle tables to emphasize that they would have no active role if included in the model. Even though electrically charged particles (electron, muon, or tau) do not engage in the weak interaction specifically, they can still interact electrically, and hence still participate in the combined electro-weak force, although with a different strengths (Y_W).

Electromagnetic Interaction

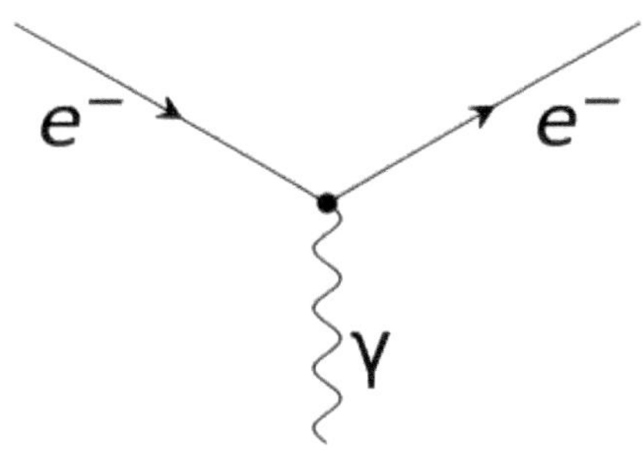

Lepton–photon interaction

One of the most prominent properties of leptons is their electric charge, Q. The electric charge determines the strength of their electromagnetic interactions. It determines the strength of the electric field generated by the particle and how strongly the particle reacts to an external electric or magnetic field. Each generation contains one lepton with $Q = -e$ (conventionally the charge of a particle is expressed in units of the elementary charge) and one lepton with zero electric charge. The lepton with electric charge is commonly simply referred to as a 'charged lepton' while the neutral lepton is called a neutrino. For example, the first generation consists of the electron e^- with a negative electric charge and the electrically neutral electron neutrino ν_e.

In the language of quantum field theory, the electromagnetic interaction of the charged leptons is expressed by the fact that the particles interact with the quantum of the electromagnetic field, the photon.

Because leptons possess an intrinsic rotation in the form of their spin, charged leptons generate a magnetic field. The size of their magnetic dipole moment μ is given by

$$\mu = g\frac{Q\hbar}{4m},$$

where m is the mass of the lepton and g is the so-called g-factor for the lepton. First order approximation quantum mechanics predicts that the g-factor is 2 for all leptons. However, higher order quantum effects caused by loops in Feynman diagrams introduce corrections to this value. These corrections, referred to as the anomalous magnetic dipole moment, are very sensitive to the details of a quantum field theory model and thus provide the opportunity for precision tests of the standard model. The theoretical and measured values for the electron anomalous magnetic dipole moment are within agreement within eight significant figures.

Weak Interaction

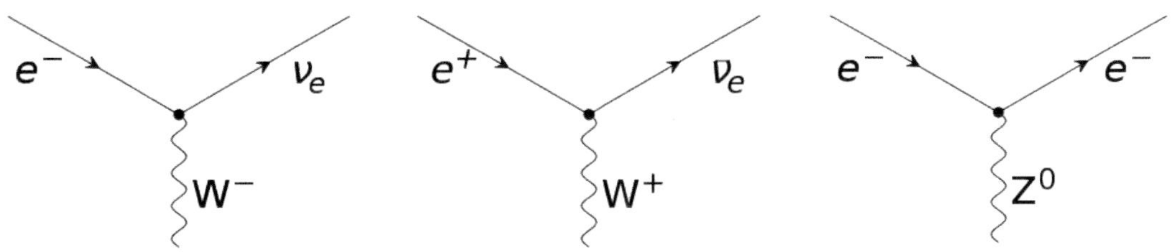

The weak interactions of the first generation leptons.

In the Standard Model, the left-handed charged lepton and the left-handed neutrino are arranged in doublet (ν_{e^L} ,e^-_L) that transforms in the spinor representation (T = ½) of the weak isospin SU(2) gauge symmetry. This means that these particles are eigenstates of the isospin projection T_3 with eigenvalues ½ and −½ respectively. In the meantime, the right-handed charged lepton transforms as a weak isospin scalar (T = 0) and thus does not participate in the weak interaction, while there is no evidence that a right-handed neutrino exists at all.

The Higgs mechanism recombines the gauge fields of the weak isospin SU(2) and the weak hypercharge U(1) symmetries to three massive vector bosons (W^+, W^-, Z^0) mediating the weak interaction, and one massless vector boson, the photon, responsible for the electromagnetic interaction. The electric charge Q can be calculated from the isospin projection T_3 and weak hypercharge Y_W through the Gell-Mann–Nishijima formula,

$$Q = T_3 + \tfrac{1}{2} Y_W$$

To recover the observed electric charges for all particles, the left-handed weak isospin doublet (ν_{e^L} , e^-_L) must thus have Y_W = −1, while the right-handed isospin scalar e^-_R must have Y_W = −2. The interaction of the leptons with the massive weak interaction vector bosons is shown in the figure.

Mass

In the Standard Model, each lepton starts out with no intrinsic mass. The charged leptons (i.e. the electron, muon, and tau) obtain an effective mass through interaction with the Higgs field, but the neutrinos remain massless. For technical reasons, the masslessness of the neutrinos implies that there is no mixing of the different generations of charged leptons as there is for quarks. This is in close agreement with current experimental observations.

However, it is known from experiments—most prominently from observed neutrino oscillations—that neutrinos do in fact have some very small mass, probably less than 2 eV/c^2. This implies the existence of physics beyond the Standard Model. The currently most favoured extension is the so-called seesaw mechanism, which would explain both why the left-handed neutrinos are so light compared to the corresponding charged leptons, and why we have not yet seen any right-handed neutrinos.

Leptonic Numbers

The members of each generation's weak isospin doublet are assigned leptonic numbers that are conserved under the Standard Model. Electrons and electron neutrinos have an *electronic number* of $L_e = 1$, while muons and muon neutrinos have a *muonic number* of $L_\mu = 1$, while tau particles and tau neutrinos have a *tauonic number* of $L_\tau = 1$. The antileptons have their respective generation›s leptonic numbers of −1.

Conservation of the leptonic numbers means that the number of leptons of the same type remains the same, when particles interact. This implies that leptons and antileptons must be created in pairs of a single generation. For example, the following processes are allowed under conservation of leptonic numbers:

Each generation forms a weak isospin doublet.

$$e^- + e^+ \rightarrow \gamma + \gamma,$$

$$\tau^- + \tau^+ \rightarrow Z^0 + Z^0,$$

but not these:

$$\gamma \rightarrow e^- + \mu^+,$$

$$W^- \rightarrow e^- + \nu_\tau,$$

$$Z^0 \rightarrow \mu^- + \tau^+.$$

$$\begin{pmatrix} \nu_e \\ e^- \end{pmatrix}, \begin{pmatrix} \nu_\mu \\ \mu^- \end{pmatrix}, \begin{pmatrix} \nu_\tau \\ \tau^- \end{pmatrix}$$

However, neutrino oscillations are known to violate the conservation of the individual leptonic numbers. Such a violation is considered to be smoking gun evidence for physics beyond the Standard Model. A much stronger conservation law is the conservation of the total number of leptons (*L*), conserved even in the case of neutrino oscillations, but even it is still violated by a tiny amount by the chiral anomaly.

Universality

The coupling of the leptons to gauge bosons are flavour-independent (i.e., the interactions between leptons and gauge bosons are the same for all leptons). This property is called lepton universality and has been tested in measurements of the tau and muon lifetimes and of Z boson partial decay widths, particularly at the Stanford Linear Collider (SLC) and Large Electron-Positron Collider (LEP) experiments.

The decay rate (Γ) of muons through the process $\mu^- \to e^- + \nu_e + \nu_\mu$ is approximately given by an expression of the form

$$\Gamma\left(\mu^- \to e^- + \overline{\nu_e} + \nu_\mu\right) = K_1 G_F^2 m_\mu^5,$$

where K_1 is some constant, and G_F is the Fermi coupling constant. The decay rate of tau particles through the process $\tau^- \to e^- + \nu_e + \nu_\tau$ is given by an expression of the same form

$$\Gamma\left(\tau^- \to e^- + \overline{\nu_e} + \nu_\tau\right) = K_2 G_F^2 m_\tau^5,$$

where K_2 is some constant. Muon–Tauon universality implies that $K_1 = K_2$. On the other hand, electron–muon universality implies

$$\Gamma\left(\tau^- \to e^- + \overline{\nu_e} + \nu_\tau\right) = \Gamma\left(\tau^- \to \mu^- + \overline{\nu_\mu} + \nu_\tau\right).$$

This explains why the branching ratios for the electronic mode (17.85%) and muonic (17.36%) mode of tau decay are equal (within error).

Universality also accounts for the ratio of muon and tau lifetimes. The lifetime of a lepton (τ_ℓ) is related to the decay rate by

$$\tau_\ell = \frac{B\left(\ell^- \to e^- + \overline{\nu_e} + \nu_\ell\right)}{\Gamma\left(\ell^- \to e^- + \overline{\nu_e} + \nu_\ell\right)},$$

where $B(x \to y)$ and Γ $(x \to y)$ denotes the branching ratios and the resonance width of the process $x \to y$.

The ratio of tau and muon lifetime is thus given by

$$\frac{\tau_\tau}{\tau_\mu} = \frac{B\left(\tau^- \to e^- + \overline{\nu}_e + \nu_\tau\right)}{B\left(\mu^- \to e^- + \overline{\nu}_e + \nu_\mu\right)} \left(\frac{m_\mu}{m_\tau}\right)^5.$$

Using the values of the 2008 *Review of Particle Physics* for the branching ratios of muons and tau yields a lifetime ratio of ~1.29×10^{-7}, comparable to the measured lifetime ratio of ~1.32×10^{-7}. The difference is due to K_1 and K_2 not actually being constants: They depend on the mass of leptons.

Recent tests of lepton universality in B meson decays, performed by the LHCb, BaBar and Belle experiments, have shown consistent deviations from the Standard Model predictions. However the statistical significance is not yet high enough to claim an observation of new physics.

Table of Leptons

Properties of leptons								
Particle / Antiparticle name	**Symbol**	**Charge Q (e)**	**Spin J**	**L_e**	**L_μ**	**L_τ**	**Mass (MeV/c²)**	**Lifetime (seconds)**
Electron	e^-	−1	½	+1	0	0	0.510 998 910 (±13)	Stable
Positron	e^+	+1		−1				
Muon	μ^-	−1	½	0	+1	0	105.658 366 8 (±3 8)	2.197 019×10^{-6} (±21)
Antimuon	μ^+	+1			−1			
Tau	τ^-	−1	½	0	0	+1	1 776.84 (±.17)	2.906×10^{-13} (±.010)
Antitau	τ^+	+1				−1		
Electron neutrino	ν_e	0	½	+1	0	0	< 0.000 002 2	Unknown
Electron antineutrino	$\overline{\nu}_e$			−1				
Muon neutrino	ν_μ	0	½	0	+1	0	< 0.17	Unknown
Muon antineutrino	$\overline{\nu}_\mu$				−1			

Tau neutrino	ν_τ	0	½	0	0	+1	< 15.5	Unknown
Tau antineutrino	$\overline{\nu}_\tau$							

Neutrino

Neutrino, elementary subatomic particle with no electric charge, very little mass, and 1/2 unit of spin. Neutrinos belong to the family of particles called leptons, which are not subject to the strong force. Rather, neutrinos are subject to the weak force that underlies certain processes of radioactive decay. There are three types of neutrino, each associated with a charged lepton—i.e., the electron, the muon, and the tau—and therefore given the corresponding names electron-neutrino, muon-neutrino, and tau-neutrino. Each type of neutrino also has an antimatter component, called an antineutrino; the term neutrino is sometimes used in a general sense to refer to both the neutrino and its antiparticle.

The basic properties of the electron-neutrino—no electric charge and little mass—were predicted in 1930 by the Austrian physicist Wolfgang Pauli to explain the apparent loss of energy in the process of radioactive beta decay. The Italian-born physicist Enrico Fermi further elaborated (1934) the theory of beta decay and gave the "ghost" particle its name. An electron-neutrino is emitted along with a positron in positive beta decay, while an electron-antineutrino is emitted with an electron in negative beta decay.

Despite such predictions, neutrinos were not detected experimentally for 20 years, owing to the weakness of their interactions with matter. Because they are not electrically charged, neutrinos do not experience the electromagnetic force and thus do not cause ionization of matter. Furthermore, they react with matter only through the very weak interaction of the weak force. Neutrinos are therefore the most penetrating of subatomic particles, capable of passing through an enormous number of atoms without causing any reaction. Only 1 in 10 billion of these particles, traveling through matter for a distance equal to Earth's diameter, reacts with a proton or a neutron. Finally, in 1956 a team of American physicists led by Frederick Reines reported the discovery of the electron-antineutrino. In their experiments antineutrinos emitted in a nuclear reactor were allowed to react withprotons to produce neutrons and positrons. The unique (and rare) energy signatures of the fates of these latter by-products provided the evidence for the existence of the elelectron-antineutrino.

The discovery of the second type of charged lepton, the muon, became the starting point for the eventual identification of a second type of neutrino, the muon-neutrino. Identification of the muon-neutrino as distinct from the electron-neutrino was accomplished in 1962 on the basis of the results of a particle-accelerator experiment.

High-energy muon-neutrinos were produced by decay of pi-mesons and were directed to a detector so that their reactions with matter could be studied. Although they are as unreactive as the other neutrinos, muon-neutrinos were found to produce muons but never electrons on the rare occasions when they reacted with protons or neutrons. The American physicists Leon Lederman, Melvin Schwartz, and Jack Steinberger received the 1988 Nobel Prize for Physics for having established the identity of muon-neutrinos.

In the mid-1970s particle physicists discovered yet another variety of charged lepton, the tau. A tau-neutrino and tau-antineutrino are associated with this third charged lepton as well. In 2000 physicists at the Fermi National Accelerator Laboratory reported the first experimental evidence for the existence of the tau-neutrino.

All types of neutrino have masses much smaller than those of their charged partners. For example, experiments show that the mass of the electron-neutrino must be less than 0.002 percent that of the electron and that the sum of the masses of the three types of neutrinos must be less than 0.48 electron volt. For many years it seemed that neutrinos' masses might be exactly zero, although there was no compelling theoretical reason why this should be so. Then in 2002 the Sudbury Neutrino Observatory (SNO), in Ontario, Canada, found the first direct evidence that electron-neutrinos emitted by nuclear reactions in the core of the Sun change type as they travel through the Sun. Such neutrino "oscillations" are possible only if one or more of the neutrino types has some small mass. Studies of neutrinos produced in the interactions of cosmic rays in Earth's atmosphere also indicate that neutrinos have mass, but further experiments are needed to understand the exact masses involved.

Unlike the charged leptons, the electrically neutral leptons, the neutrinos, do not come under the influence of the electromagnetic force. They experience only the weakest two of nature's forces, the weak force and gravity. For this reason neutrinos react extremely weakly with matter. They can, for example, pass through Earth without interacting, which makes it difficult to detect neutrinos and to measure their properties.

Although electrically neutral, the neutrinos seem to carry an identifying property that associates them specifically with one type of charged lepton. In the example of the muon's decay, the antineutrino produced is not simply the antiparticle of the neutrino that appears with it. The neutrino carries a muon-type hallmark, while the antineutrino, like the antineutrino emitted when a neutron decays, is always an electron-antineutrino. In interactions with matter, such electron-neutrinos and antineutrinos never produce muons, only electrons. Likewise, muon-neutrinos give rise to muons only, never to electrons.

Theory does not require the mass of neutrinos to be any specific amount, and in the past it was assumed to be zero. Observations of the cosmic microwave background and gravitational lensing of distant galaxies indicate the three flavours of neutrino have a combined mass of 0.32 eV, or less than $^{1}/_{100,000,000}$ the mass of an electron. Neutrinos can change from one type to another, or "oscillate." This can happen only if the neutrino types in question have small differences in mass—and hence must have mass.

Three types of neutrinos are known; there is strong evidence that no additional neutrinos exist, unless their properties are unexpectedly very different from the known types. Each type or "flavor" of neutrino is related to a charged particle (which gives the corresponding neutrino its name). Hence, the "electron neutrino" is associated with the electron, and two other neutrinos are associated with heavier versions of the electron called the muon and the tau (elementary particles are frequently labelled with Greek letters, to confuse the layman). The table below lists the known types of neutrinos (and their electrically charged partners).

Neutrino	n_e	n_m	n_t
Charged Partner	electron (e)	muon (m)	tau (t)

Electron Neutrino

The electron neutrino (a lepton) was first postulated in 1930 by Wolfgang Pauli to explain why the electrons in beta decay were not emitted with the full reaction energy of the nuclear transition. The apparent violation of conservation of energy and momentum was most easily avoided by postulating another particle. Enrico Fermi called the particle a neutrino and developed a theory of beta decay based on it, but it was not experimentally observed until 1956. This elusive particle, with no charge and almost no mass, could penetrate vast thicknesses of material without interaction. The mean free path of a neutrino in water would be on the order of 10x the distance from the Earth to the Sun. In the standard Big Bang model, the neutrinos left over from the creation of the universe are the most abundant particles in the universe. This remnant neutrino density is put at 100 per cubic centimeter at an effective temperature of 2K (Simpson). The background temperature for neutrinos is lower than that for the microwave background (2.7K) because the neutrino transparency point came earlier. The sun emits vast numbers of neutrinos which can pass through the earth with little or no interaction. This leads to the statement "Solar neutrinos shine down on us during the day, and shine up on us during the night!". Bahcall's modeling of the solar neutrino flux led to the prediction of about 5×10^6 neutrinos/cm^2s.

A remarkable opportunity for observing neutrinos came with Supernova 1987A when the Japanese observing team detected neutrinos almost coincident with the discovery of the light from the supernova.

Neutrinos interact only by the weak interaction. Their interactions are usually represented in terms of Feynman diagrams.

The electron neutrino (ν_e) is a subatomic lepton elementary particle which has no net electric charge. Together with the electron it forms the first generation of leptons, hence the name electron neutrino. It was first hypothesized by Wolfgang Pauli in 1930, to account for missing momentum and missing energy in beta decay, and was discovered in 1956 by a team led by Clyde Cowan and Frederick Reines.

Proposal

In the early 1900s, theories predicted that the electrons resulting from beta decay should have been emitted at a specific energy. However, in 1914, James Chadwick showed that electrons were instead emitted in a continuous spectrum.

$$n^0 \rightarrow p^+ + e^-$$

The early understanding of beta decay

In 1930, Wolfgang Pauli theorized that an undetected particle was carrying away the observed difference between the energy, momentum, and angular momentum of the initial and final particles.

$$n^0 \rightarrow p^+ + e^- + \nu_e^{-0}$$

Pauli's version of beta decay

Discovery

The electron neutrino was discovered by Clyde Cowan and Frederick Reines in 1956.

Name

Pauli originally named his proposed light particle a neutron. When James Chadwick discovered a much more massive nuclear particle in 1932 and also named it a neutron, this left the two particles with the same name. Enrico Fermi, who developed the theory of beta decay, introduced the term neutrino in 1934 (it was jokingly coined by Edoardo Amaldi during a conversation with Fermi at the Institute of physics of via Panisperna in Rome, in order to distinguish this light neutral particle from Chadwick's neutron) to resolve the confusion. It was a pun on neutrone, the Italian equivalent of neutron: the -one ending can be an augmentative in Italian, so neutrone could be read as the "large neutral thing"; -ino replaces the augmentative suffix with a diminutive one.

Upon the prediction and discovery of a second neutrino, it became important to distinguish between different types of neutrinos. Pauli's neutrino is now identified as the electron neutrino, while the second neutrino is identified as the muon neutrino.

Electron Antineutrino

The electron neutrino has a corresponding antiparticle, the electron antineutrino ($\overline{\nu}_e$), which differs only in that some of its properties have equal magnitude but opposite sign. One open question of particle physics is whether or not neutrinos and anti-neutrinos are the same particle in which case it would be a Majorana fermion or whether they are different particles in which case they would be Dirac fermions. They are produced in beta decay and other types weak interactions.

Muon Neutrino

The muon neutrino is a lepton, an elementary subatomic particle which has the symbol ν_μ and no net electric charge. Together with the muon it forms the second generation of leptons, hence the name muon neutrino. It was first hypothesized in the early 1940s by several people, and was discovered in 1962 by Leon Lederman, Melvin Schwartz and Jack Steinberger. The discovery was rewarded with the 1988 Nobel Prize in Physics.

Speed

In September 2011 OPERA researchers reported that muon neutrinos were apparently traveling at faster than light speed. This result was confirmed again in a second experiment in November 2011. These results have been viewed skeptically by the scientific community at large, and more experiments have/are investigating the phenomenon. In March 2012 the ICARUS team published results directly contradicting the results of OPERA.

Later, in July 2012, the apparent anomalous super-luminous propagation of neutrinos was traced to a faulty element of the fibre optic timing system in Gran-Sasso. After it was corrected the neutrinos appeared to travel with the speed of light within the errors of the experiment.

Discovery of Muon Neutrino

1962 - Lederman, Schwarts, Steinberger (Brookhaven) make an experiment to check whether muon and electron neutrino indeed different. One has to start from muon neutrinos and see whether all they can make are muons:

$$\bar{\nu}_\mu + p \rightarrow n + \mu^+$$

$$\nu_\mu + n \rightarrow p + \mu^-$$

Pions is an excellent source of muon neutrinos...

The idea of neutrino beams actually belongs to Pontecorvo...

- 15 GeV protons → Be target

  → lots of pions: π^+ , π^- , π^0
- 20 m decay distance: not too short (good fraction of pions decay), not too long (muons do not decay)

  $$\pi^+ \to \mu^+ \ \nu_\mu \left(10^{-8} \ s\right), \ \ \mu^+ \to e^+ \ \nu_e \ \overline{\nu}_\mu \left(10^{-6} s\right)$$

  $$\pi^- \ \to \ \mu^- \overline{\nu}_\mu \left(10^{-8} \ s\right), \ \ \mu^- \to e^- \ \nu_\mu \ \overline{\nu}_e \left(10^{-6} s\right),$$

  → lots of photons, pions, muons, and high energy muon neutrinos and antineutrinos
- 13 m of steel to absorb all particles but neutrinos;
- spark chamber with Al sheets
- cover detectors with scintillator to veto cosmic rays

  → 10^{14} muon neutrinos and antineutrinos muons should produce nice tracks running throughout the entire detector (29 events) electrons should produce showers as they cross Al sheets (none)

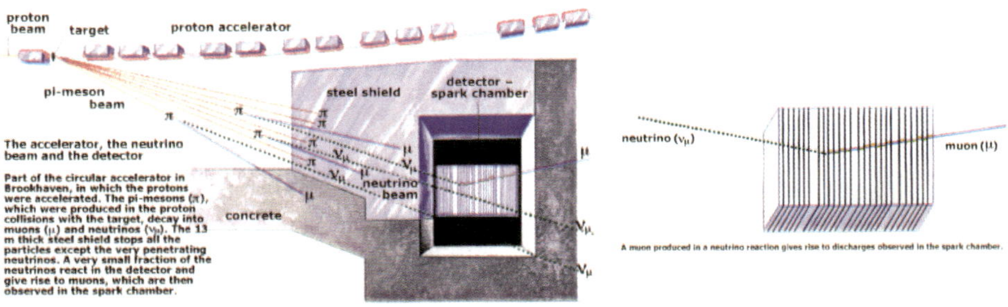

Mass limit from direct mass measurements <200 keV

1988 - Lederman, Schwarts, Steinberger receive the Nobel Prize "for the neutrino beam method and the demonstration of the doublet structure of the leptons through the discovery of the muon neutrino"

Tau Neutrino

The tau neutrino or tauon neutrino is a subatomic elementary particle which has the symbol ν_T and no net electric charge. Together with the tau, it forms the third generation of leptons, hence the name tau neutrino. Its existence was immediately implied after the tau particle was detected in a series of experiments between 1974 and 1977 by

Martin Lewis Perl with his colleagues at the SLAC–LBL group. The discovery of the tau neutrino was announced in July 2000 by the DONUT collaboration (Direct Observation of the Nu Tau).

Lepton Number

In particle physics, the lepton number is used to denote which particles are leptons and which particles are not. Each lepton has a lepton number of 1 and each antilepton has a lepton number of -1. Other non-leptonic particles have a lepton number of 0. The lepton number is a conserved quantum number in all particle reactions. A slight asymmetry in the laws of physics allowed leptons to be created in the Big Bang.

Q/e	$L_e=-1$	$L_\mu=-1$	$L_\tau=-1$
0 +1	$\begin{pmatrix}\overline{\nu}_e \\ e^+\end{pmatrix}$	$\begin{pmatrix}\overline{\nu}_\mu \\ \mu^+\end{pmatrix}$	$\begin{pmatrix}\overline{\nu}_\tau \\ \tau^+\end{pmatrix}$
Q/e	$L_e=1$	$L_\mu=1$	$L_\tau=1$
0 −1	$\begin{pmatrix}\nu_e \\ e^-\end{pmatrix}$	$\begin{pmatrix}\nu_\mu \\ \mu^-\end{pmatrix}$	$\begin{pmatrix}\nu_\tau \\ \tau^-\end{pmatrix}$

Lepton Flavor Conservation

In addition to lepton number, lepton family numbers are defined as

- L_e , the electron number, for the electron and the electron neutrino;
- L_μ, the muon number, for the muon and the muon neutrino; and
- L_τ , the tau number, for the tau and the tau neutrino.

Prominent examples of lepton flavor conservation are the muon decays $\mu^- \rightarrow e^- + \overline{\nu}_e + \nu_\mu$ and $\mu^+ \rightarrow e^+ + \nu_e + \overline{\nu}_\mu$. In these, the creation of an electron is accompanied by the creation of an electron antineutrino, and the creation of a positron is accompanied by the creation of an electron neutrino. Likewise, a decaying negative muon results in the creation of a muon neutrino, while a decaying positive muon results in the creation of a muon antineutrino.

Violations of the Lepton Number Conservation Laws

Lepton flavor is only approximately conserved, and is notably not conserved in neutrino oscillation. However, total lepton number is still conserved in the Standard Model.

Numerous searches for physics beyond the Standard Model incorporate searches for lepton number or lepton flavor violation, such as the decays $\mu^- \rightarrow e^- + \gamma$. Experiments such as MEGA and SINDRUM have searched for lepton number violation in muon decays to electrons; MEG set the current branching limit of order 10^{-13} and plans to lower to limit to 10^{-14} after 2016. Some theories beyond the Standard Model, such as supersymmetry, predict branching ratios of order 10^{-12} to 10^{-14}. The Mu2e experiment, in construction as of 2017, has a planned sensitivity of order 10^{-17}.

Because the lepton number conservation law in fact is violated by chiral anomalies, there are problems applying this symmetry universally over all energy scales. However, the quantum number B − L is commonly conserved in Grand Unified Theory models.

Law of Conservation of Lepton Number

The conservation of lepton number means that whenever a lepton of a certain generation is created or destroyed in a reaction, a corresponding antilepton from the same generation must be created or destroyed. It must be added, there is a separate requirement for each of the three generations of leptons, the electron, muon and tau and their associated neutrinos.

References

- R. Shankar (1994). "Chapter 2: Rotational Invariance and Angular Momentum". Principles of Quantum Mechanics (2nd ed.). Springer. pp. 305–352. ISBN 978-0-306-44790-7
- G. Danby; J.-M. Gaillard; K. Goulianos; L. M. Lederman; N. B. Mistry; M. Schwartz; J. Steinberger (1962). "Observation of high-energy neutrino reactions and the existence of two kinds of neutrinos". Physical Review Letters. 9: 36. Bibcode:1962PhRvL...9...36D. doi:10.1103/PhysRevLett.9.36
- L.M. Brown (1978). "The idea of the neutrino". Physics Today. 31 (9): 23. Bibcode:1978PhT....31i..23B. doi:10.1063/1.2995181
- Griffiths, David J. (1987). Introduction to Elementary Particles. Wiley, John & Sons, Inc. ISBN 0-471-60386-4. ; Tipler, Paul; Llewellyn, Ralph (2002). Modern Physics (4th ed.). W. H. Freeman. ISBN 0-7167-4345-0
- R. Wilson (1997). Astronomy Through the Ages: The Story of the Human Attempt to Understand the Universe. CRC Press. p. 138. ISBN 0-7484-0748-0
- Gribov, V.; Pontecorvo, B. (1969-01-20). "Neutrino astronomy and lepton charge". Physics Letters B. 28 (7): 493–496. Bibcode:1969PhLB...28..493G. doi:10.1016/0370-2693(69)90525-5. ISSN 0370-2693
- "OPERA experiment reports anomaly in flight time of neutrinos from CERN to Gran Sasso (UPDATE 8 June 2012)". CERN press office. 8 June 2012. Retrieved 19 April 2013

5

Particle Physics: Models and Theories

Fundamental particles and their interactions are understood from a quantum field theoretic model called the Standard Model. This chapter includes a detailed explanation of the Standard Model and various other theories associated to particle physics like the Grand Unified Theory and String theory. It also closely examines some fundamental particles as given by the Standard Model such as Gauge bosons and the Higgs boson, in order to provide an extensive understanding of the subject.

Standard Model

The Standard Model is a kind of periodic table of the elements for particle physics. But instead of listing the chemical elements, it lists the fundamental particles that make up the atoms that make up the chemical elements, along with any other particles that cannot be broken down into any smaller pieces.

The standard model describes the types of particles that make up matter and the possible interactions between these particles.

Types of Particle

All particles belong to one of two categories: fermions or bosons. The classification of which category a particle belongs to is based on the particle's spin. Spin is an important quantum mechanical property that is intrinsic to the particle and nothing to do with the particle actually physically spinning.

A fermion is any particle that has a half-integer spin (1/2, 3/2, 5/2, etc.). The consequence of this, called the Pauli exclusion principle, is that fermions can't occupy the same quantum state; a quantum state simply being all the values of the particle's quantum properties (energy, spin etc.). A boson is any particle that has an integer spin (0, 1, 2, etc.). There is no restriction on bosons, any number of them can occupy the same quantum state. The categorisation of boson or fermion determines the particle's behaviour. Elementary fermions are the particles that make up matter and elementary bosons carry forces between these matter particles.

There are only four fundamental interactions that have been observed within our universe: gravity, electromagnetism, the strong interaction and the weak interaction.

Gravity

Gravity is the only fundamental interaction unexplained within the standard model. Our current model of gravity is Einstein's theory of general relativity. However, general relativity hasn't been reconciled with QFT. This isn't a large practical problem because gravity is the weakest force by a significant margin and gravitational forces can be assumed as negligible for interactions between particles.

Electromagnetism

The electromagnetic interaction affects all particles that carry an electrical charge. It causes like charged particles to repel each other and oppositely charged particles to attract each other. The force carrier for electromagnetism is the photon (γ). The photon is electrically neutral and massless, hence the electromagnetic interaction has an infinite range (although it does get significantly weaker with distance). The characteristic time for electromagnetic interactions is somewhere between 10^{-20} and 10^{-15} seconds.

Examples of the electromagnetic interaction include atoms being held together, the force preventing stuff from falling through the floor, friction between objects and the interactions with electrons that form the basis of all electrical technology. All everyday phenomena are either produced by electromagnetic or gravitational interactions.

Strong Interaction

Analogous to how the electromagnetic interaction couples to electrical charge, the strong interaction couples to a different kind of charge, the colour charge. All quarks carry a colour, which is either red, blue or green. Antiquarks can carry anti-red, anti-blue or anti-green. This has nothing to do with the quark's visible colour, it is simply another quantum number. All observable particles (i.e. hadrons) have to be colour neutral. Mesons can obviously meet this requirement e.g. red + anti-red is colour neutral. Baryons can also be colour neutral if each quark is a different colour (an RGB combination), this is a labelling concept that draws from white light being a sum of different coloured lights.

The force carrier for strong interactions is the gluon (g). Gluons are massless and they also carry colour charges, meaning gluons can strongly interact with each other, unlike photons. A gluon carries both a colour and an anti-colour, this leads to there being eight different gluon types. The strong interaction is attractive but it gets stronger with distance. This effect leads to the phenomena of colour confinement. When attempting to pull apart quarks, the increasing distance between them increases the energy of the colour field between them. Eventually, it becomes a high enough energy that a quark-antiquark pair is created. This phenomenon prevents individual quarks from ever being observed.

Even though gluons are massless, the effect of colour confinement limits the range of the strong interaction to about 10^{-15} metres, roughly the size of the nucleus within an atom. The characteristic time for strong interactions is less than 10^{-20} seconds. At the atomic scale, the strong interaction is the dominant interaction and hence it is responsible for binding together quarks into hadrons and holding the nucleus together.

Weak Interaction

The weak interaction causes massive particles to decay into lighter particles. Unlike the electromagnetic and strong interactions, the weak interaction is allowed to break the conservation of some quantum numbers such as strangeness. Breaking these conservation laws allows the weak interaction to change the flavour of quarks and so it can cause quarks to decay between generations. However, electrical charge, baryon number and lepton numbers must still be conserved.

The weak interaction has three force carriers: the W^+, the W^- and the Z^o. All three particles have heavy masses (around 80 GeV/c^2) which limits the range of the interaction to about 10^{-18} metres, significantly smaller than the size of a proton. The characteristic time for weak interactions is typically longer than 10^{-15} seconds. The weak nuclear interaction is responsible for nuclear decays and hence it is the fundamental interaction behind radioactivity.

Within the standard model, the weak interaction has been successfully unified with the electromagnetic interaction. Essentially, electromagnetism and the weak interaction are actually manifestations of the same interaction, the electroweak interaction. At our universe's current low energy they appear to be very different but at a high enough energy (around 246 GeV) they merge to become the same interaction.

Construction of the Standard Model Lagrangian

Parameters of the Standard Model			
Symbol	**Description**	**Renormalization scheme (point)**	**Value**
m_e	Electron mass		511 keV
m_μ	Muon mass		105.7 MeV
m_τ	Tau mass		1.78 GeV
m_u	Up quark mass	μ $\overline{MS} = 2GeV$	1.9 MeV
m_d	Down quark mass	μ $\overline{MS} = 2GeV$	4.4 MeV
m_s	Strange quark mass	μ $\overline{MS} = 2GeV$	87 MeV
m_c	Charm quark mass	μ $\overline{MS} = M_c$	1.32 GeV
m_b	Bottom quark mass	μ $\overline{MS} = M_b$	4.24 GeV

m_t	Top quark mass	On shell scheme	173.5 GeV
θ_{12}	CKM 12-mixing angle		13.1°
θ_{23}	CKM 23-mixing angle		2.4°
θ_{13}	CKM 13-mixing angle		0.2°
δ	CKM CP violation Phase		0.995
g_1 or g'	U(1) gauge coupling	$\mu_{\overline{MS}} = M_Z$	0.357
g_2 or g	SU(2) gauge coupling	$\mu_{\overline{MS}} = M_Z$	0.652
g_3 or g_s	SU(3) gauge coupling	$\mu_{\overline{MS}} = M_Z$	1.221
θ_{QCD}	QCD vacuum angle		~0
v	Higgs vacuum expectation value		246 GeV
m_H	Higgs mass		125.09±0.24 GeV

Technically, quantum field theory provides the mathematical framework for the Standard Model, in which a Lagrangian controls the dynamics and kinematics of the theory. Each kind of particle is described in terms of a dynamical field that pervades space-time. The construction of the Standard Model proceeds following the modern method of constructing most field theories: by first postulating a set of symmetries of the system, and then by writing down the most general renormalizable Lagrangian from its particle (field) content that observes these symmetries.

The global Poincaré symmetry is postulated for all relativistic quantum field theories. It consists of the familiar translational symmetry, rotational symmetry and the inertial reference frame invariance central to the theory of special relativity. The local SU(3)×SU(2)×U(1) gauge symmetry is an internal symmetry that essentially defines the Standard Model. Roughly, the three factors of the gauge symmetry give rise to the three fundamental interactions. The fields fall into different representations of the various symmetry groups of the Standard Model. Upon writing the most general Lagrangian, one finds that the dynamics depends on 19 parameters, whose numerical values are established by experiment. The parameters are summarized in the table above (note: the Higgs mass is at 125 GeV, the Higgs self-coupling strength $\lambda \sim 1/8$).

Quantum Chromodynamics Sector

The quantum chromodynamics (QCD) sector defines the interactions between quarks and gluons, with SU(3) symmetry, generated by T^a. Since leptons do not interact with gluons, they are not affected by this sector. The Dirac Lagrangian of the quarks coupled to the gluon fields is given by

$$\mathcal{L}_{QCD} = \sum_{\psi} \bar{\psi}_i \left(i\gamma^{\mu} (\partial_{\mu}\delta_{ij} - i g_s G_{\mu}^{a} T_{ij}^{a}) - m_{\psi}\delta_{ij} \right) \psi_j - \frac{1}{4} G_{\mu\nu}^{a} G_{a}^{\mu\nu},$$

Where

ψ_i is the Dirac spinor of the quark field, where $i = \{r, g, b\}$ represents color,

γ^{μ} are the Dirac matrices,

G^{a}_{μ} is the 8-component ($a = 1,2,\ldots,8$) SU(3) gauge field,

T^{a}_{ij} are the 3 × 3 Gell-Mann matrices, generators of the SU(3) color group,

$G^{a}_{\mu\nu}$ are the field strength tensors for the gluons,

g_s is the strong coupling constant.

Electroweak Sector

The electroweak sector is a Yang–Mills gauge theory with the simple symmetry group U(1) × SU(2)$_L$,

$$\mathcal{L}_{EW} = \sum_{\psi} \bar{\psi} \gamma^{\mu} \left(i\partial_{\mu} - g' \tfrac{1}{2} Y_W B_{\mu} - g \tfrac{1}{2} \vec{\tau}_L \vec{W}_{\mu} \right) \psi - \tfrac{1}{4} W_{a}^{\mu\nu} W_{\mu\nu}^{a} - \tfrac{1}{4} B^{\mu\nu} B_{\mu\nu},$$

where

B_{μ} is the U(1) gauge field,

Y_W is the weak hypercharge – the generator of the U(1) group,

$\vec{W}_{\mu}$ is the 3-component SU(2) gauge field,

$\vec{\tau}_L$ are the Pauli matrices – infinitesimal generators of the SU(2) group – with subscript L to indicate that they only act on *left*-chiral fermions,

g' and *g* are the U(1) and SU(2) coupling constants respectively,

$W^{a\mu\nu}$ ($a = 1,2,3$) and $B^{\mu\nu}$ are the field strength tensors for the weak isospin and weak hypercharge fields.

Notice that the addition of fermion mass terms into the electroweak lagrangian is forbidden, since terms of the form $\bar{\psi}\psi$ do not respect U(1) × SU(2)$_L$ gauge invariance. Neither is it possible to add explicit mass terms for the U(1) and SU(2) gauge fields. The Higgs mechanism is responsible for the generation of the gauge boson masses, and the fermion masses result from Yukawa-type interactions with the Higgs field.

Higgs Sector

In the Standard Model, the Higgs field is a complex scalar of the group $SU(2)_L$:

$$\varphi = \frac{1}{\sqrt{2}}\begin{pmatrix}\varphi^+ \\ \varphi^0\end{pmatrix},$$

where the superscripts + and 0 indicate the electric charge (Q) of the components. The weak isospin (Y_W) of both components is 1.

Before symmetry breaking, the Higgs Lagrangian is

$$\mathcal{L}_H = \varphi^\dagger\left(\partial^\mu - \frac{i}{2}\left(g'Y_W B^\mu + g\vec{\tau}\vec{W}^\mu\right)\right)\left(\partial_\mu + \frac{i}{2}\left(g'Y_W B_\mu + g\vec{\tau}\vec{W}_\mu\right)\right)\varphi - \frac{\lambda^2}{4}\left(\varphi^\dagger\varphi - v^2\right)^2,$$

which can also be written as

$$\mathcal{L}_H = \left|\left(\partial_\mu + \frac{i}{2}\left(g'Y_W B_\mu + g\vec{\tau}\vec{W}_\mu\right)\right)\varphi\right|^2 - \frac{\lambda^2}{4}\left(\varphi^\dagger\varphi - v^2\right)^2.$$

Yukawa Sector

The Yukawa interaction terms are

$$\mathcal{L}_{Yukawa} = \bar{U}_L G_u U_R \phi^0 - \bar{D}_L G_u U_R \phi^- + \bar{U}_L G_d D_R \phi^+ + \bar{D}_L G_d D_R \phi^0 + hc,$$

where $G_{u,d}$ are 3 × 3 matrices of Yukawa couplings, with the ij term giving the coupling of the generations i and j.

Challenges

Self-consistency of the Standard Model (currently formulated as a non-abelian gauge theory quantized through path-integrals) has not been mathematically proven. While regularized versions useful for approximate computations (for example lattice gauge theory) exist, it is not known whether they converge (in the sense of S-matrix elements) in the limit that the regulator is removed. A key question related to the consistency is the Yang–Mills existence and mass gap problem.

Experiments indicate that neutrinos have mass, which the classic Standard Model did not allow. To accommodate this finding, the classic Standard Model can be modified to include neutrino mass.

If one insists on using only Standard Model particles, this can be achieved by adding a non-renormalizable interaction of leptons with the Higgs boson. On a fundamental level, such an interaction emerges in the seesaw mechanism where heavy right-handed neutrinos are added to the theory. This is natural in the left-right symmetric extension of the Standard Model and in certain grand unified theories. As long as new physics appears below or around 10^{14} GeV, the neutrino masses can be of the right order of magnitude.

Theoretical and experimental research has attempted to extend the Standard Model into a Unified field theory or a Theory of everything, a complete theory explaining all physical phenomena including constants. Inadequacies of the Standard Model that motivate such research include:

- The model does not explain gravitation, although physical confirmation of a theoretical particle known as a graviton would account for it to a degree. Though it addresses strong and electroweak interactions, the Standard Model does not consistently explain the canonical theory of gravitation, general relativity, in terms of quantum field theory. The reason for this is, among other things, that quantum field theories of gravity generally break down before reaching the Planck scale. As a consequence, we have no reliable theory for the very early universe.
- Some physicists consider it to be *ad hoc* and inelegant, requiring 19 numerical constants whose values are unrelated and arbitrary. Although the Standard Model, as it now stands, can explain why neutrinos have masses, the specifics of neutrino mass are still unclear. It is believed that explaining neutrino mass will require an additional 7 or 8 constants, which are also arbitrary parameters.
- The Higgs mechanism gives rise to the hierarchy problem if some new physics (coupled to the Higgs) is present at high energy scales. In these cases, in order for the weak scale to be much smaller than the Planck scale, severe fine tuning of the parameters is required; there are, however, other scenarios that include quantum gravity in which such fine tuning can be avoided. There are also issues of quantum triviality, which suggests that it may not be possible to create a consistent quantum field theory involving elementary scalar particles.
- The model is inconsistent with the emerging "Standard Model of cosmology". More common contentions include the absence of an explanation in the Standard Model of particle physics for the observed amount of cold dark matter (CDM) and its contributions to dark energy, which are many orders of magnitude too large. It is also difficult to accommodate the observed predominance of matter over antimatter (matter/antimatter asymmetry). The isotropy and homogeneity of the visible universe over large distances seems to require a mechanism like cosmic inflation, which would also constitute an extension of the Standard Model.

Currently, no proposed Theory of Everything has been widely accepted or verified.

Limitations of the Standard Model

Despite being a successful, thoroughly tested theory we know that there are problems the theory can't explain. These gaps in our understanding indicate that the standard model is only an approximation (at low energies) of a more complete theory.

These are currently the major unsolved problems within the standard model:

- Neutrino mass - The standard model initially assumed neutrinos were massless, until it was experimentally determined that they had some tiny mass. Their exact masses and the mechanism behind is not yet understood.
- Interaction unification - Two of the four fundamental interactions have already been unified into the electroweak interaction. The next step is to try and unify this with the strong force, at an even higher unification energy. The aim would then be to unify this with gravity and have one fundamental unified interaction.
- Gravity - Gravity isn't explained within the standard model. The properties of a graviton, the name coined for a potential force carrier, are still debated let alone observed by experiment. Most likely, a new model is needed to reconcile quantum mechanics with gravity; one of many potential candidates is string theory.
- Dark matter and dark energy - Only about 5% of our universe consists of 'ordinary matter' as explained by the standard model. The rest is made up of completely unexplained kinds of matter and energy, which we currently call dark matter and dark energy.
- Matter-antimatter asymmetry - We observe that our universe is almost exclusively matter, where is all the antimatter? The standard model gives no reasons why antimatter and matter shouldn't be produced in equal amounts. Therefore, it is a question of cosmology and particle physics as to why our universe has developed into such an asymmetrical state.

Gauge Boson

Have you ever wondered how a magnet "knows" to stick to a refrigerator? If you hold a magnet sufficiently close, it will actually jump out of your hand and attach itself to the metal surface. But how does it know that it is close to a refrigerator or which way to jump? It is not as if the magnet has eyes to see which direction to go. Somehow, it can sense the metal even though there is some distance between the two. How is this action at a distance possible? Physicists believe the answer involves subatomic particles called *gauge bosons*.

Gauge bosons are carrier particles for three of the four fundamental forces. In the Standard Model of particle physics, there are four kinds of gauge bosons. The three gauge bosons are as follows:

1. W and Z bosons, which carry the weak force
2. Gluons, which carry the strong force
3. Photons, which carry the electromagnetic force

The only remaining fundamental force that has no known gauge boson is gravity. The theoretical gauge boson for gravity is called a graviton.

All gauge bosons are bosons, which simply means that two or more of them can exist in the same place at the same time, unlike fermions. It also means that gauge bosons have a spin of 0 or 1.

Gauge bosons are thought to interact with the Higgs field. This theoretical field is believed by many scientists to be responsible for why some gauge bosons – like W and Z bosons – have mass, while others – such as photons – do not have mass.

Photons

The electromagnetic force is responsible for keeping a magnet stuck to a metal refrigerator. It also keeps electrons bound in orbitals around the nucleus of an atom. So, without the electromagnetic force, matter could not exist. The photon (γ) is a particle of light and is also the messenger particle of the electromagnetic force. So, the particle that allows our eyes to see the world is the same particle that holds matter together.

Most physicists believe that the process works essentially as follows. A charged particle emits a low-energy photon toward another charged particle with the message "come a bit closer" or "move farther away from here" depending on the relative charges. When the other particle receives the photon, it responds by moving closer or moving away. Charged particles therefore "know" the location of other charged particles due to the exchange of virtual photons that tell them how to move. The term "virtual" means that we cannot directly observe the photon exchange; it is inferred based on our observations of how particles move. But the model does make correct predictions.

The photon itself is a massless, uncharged boson with spin 1. Recall that bosons are integer-spin particles that do not obey the Pauli Exclusion Principle. Photons are stable and are their own antiparticle.

Gluons

Just as photons are the messengers of the electromagnetic force, so *gluons* (g) are the messengers of the strong nuclear force. Recall that the strong force has six types of charge, which we call red, green, blue, antired, antigreen, and antiblue. (These terms are unrelated to literal color.) Since there are six charges rather than two, gluon behavior is a bit more complicated than photon behavior. But the same basic principle applies: a quark will emit a gluon that is absorbed by another quark, telling it how to move. Gluons possess a color charge and simultaneously an anticolor charge.

Here is how it works. Recall that the three quarks of a baryon have color charge. One will be red, one green, and one blue (Figure A below). The red quark emits a gluon. This gluon has color and anticolor. Let's suppose the gluon is red and antiblue (Figure B below). Since the baryon's color charge is conserved, the red quark must change to a blue quark, with the gluon carrying away its redness and antiblue to cancel out the blue. This red-antiblue gluon is then absorbed by the blue quark, which turns it red (Figure C below). The reaction causes the red and blue quarks to attract a bit, but it also swaps their colors. The same kind of process happens between the red and green quarks, and between green and blue quarks; they are constantly swapping colors by gluon exchange. At any given moment, the total color charge of the baryon is always white, because this is the sum of all the quark and gluon colors.

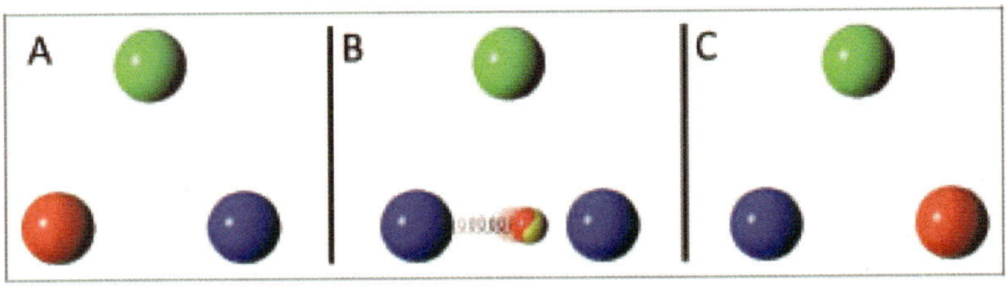

Figure, Gluon exchange between quarks within a baryon.

Like photons, gluons are massless, uncharged, spin 1 bosons. But unlike photons, gluons have not been directly observed. However, they do account nicely for the properties of the strong nuclear force. Furthermore, since gluons themselves possess color charge, they can interact with other gluons.Since gluons possess any one of three colors and any one of three anticolors, there are actually eight unique color combinations of gluons.Antigluons are also gluons because reversing all colors simply results in another gluon.

W and Z Bosons

The weak nuclear force is able to change one type of particle into another at short distances. It is responsible for phenomena such as the radioactive decay in unstable atoms. The weak force is mediated by three different bosons: the W^+, the W^-, and the Z^o. All three are spin 1 bosons. Unlike all other gauge bosons, these have mass and are unstable, which limits the range of the weak force. Since the W bosons have electrical charge (+1 or -1), a weak force interaction can change not only the type of particle but also the charge by ±1. For example, a down quark can decay into an up quark by emitting a W^- or by absorbing a W^+. Weak decays that do not alter the charge of the particle involve the Z^o. The W^- is the antiparticle of the W^+ and vice versa. The Z^o is its own antiparticle.

Higgs Boson

"Over the past few decades, particle physicists have developed an elegant theoretical model (the Standard Model) that gives a framework for our current understanding of the fundamental particles and forces of nature. One major ingredient in this model is a hypothetical, ubiquitous quantum field that is supposed to be responsible for giving particles their masses (this field would answer the basic question of why particles have the masses they do--or indeed, why they have any mass at all). This field is called the Higgs field. As a consequence of wave-particle duality, all quantum fields have a fundamental particle associated with them. The particle associated with the Higgs field is called the Higgs boson.

"Because the Higgs field would be responsible for mass, the very fact that the fundamental particles do have mass is regarded by many physicists as an indication of the existence of the Higgs field. We can even take all our data on particle physics data and interpret them in terms of the mass of a hypothetical Higgs boson."

Most physicists believe that particles that strongly "feel" the Higgs field have more mass than those that feel it weakly. Massless particles, like photons and gluons, don't interact with the Higgs field at all. The Higgs boson has a spin of zero, a charge of zero, and is extremely massive, more than a hundred times heavier than a proton. Some people have referred to the Higgs boson as the *God particle* because it supposedly "rules" over all other particles by setting their mass.

Why has the Higgs been the subject of so much hype, funding, and (mis)information? For two reasons. One, it was the last hold-out particle remaining hidden during the quest to check the accuracy of the Standard Model of Physics. This meant its discovery would validate more than a generation of scientific publication. Two, the Higgs is the particle which gives other particles their mass, making it both centrally important and seemingly magical. We tend to think of mass as an intrinsic property of all things, yet physicists believe that without the Higgs boson, mass fundamentally doesn't exist.

Non-technical Summary

The Standard Model

Physicists explain the properties and forces between elementary particles in terms of the Standard Model – a widely accepted framework for understanding almost everything in the known universe, other than gravity. (A separate theory, General Relativity, is used for gravity.) In this model, the fundamental forces in nature arise from properties of our universe called gauge invariance and symmetries. The forces are transmitted by particles known as gauge bosons.

The Problem of Gauge Boson Mass

Field theories had been used with great success in understanding the electromagnetic field and the strong force, but by around 1960 all attempts to create a *gauge invariant* theory for the weak force (and its combination with fundamental force electromagnetism, the electroweak interaction) had consistently failed, with gauge theories thereby starting to fall into disrepute as a result. The problem was that the symmetry requirements in gauge theory predicted that both electromagnetism's gauge boson (the photon) and the weak force's gauge bosons (W and Z) should have zero mass. Although the photon is indeed massless, experiments show that the weak force's bosons have mass. This meant that either gauge invariance was an incorrect approach, or something else – unknown – was giving these particles their mass, but all attempts to suggest a theory able to solve this problem just seemed to create new theoretical issues.

In the late 1950s, physicists had "no idea" how to resolve these issues, which were significant obstacles to developing a full-fledged theory for particle physics.

Symmetry Breaking

By the early 1960s, physicists had realised that a given symmetry law might not always be followed under certain conditions, at least in some areas of physics. This is called symmetry breaking and was recognised in the late 1950s by Yoichiro Nambu. Symmetry breaking can lead to surprising and unexpected results. In 1962 physicist Philip Anderson – an expert in superconductivity – wrote a paper that considered symmetry breaking in particle physics, and suggested that perhaps symmetry breaking might be the missing piece needed to solve the problems of gauge invariance in particle physics. If electroweak symmetry was somehow being broken, it might explain why electromagnetism's boson is massless, yet the weak force bosons have mass, and solve the problems. Shortly afterwards, in 1963, this was shown to be theoretically possible, at least for some limited cases.

Higgs Mechanism

Following the 1962 and 1963 papers, three groups of researchers independently published papers in 1964 with similar conclusions: that the conditions for electroweak symmetry would be "broken" if an unusual type of field existed throughout the universe, and indeed, some fundamental particles would acquire mass. The field required for this to happen (which was purely hypothetical at the time) became known as the *Higgs field* (after Peter Higgs, one of the researchers) and the mechanism by which it led to symmetry breaking, known as the *Higgs mechanism*. A key feature of the necessary field is that it would take *less* energy for the field to have a non-zero value than a zero value, unlike all other known fields, therefore, the Higgs field has a non-zero value (or *vacuum expectation*) *everywhere*. It was the first proposal capable of showing how the weak force gauge bosons could have mass despite their governing symmetry, within a gauge invariant theory.

Although these ideas did not gain much initial support or attention, by 1972 they had been developed into a comprehensive theory and proved capable of giving "sensible" results that accurately described particles known at the time, and which, with exceptional accuracy, predicted several other particles discovered during the following years. During the 1970s these theories rapidly became the Standard Model of particle physics. There was not yet any direct evidence that the Higgs field existed, but even without proof of the field, the accuracy of its predictions led scientists to believe the theory might be true. By the 1980s the question of whether or not the Higgs field existed, and therefore whether or not the entire Standard Model was correct, had come to be regarded as one of the most important unanswered questions in particle physics.

Higgs Field

According to the Standard Model, a field of the necessary kind (the *Higgs field*) exists throughout space and breaks certain symmetry laws of the electroweak interaction. Via the Higgs mechanism, this field causes the gauge bosons of the weak force to be massive at all temperatures below an extreme high value. When the weak force bosons acquire mass, this affects their range, which becomes very small. Furthermore, it was later realised that the same field would also explain, in a different way, why other fundamental constituents of matter (including electrons and quarks) have mass.

For many decades, scientists had no way to determine whether or not the Higgs field existed, because the technology needed for its detection did not exist at that time. If the Higgs field did exist, then it would be unlike any other known fundamental field, but it also was possible that these key ideas, or even the entire Standard Model, were somehow incorrect. Only discovering that the Higgs boson and therefore the Higgs field existed solved the problem.

Higgs Boson

The hypothesised Higgs mechanism made several accurate predictions, however to confirm its existence there was an extensive search for a matching particle associated with it — the "Higgs boson". Detecting Higgs bosons was difficult due to the energy required to produce them and their very rare production even if the energy is sufficient. It was therefore several decades before the first evidence of the Higgs boson was found. Particle colliders, detectors, and computers capable of looking for Higgs bosons took more than 30 years (c. 1980–2010) to develop.

By March 2013, the existence of the Higgs boson was confirmed, and therefore, the concept of some type of Higgs field throughout space is strongly supported. The nature and properties of this field are now being investigated further, using more data collected at the LHC.

Interpretation

Various analogies have been used to describe the Higgs field and boson, including analogies with well-known symmetry-breaking effects such as the rainbow and prism, electric fields, ripples, and resistance of macro objects moving through media (such as people moving through crowds or some objects moving through syrup or molasses). However, analogies based on simple resistance to motion are inaccurate, as the Higgs field does not work by resisting motion.

Significance

Evidence of the Higgs field and its properties has been extremely significant for many reasons. The importance of the Higgs boson is largely that it is able to be examined using existing knowledge and experimental technology, as a way to confirm and study the entire Higgs field theory. Conversely, proof that the Higgs field and boson do *not* exist would have also been significant.

Validation of the Standard Model

The Higgs boson validates the Standard Model through the mechanism of mass generation. As more precise measurements of its properties are made, more advanced extensions may be suggested or excluded. As experimental means to measure the field's behaviours and interactions are developed, this fundamental field may be better understood. If the Higgs field had not been discovered, the Standard Model would have needed to be modified or superseded.

Related to this, a belief generally exists among physicists that there is likely to be "new" physics beyond the Standard Model, and the Standard Model will at some point be extended or superseded. The Higgs discovery, as well as the many measured collisions occurring at the LHC, provide physicists a sensitive tool to parse data for where the Standard Model fails, and could provide considerable evidence guiding researchers into future theoretical developments.

Symmetry Breaking of the Electroweak Interaction

Below an extremely high temperature, electroweak symmetry breaking causes the electroweak interaction to manifest in part as the short-ranged weak force, which is carried by massive gauge bosons. This symmetry breaking is required for atoms and other structures to form, as well as for nuclear reactions in stars, such as our Sun. The Higgs field is responsible for this symmetry breaking.

Particle Mass Acquisition

The Higgs field is pivotal in generating the masses of quarks and charged leptons (through Yukawa coupling) and the W and Z gauge bosons (through the Higgs mechanism).

It is worth noting that the Higgs field does not "create" mass out of nothing (which would violate the law of conservation of energy), nor is the Higgs field responsible for the mass of all particles. For example, approximately 99% of the mass of baryons (composite particles such as the proton and neutron), is due instead to quantum chromodynamics binding energy, which is the sum of the kinetic energies of quarks and the energies of the massless gluons mediating the strong interaction inside the baryons. In Higgs-based theories, the property of "mass" is a manifestation of potential energy transferred to fundamental particles when they interact ("couple") with the Higgs field, which had contained that mass in the form of energy.

Scalar Fields and Extension of the Standard Model

The Higgs field is the only scalar (spin 0) field to be detected; all the other fields in the Standard Model are spin ½ fermions or spin 1 bosons. According to Rolf-Dieter Heuer, director general of CERN when the Higgs boson was discovered, this existence proof of a scalar field is almost as important the Higgs' role in determining the mass of other particles. It suggests that other hypothetical scalar fields suggested by other theories, from the inflaton to quintessence, could perhaps exist as well.

Cosmology

Inflaton

There has been considerable scientific research on possible links between the Higgs field and the inflaton – a hypothetical field suggested as the explanation for the expansion of space during the first fraction of a second of the universe (known as the "inflationary epoch"). Some theories suggest that a fundamental scalar field might be responsible for this phenomenon; the Higgs field is such a field, and its existence has led to papers analysing whether it could also be the *inflaton* responsible for this exponential expansion of the universe during the Big Bang. Such theories are highly tentative and face significant problems related to unitarity, but may be viable if combined with additional features such as large non-minimal coupling, a Brans–Dicke scalar, or other "new" physics, and they have received treatments suggesting that Higgs inflation models are still of interest theoretically.

Nature of the Universe and its Possible Fates

In the Standard Model, there exists the possibility that the underlying state of our universe -known as the "vacuum" - is long-lived, but not completely stable. In this scenario, the universe as we know it could effectively be destroyed by collapsing into a more stable vacuum state. This was sometimes misreported as the Higgs boson "ending" the universe. If the masses of the Higgs boson and top quark are known more precisely, and the Standard Model provides an accurate description of particle physics up to extreme energies of the Planck scale, then it is possible to calculate whether the vacuum is stable or merely long-lived. A 125 – 127 GeV Higgs mass seems to be extremely close to the boundary for stability, but a definitive answer requires much more precise measurements of the pole mass of the top quark. New physics can change this picture.

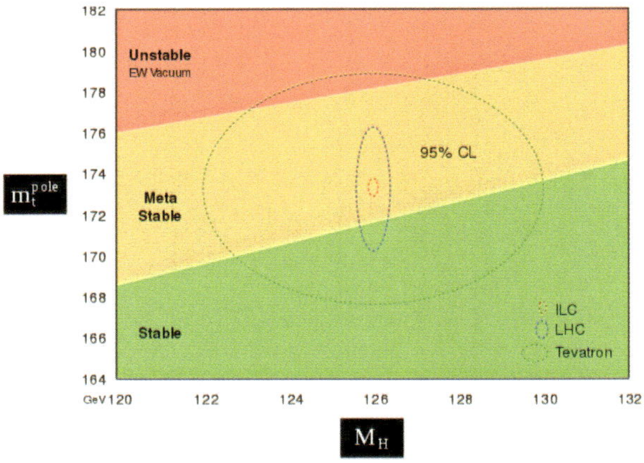

Diagram showing the Higgs boson and top quark masses, which could indicate whether our universe is stable, or a long-lived 'bubble'. As of 2012, the 2σ ellipse based on Tevatron and LHC data still allows for both possibilities.

If measurements of the Higgs boson suggest that our universe lies within a false vacuum of this kind, then it would imply – more than likely in many billions of years – that the universe's forces, particles, and structures could cease to exist as we know them (and be replaced by different ones), if a true vacuum happened to nucleate. It also suggests that the Higgs self-coupling λ and its β_λ function could be very close to zero at the Planck scale, with "intriguing" implications, including theories of gravity and Higgs-based inflation. A future electron–positron collider would be able to provide the precise measurements of the top quark needed for such calculations.

Vacuum Energy and the Cosmological Constant

More speculatively, the Higgs field has also been proposed as the energy of the vacuum, which at the extreme energies of the first moments of the Big Bang caused the universe to be a kind of featureless symmetry of undifferentiated, extremely high energy. In this kind of speculation, the single unified field of a Grand Unified Theory is identified as (or modelled upon) the Higgs field, and it is through successive symmetry breakings of the Higgs field, or some similar field, at phase transitions that the presently known forces and fields of the universe arise.

The relationship (if any) between the Higgs field and the presently observed vacuum energy density of the universe has also come under scientific study. As observed, the present vacuum energy density is extremely close to zero, but the energy density expected from the Higgs field, supersymmetry, and other current theories are typically many orders of magnitude larger. It is unclear how these should be reconciled. This cosmological constant problem remains a further major unanswered problem in physics.

Practical and Technological Impact

As yet, there are no known immediate technological benefits of finding the Higgs particle. However, a common pattern for fundamental discoveries is for practical applications to follow later, and once the discovery has been explored further, perhaps becoming the basis for new technologies of importance to society.

The challenges in particle physics have furthered major technological progress of widespread importance. For example, the World Wide Web began as a project to improve CERN's communication system. CERN's requirement to process massive amounts of data produced by the Large Hadron Collider also led to contributions to the fields of distributed and cloud computing.

Theoretical Properties

Theoretical Need for the Higgs

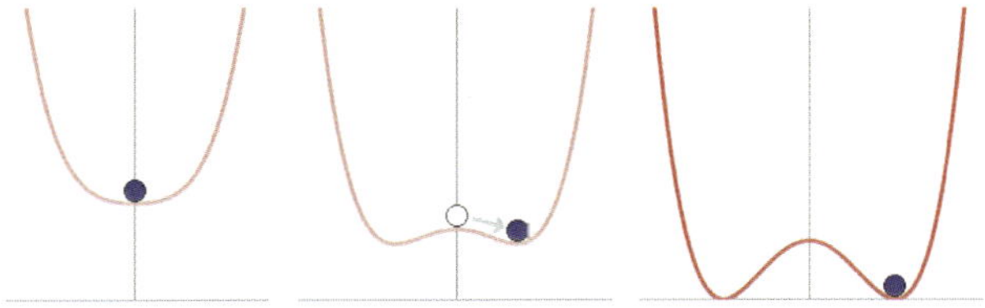

"Symmetry breaking illustrated": – At high energy levels *(left)* the ball settles in the centre, and the result is symmetrical. At lower energy levels *(right)*, the overall "rules" remain symmetrical, but the "Mexican hat" potential comes into effect: "local" symmetry inevitably becomes broken since eventually the ball must at random roll one way or another.

Gauge invariance is an important property of modern particle theories such as the Standard Model, partly due to its success in other areas of fundamental physics such as electromagnetism and the strong interaction (quantum chromodynamics). However, there were great difficulties in developing gauge theories for the weak nuclear force or a possible unified electroweak interaction. Fermions with a mass term would violate gauge symmetry and therefore cannot be gauge invariant. (This can be seen by examining the Dirac Lagrangian for a fermion in terms of left and right handed components; we find none of the spin-half particles could ever flip helicity as required for mass, so they must be massless.) W and Z bosons are observed to have mass, but a boson mass term contains terms which clearly depend on the choice of gauge, and therefore these masses too cannot be gauge invariant. Therefore, it seems that *none* of the standard model fermions *or* bosons could "begin" with mass as an inbuilt property except by abandoning gauge invariance. If gauge invariance were to be retained, then these particles had to be acquiring their mass by some other mechanism or interaction. Additionally, whatever was giving these particles their mass had to not "break" gauge invariance as the basis for other parts of the theories where it worked well, *and* had to not require or predict unexpected massless particles or long-range forces (seemingly an inevitable consequence of Goldstone's theorem) which did not actually seem to exist in nature.

A solution to all of these overlapping problems came from the discovery of a previously unnoticed borderline case hidden in the mathematics of Goldstone's theorem, that under certain conditions it *might* theoretically be possible for a symmetry to be broken *without* disrupting gauge invariance and *without* any new massless particles or forces, and having "sensible" (renormalisable) results mathematically. This became known as the Higgs mechanism.

The Standard Model hypothesises a field which is responsible for this effect, called the Higgs field (symbol: ϕ), which has the unusual property of a non-zero amplitude in its ground state; i.e., a non-zero vacuum expectation value. It can have this effect because of its unusual "Mexican hat" shaped potential whose lowest "point" is not at its "centre". In simple terms, unlike all other known fields, the Higgs field requires *less* energy to have a non-zero value than a zero value, so it ends up having a non-zero value *everywhere*. Below a certain extremely high energy level the existence of this non-zero vacuum expectation spontaneously breaks electroweak gauge symmetry which in turn gives rise to the Higgs mechanism and triggers the acquisition of mass by those particles interacting with the field. This effect occurs because scalar field components of the Higgs field are "absorbed" by the massive bosons as degrees of freedom, and couple to the fermions via Yukawa coupling, thereby producing the expected mass terms. When symmetry breaks under these conditions, the Goldstone bosons that arise *interact* with the Higgs field (and with other particles capable of interacting with the Higgs field) instead of becoming new massless particles. The intractable problems of both underlying theories "neutralise" each other, and the residual outcome is that elementary particles acquire a consistent mass based on how strongly they interact with the Higgs field. It is the simplest known process capable of giving mass to the gauge bosons while remaining compatible with gauge theories. Its quantum would be a scalar boson, known as the Higgs boson.

Properties of the Higgs Field

In the Standard Model, the Higgs field is a scalar tachyonic field – *scalar* meaning it does not transform under Lorentz transformations, and *tachyonic* meaning the field (but not the particle) has imaginary mass, and in certain configurations must undergo symmetry breaking. It consists of four components: two neutral ones and two charged component fields. Both of the charged components and one of the neutral fields are Goldstone bosons, which act as the longitudinal third-polarisation components of the massive W^+, W^-, and Z bosons. The quantum of the remaining neutral component corresponds to (and is theoretically realised as) the massive Higgs boson, this component can also interact with fermions via Yukawa coupling to give them mass, as well.

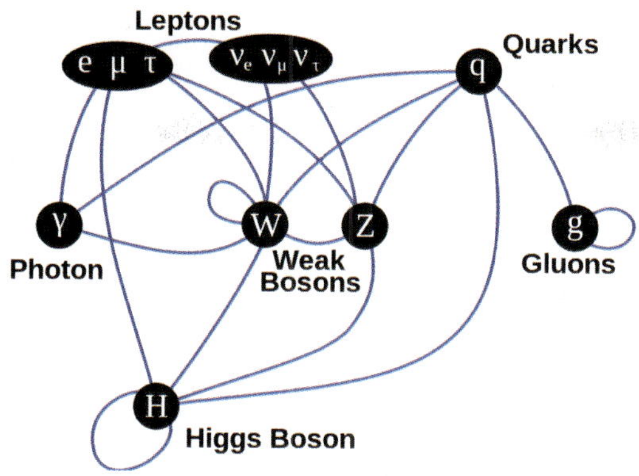

Summary of interactions between certain particles described by the Standard Model.

Mathematically, the Higgs field has imaginary mass and is therefore a tachyonic field. While tachyons (particles that move faster than light) are a purely hypothetical concept, fields with imaginary mass have come to play an important role in modern physics. Under no circumstances do any excitations ever propagate faster than light in such theories — the presence or absence of a tachyonic mass has no effect whatsoever on the maximum velocity of signals (there is no violation of causality). Instead of faster-than-light particles, the imaginary mass creates an instability: Any configuration in which one or more field excitations are tachyonic must spontaneously decay, and the resulting configuration contains no physical tachyons. This process is known as tachyon condensation, and is now believed to be the explanation for how the Higgs mechanism itself arises in nature, and therefore the reason behind electroweak symmetry breaking.

Although the notion of imaginary mass might seem troubling, it is only the field, and not the mass itself, that is quantised. Therefore, the field operators at spacelike separated points still commute (or anticommute), and information and particles still do not propagate faster than light. Tachyon condensation drives a physical system that has reached a local limit – and might naively be expected to produce physical tachyons – to an alternate stable state where no physical tachyons exist. Once a tachyonic field such as the Higgs field reaches the minimum of the potential, its quanta are not tachyons any more but rather are ordinary particles such as the Higgs boson.

Properties of the Higgs Boson

Since the Higgs field is scalar, the Higgs boson has no spin. The Higgs boson is also its own antiparticle and is CP-even, and has zero electric and colour charge.

The Standard Model does not predict the mass of the Higgs boson. If that mass is between 115 and 180 GeV/c^2 (consistent with empirical observations of 125 GeV/c^2), then the Standard Model can be valid at energy scales all the way up to the Planck scale (10^{19} GeV). Many theorists expect new physics beyond the Standard Model to emerge at the TeV-scale, based on unsatisfactory properties of the Standard Model. The highest possible mass scale allowed for the Higgs boson (or some other electroweak symmetry breaking mechanism) is 1.4 TeV; beyond this point, the Standard Model becomes inconsistent without such a mechanism, because unitarity is violated in certain scattering processes.

It is also possible, although experimentally difficult, to estimate the mass of the Higgs boson indirectly. In the Standard Model, the Higgs boson has a number of indirect effects; most notably, Higgs loops result in tiny corrections to masses of W and Z bosons. Precision measurements of electroweak parameters, such as the Fermi constant and masses of W/Z bosons, can be used to calculate constraints on the mass of the Higgs. As of July 2011, the precision electroweak measurements tell us that the mass of the Higgs boson is likely to be less than about 161 GeV/c^2 at 95% confidence level (this upper limit would increase to 185 GeV/c^2 if the lower bound of 114.4 GeV/c^2 from the LEP-2 direct search is allowed for). These indirect constraints rely on the assumption that the Standard Model is correct. It may still be possible to discover a Higgs boson above these masses if it is accompanied by other particles beyond those predicted by the Standard Model.

Production

Feynman diagrams for Higgs production

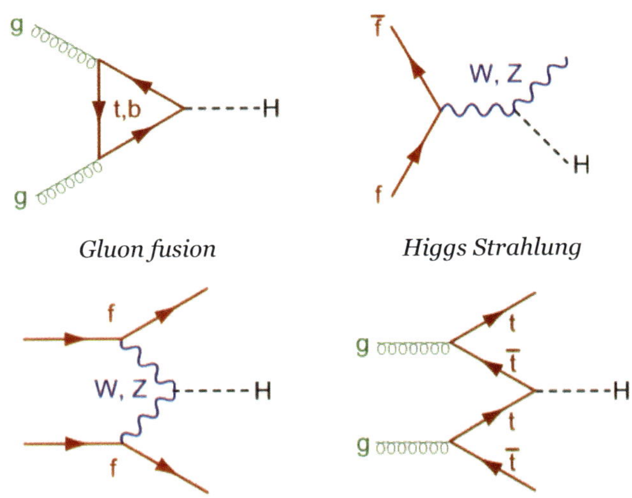

Gluon fusion *Higgs Strahlung*

Vector boson fusion *Top fusion*

If Higgs particle theories are valid, then a Higgs particle can be produced much like other particles that are studied, in a particle collider. This involves accelerating a large number of particles to extremely high energies and extremely close to the speed of light, then allowing them to smash together. Protons and lead ions (the bare nuclei of lead atoms) are used at the LHC. In the extreme energies of these collisions, the desired esoteric particles will occasionally be produced and this can be detected and studied; any absence or difference from theoretical expectations can also be used to improve the theory. The relevant particle theory (in this case the Standard Model) will determine the necessary kinds of collisions and detectors. The Standard Model predicts that Higgs bosons could be formed in a number of ways, although the probability of producing a Higgs boson in any collision is always expected to be very small—for example, only 1 Higgs boson per 10 billion collisions in the Large Hadron Collider. The most common expected processes for Higgs boson production are:

- *Gluon fusion*. If the collided particles are hadrons such as the proton or anti-proton—as is the case in the LHC and Tevatron—then it is most likely that two of the gluons binding the hadron together collide. The easiest way to produce a Higgs particle is if the two gluons combine to form a loop of virtual quarks. Since the coupling of particles to the Higgs boson is proportional to their mass, this process is more likely for heavy particles. In practice it is enough to consider the contributions of virtual top and bottom quarks (the heaviest quarks). This process is the dominant contribution at the LHC and Tevatron being about ten times more likely than any of the other processes.
- *Higgs Strahlung*. If an elementary fermion collides with an anti-fermion—e.g., a quark with an anti-quark or an electron with a positron—the two can merge to form a virtual W or Z boson which, if it carries sufficient energy, can then emit a Higgs boson. This process was the dominant production mode at the LEP, where an electron and a positron collided to form a virtual Z boson, and it was the second largest contribution for Higgs production at the Tevatron. At the LHC this process is only the third largest, because the LHC collides protons with protons, making a quark-antiquark collision less likely than at the Tevatron. Higgs Strahlung is also known as *associated production*.
- *Weak boson fusion*. Another possibility when two (anti-)fermions collide is that the two exchange a virtual W or Z boson, which emits a Higgs boson. The colliding fermions do not need to be the same type. So, for example, an up quark may exchange a Z boson with an anti-down quark. This process is the second most important for the production of Higgs particle at the LHC and LEP.
- *Top fusion*. The final process that is commonly considered is by far the least likely (by two orders of magnitude). This process involves two colliding gluons, which each decay into a heavy quark–antiquark pair. A quark and antiquark from each pair can then combine to form a Higgs particle.

Decay

Quantum mechanics predicts that if it is possible for a particle to decay into a set of lighter particles, then it will eventually do so. This is also true for the Higgs boson. The likelihood with which this happens depends on a variety of factors including: the difference in mass, the strength of the interactions, etc. Most of these factors are fixed by the Standard Model, except for the mass of the Higgs boson itself. For a Higgs boson with a mass of 125 GeV/c^2 the SM predicts a mean life time of about 1.6×10^{-22} s.

Since it interacts with all the massive elementary particles of the SM, the Higgs boson has many different processes through which it can decay. Each of these possible processes has its own probability, expressed as the *branching ratio*; the fraction of the total number decays that follows that process. The SM predicts these branching ratios as a function of the Higgs mass.

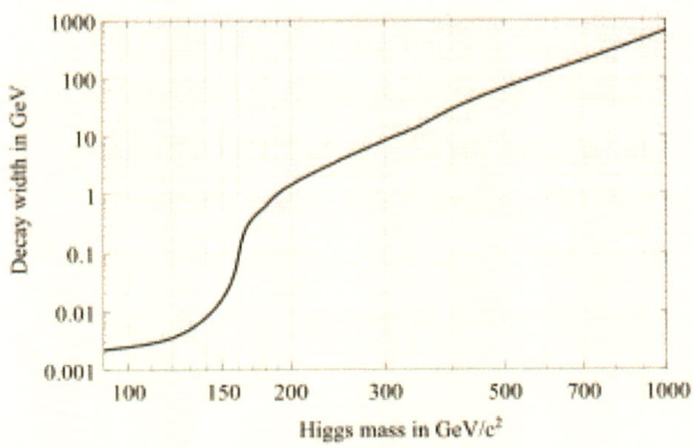

The Standard Model prediction for the decay width of the Higgs particle depends on the value of its mass.

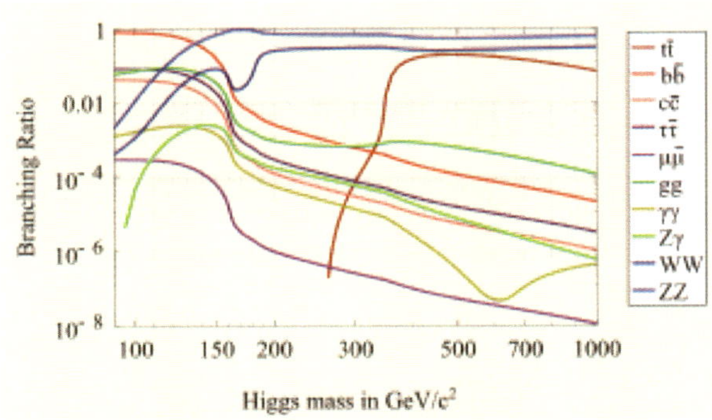

The Standard Model prediction for the branching ratios of the different decay modes of the Higgs particle depends on the value of its mass.

One way that the Higgs can decay is by splitting into a fermion–antifermion pair. As general rule, the Higgs is more likely to decay into heavy fermions than light fermions, because the mass of a fermion is proportional to the strength of its interaction with the Higgs. By this logic the most common decay should be into a top–antitop quark pair. However, such a decay would only be possible if the Higgs were heavier than ~346 GeV/c^2, twice the mass of the top quark. For a Higgs mass of 125 GeV/c^2 the SM predicts that the most common decay is into a bottom–antibottom quark pair, which happens 57.7% of the time. The second most common fermion decay at that mass is a tau–antitau pair, which happens only about 6.3% of the time.

Another possibility is for the Higgs to split into a pair of massive gauge bosons. The most likely possibility is for the Higgs to decay into a pair of W bosons (the light blue line in the plot), which happens about 21.5% of the time for a Higgs boson with a mass of 125 GeV/c^2. The W bosons can subsequently decay either into a quark and an antiquark or into a charged lepton and a neutrino. The decays of W bosons into quarks are difficult to distinguish from the background, and the decays into leptons cannot be fully reconstructed (because neutrinos are impossible to detect in particle collision experiments). A cleaner signal is given by decay into a pair of Z-bosons (which happens about 2.6% of the time for a Higgs with a mass of 125 GeV/c^2), if each of the bosons subsequently decays into a pair of easy-to-detect charged leptons (electrons or muons).

Decay into massless gauge bosons (i.e., gluons or photons) is also possible, but requires intermediate loop of virtual heavy quarks (top or bottom) or massive gauge bosons. The most common such process is the decay into a pair of gluons through a loop of virtual heavy quarks. This process, which is the reverse of the gluon fusion process mentioned above, happens approximately 8.6% of the time for a Higgs boson with a mass of 125 GeV/c^2. Much rarer is the decay into a pair of photons mediated by a loop of W bosons or heavy quarks, which happens only twice for every thousand decays. However, this process is very relevant for experimental searches for the Higgs boson, because the energy and momentum of the photons can be measured very precisely, giving an accurate reconstruction of the mass of the decaying particle.

Alternative Models

The Minimal Standard Model as described above is the simplest known model for the Higgs mechanism with just one Higgs field. However, an extended Higgs sector with additional Higgs particle doublets or triplets is also possible, and many extensions of the Standard Model have this feature. The non-minimal Higgs sector favoured by theory are the two-Higgs-doublet models (2HDM), which predict the existence of a quintet of scalar particles: two CP-even neutral Higgs bosons h^0 and H^0, a CP-odd neutral Higgs boson A^0, and two charged Higgs particles $H^{\pm}$. Supersymmetry ("SUSY") also predicts relations between the Higgs-boson masses and the masses of the gauge bosons, and could accommodate a 125 GeV/c^2 neutral Higgs boson.

The key method to distinguish between these different models involves study of the particles' interactions ("coupling") and exact decay processes ("branching ratios"), which can be measured and tested experimentally in particle collisions. In the Type-I 2HDM model one Higgs doublet couples to up and down quarks, while the second doublet does not couple to quarks. This model has two interesting limits, in which the lightest Higgs couples to just fermions ("gauge-phobic") or just gauge bosons ("fermiophobic"), but not both. In the Type-II 2HDM model, one Higgs doublet only couples to up-type quarks, the other only couples to down-type quarks. The heavily researched Minimal Supersymmetric Standard Model (MSSM) includes a Type-II 2HDM Higgs sector, so it could be disproven by evidence of a Type-I 2HDM Higgs.

In other models the Higgs scalar is a composite particle. For example, in technicolor the role of the Higgs field is played by strongly bound pairs of fermions called techniquarks.

Other models, feature pairs of top quarks. In yet other models, there is no Higgs field at all and the electroweak symmetry is broken using extra dimensions.

Further Theoretical Issues and Hierarchy Problem

The Standard Model leaves the mass of the Higgs boson as a parameter to be measured, rather than a value to be calculated. This is seen as theoretically unsatisfactory, particularly as quantum corrections (related to interactions with virtual particles) should apparently cause the Higgs particle to have a mass immensely higher than that observed, but at the same time the Standard Model requires a mass of the order of 100 to 1000 GeV to ensure unitarity (in this case, to unitarise longitudinal vector boson scattering). Reconciling these points appears to require explaining why there is an almost-perfect cancellation resulting in the visible mass of ~ 125 GeV, and it is not clear how to do this. Because the weak force is about 10^{32} times stronger than gravity, and (linked to this) the Higgs boson's mass is so much less than the Planck mass or the grand unification energy, it appears that either there is some underlying connection or reason for these observations which is unknown and not described by the Standard Model, or some unexplained and extremely precise fine-tuning of parameters – however at present neither of these explanations is proven. This is known as a hierarchy problem. More broadly, the hierarchy problem amounts to the worry that a future theory of fundamental particles and interactions should not have excessive fine-tunings or unduly delicate cancellations, and should allow masses of particles such as the Higgs boson to be calculable. The problem is in some ways unique to spin-0 particles (such as the Higgs boson), which can give rise to issues related to quantum corrections that do not affect particles with spin. A number of solutions have been proposed, including supersymmetry, conformal solutions and solutions via extra dimensions such as braneworld models.

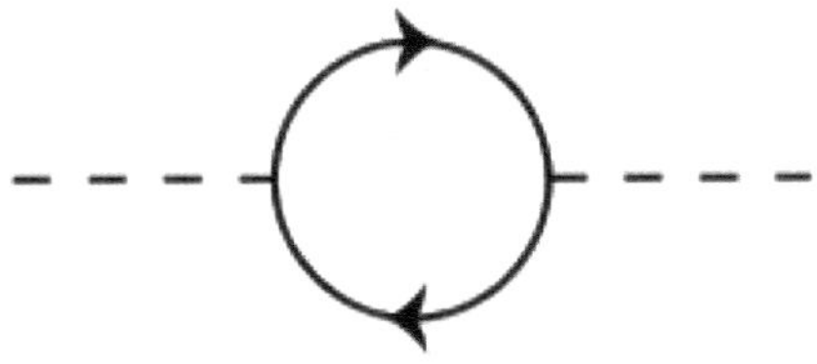

A one-loop Feynman diagram of the first-order correction to the Higgs mass. In the Standard Model the effects of these corrections are potentially enormous, giving rise to the so-called hierarchy problem.

There are also issues of quantum triviality, which suggests that it may not be possible to create a consistent quantum field theory involving elementary scalar particles. However, if quantum triviality is avoided, triviality constraints may set bounds on the Higgs Boson mass.

Experimental Search

To produce Higgs bosons, two beams of particles are accelerated to very high energies and allowed to collide within a particle detector. Occasionally, although rarely, a Higgs boson will be created fleetingly as part of the collision byproducts. Because the Higgs boson decays very quickly, particle detectors cannot detect it directly. Instead the detectors register all the decay products (the *decay signature*) and from the data the decay process is reconstructed. If the observed decay products match a possible decay process (known as a *decay channel*) of a Higgs boson, this indicates that a Higgs boson may have been created. In practice, many processes may produce similar decay signatures. Fortunately, the Standard Model precisely predicts the likelihood of each of these, and each known process, occurring. So, if the detector detects more decay signatures consistently matching a Higgs boson than would otherwise be expected if Higgs bosons did not exist, then this would be strong evidence that the Higgs boson exists.

Because Higgs boson production in a particle collision is likely to be very rare (1 in 10 billion at the LHC), and many other possible collision events can have similar decay signatures, the data of hundreds of trillions of collisions needs to be analysed and must "show the same picture" before a conclusion about the existence of the Higgs boson can be reached. To conclude that a new particle has been found, particle physicists require that the statistical analysis of two independent particle detectors each indicate that there is lesser than a one-in-a-million chance that the observed decay signatures are due to just background random Standard Model events—i.e., that the observed number of events is more than 5 standard deviations (sigma) different from that expected if there was no new particle. More collision data allows better confirmation of the physical properties of any new particle observed, and allows physicists to decide whether it is indeed a Higgs boson as described by the Standard Model or some other hypothetical new particle.

To find the Higgs boson, a powerful particle accelerator was needed, because Higgs bosons might not be seen in lower-energy experiments. The collider needed to have a high luminosity in order to ensure enough collisions were seen for conclusions to be drawn. Finally, advanced computing facilities were needed to process the vast amount of data (25 petabytes per year as of 2012) produced by the collisions. For the announcement of 4 July 2012, a new collider known as the Large Hadron Collider was constructed at CERN with a planned eventual collision energy of 14 TeV—over seven times any previous collider—and over 300 trillion (3×10^{14}) LHC proton–proton collisions were analysed by the LHC Computing Grid, the world's largest computing grid (as of 2012), comprising over 170 computing facilities in a worldwide network across 36 countries.

Search before 4 July 2012

The first extensive search for the Higgs boson was conducted at the Large Electron–Positron Collider (LEP) at CERN in the 1990s. At the end of its service in 2000, LEP had found no conclusive evidence for the Higgs. This implied that if the Higgs boson were to exist it would have to be heavier than 114.4 GeV/c^2.

The search continued at Fermilab in the United States, where the Tevatron—the collider that discovered the top quark in 1995—had been upgraded for this purpose. There was no guarantee that the Tevatron would be able to find the Higgs, but it was the only supercollider that was operational since the Large Hadron Collider (LHC) was still under construction and the planned Superconducting Super Collider had been cancelled in 1993 and never completed. The Tevatron was only able to exclude further ranges for the Higgs mass, and was shut down on 30 September 2011 because it no longer could keep up with the LHC. The final analysis of the data excluded the possibility of a Higgs boson with a mass between 147 GeV/c^2 and 180 GeV/c^2. In addition, there was a small (but not significant) excess of events possibly indicating a Higgs boson with a mass between 115 GeV/c^2 and 140 GeV/c^2.

The Large Hadron Collider at CERN in Switzerland, was designed specifically to be able to either confirm or exclude the existence of the Higgs boson. Built in a 27 km tunnel under the ground near Geneva originally inhabited by LEP, it was designed to collide two beams of protons, initially at energies of 3.5 TeV per beam (7 TeV total), or almost 3.6 times that of the Tevatron, and upgradeable to 2 × 7 TeV (14 TeV total) in future. Theory suggested if the Higgs boson existed, collisions at these energy levels should be able to reveal it. As one of the most complicated scientific instruments ever built, its operational readiness was delayed for 14 months by a magnet quench event nine days after its inaugural tests, caused by a faulty electrical connection that damaged over 50 superconducting magnets and contaminated the vacuum system.

Data collection at the LHC finally commenced in March 2010. By December 2011 the two main particle detectors at the LHC, ATLAS and CMS, had narrowed down the mass range where the Higgs could exist to around 116-130 GeV (ATLAS) and 115-127 GeV (CMS). There had also already been a number of promising event excesses that had "evaporated" and proven to be nothing but random fluctuations. However, from around May 2011, both experiments had seen among their results, the slow emergence of a small yet consistent excess of gamma and 4-lepton decay signatures and several other particle decays, all hinting at a new particle at a mass around 125 GeV. By around November 2011, the anomalous data at 125 GeV was becoming "too large to ignore" (although still far from conclusive), and the team leaders at both ATLAS and CMS each privately suspected they might have found the Higgs. On November 28, 2011, at an internal meeting of the two team leaders and the director general of CERN, the latest analyses were discussed outside their teams for the first time, suggesting both ATLAS and CMS might be converging on a possible shared result at 125 GeV, and initial preparations commenced in case of a successful finding. While this information was not known publicly at the time, the narrowing of the possible Higgs range to around 115–130 GeV and the repeated observation of small but consistent event excesses across multiple channels at both ATLAS and CMS in the 124-126 GeV region (described as "tantalising hints" of around 2-3 sigma) were public knowledge with "a lot of interest". It was therefore widely anticipated around the end of 2011, that the LHC would provide sufficient data to either exclude or confirm the finding of a Higgs boson by the end of 2012, when their 2012 collision data (with slightly higher 8 TeV collision energy) had been examined.

Discovery of Candidate Boson at CERN

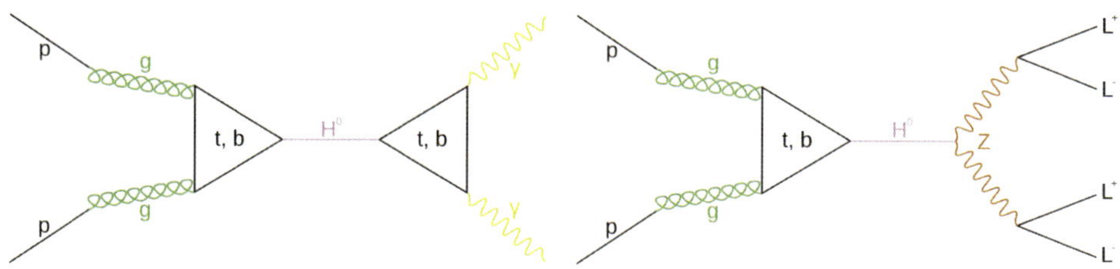

Feynman diagrams

Feynman diagrams showing the cleanest channels associated with the low-mass (~125 GeV) Higgs boson candidate observed by ATLAS and CMS at the LHC. The dominant production mechanism at this mass involves two gluons from each proton fusing to a Top-quark Loop, which couples strongly to the Higgs field to produce a Higgs boson.

Left: Diphoton channel: Boson subsequently decays into 2 gamma ray photons by virtual interaction with a W boson loop or top quark loop.

Right: 4-Lepton "golden channel": Boson emits 2 Z bosons, which each decay into 2 leptons (electrons, muons).

Experimental analysis of these channels reached a significance of more than 5 sigma in both experiments.

On 22 June 2012, CERN announced an upcoming seminar covering tentative findings for 2012, and shortly afterwards (from around 1 July 2012 according to an analysis of the spreading rumour in social media) rumours began to spread in the media that this would include a major announcement, but it was unclear whether this would be a stronger signal or a formal discovery. Speculation escalated to a "fevered" pitch when reports emerged that Peter Higgs, who proposed the particle, was to be attending the seminar, and that "five leading physicists" had been invited – generally believed to signify the five living 1964 authors – with Higgs, Englert, Guralnik, Hagen attending and Kibble confirming his invitation (Brout having died in 2011).

On 4 July 2012, both of the CERN experiments announced they had independently made the same discovery: CMS of a previously unknown boson with mass 125.3 ± 0.6 GeV/c^2 and ATLAS of a boson with mass 126.0 ± 0.6 GeV/c^2. Using the combined analysis of two interaction types (known as 'channels'), both experiments independently reached a local significance of 5 sigma — implying that the probability of getting at least as strong a result by chance alone is less than 1 in 3 million. When additional channels were taken into account, the CMS significance was reduced to 4.9 sigma.

The two teams had been working 'blinded' from each other from around late 2011 or early 2012, meaning they did not discuss their results with each other, providing additional certainty that any common finding was genuine validation of a particle. This level of evidence, confirmed independently by two separate teams and experiments, meets the formal level of proof required to announce a confirmed discovery.

On 31 July 2012, the ATLAS collaboration presented additional data analysis on the "observation of a new particle", including data from a third channel, which improved the significance to 5.9 sigma (1 in 588 million chance of obtaining at least as strong evidence by random background effects alone) and mass 126.0 ± 0.4 (stat) ± 0.4 (sys) GeV/c^2, and CMS improved the significance to 5-sigma and mass 125.3 ± 0.4 (stat) ± 0.5 (sys) GeV/c^2.

The New Particle Tested as a Possible Higgs Boson

Following the 2012 discovery, it was still unconfirmed whether or not the 125 GeV/c^2 particle was a Higgs boson. On one hand, observations remained consistent with the observed particle being the Standard Model Higgs boson, and the particle decayed into at least some of the predicted channels. Moreover, the production rates and branching ratios for the observed channels broadly matched the predictions by the Standard Model within the experimental uncertainties. However, the experimental uncertainties currently still left room for alternative explanations, meaning an announcement of the discovery of a Higgs boson would have been premature. To allow more opportunity for data collection, the LHC's proposed 2012 shutdown and 2013–14 upgrade were postponed by 7 weeks into 2013.

In November 2012, in a conference in Kyoto researchers said evidence gathered since July was falling into line with the basic Standard Model more than its alternatives, with a range of results for several interactions matching that theory's predictions. Physicist Matt Strassler highlighted "considerable" evidence that the new particle is not a pseudoscalar negative parity particle (consistent with this required finding for a Higgs boson), "evaporation" or lack of increased significance for previous hints of non-Standard Model findings, expected Standard Model interactions with W and Z bosons, absence of "significant new implications" for or against supersymmetry, and in general no significant deviations to date from the results expected of a Standard Model Higgs boson. However some kinds of extensions to the Standard Model would also show very similar results; so commentators noted that based on other particles that are still being understood long after their discovery, it may take years to be sure, and decades to fully understand the particle that has been found.

These findings meant that as of January 2013, scientists were very sure they had found an unknown particle of mass ~ 125 GeV/c^2, and had not been misled by experimental error or a chance result. They were also sure, from initial observations, that the new particle was some kind of boson. The behaviours and properties of the particle, so far as examined since July 2012, also seemed quite close to the behaviours expected of a Higgs boson. Even so, it could still have been a Higgs boson or some other unknown boson, since future tests could show behaviours that do not match a Higgs boson, so as of December 2012 CERN still only stated that the new particle was "consistent with" the Higgs boson, and scientists did not yet positively say it was the Higgs boson. Despite this, in late 2012, widespread media reports announced (incorrectly) that a Higgs boson had been confirmed during the year.

In January 2013, CERN director-general Rolf-Dieter Heuer stated that based on data analysis to date, an answer could be possible 'towards' mid-2013, and the deputy chair of physics at Brookhaven National Laboratory stated in February 2013 that a "definitive" answer might require "another few years" after the collider's 2015 restart. In early March 2013, CERN Research Director Sergio Bertolucci stated that confirming spin-0 was the major remaining requirement to determine whether the particle is at least some kind of Higgs boson.

Confirmation of Existence and Current Status

On 14 March 2013 CERN confirmed that:

> "CMS and ATLAS have compared a number of options for the spin-parity of this particle, and these all prefer no spin and even parity [two fundamental criteria of a Higgs boson consistent with the Standard Model]. This, coupled with the measured interactions of the new particle with other particles, strongly indicates that it is a Higgs boson."

This also makes the particle the first elementary scalar particle to be discovered in nature.

Examples of tests used to validate that the discovered particle is the Higgs boson:

Requirement	How tested / explanation	Current status (July 2017)
Zero spin	Examining decay patterns. Spin-1 had been ruled out at the time of initial discovery by the observed decay to two photons ($\gamma\gamma$), leaving spin-0 and spin-2 as remaining candidates.	Spin-0 confirmed. The spin-2 hypothesis is excluded with a confidence level exceeding 99.9%.
Even (Positive) parity	Studying the angles at which decay products fly apart. Negative parity was also disfavoured if spin-0 was confirmed.	Even parity tentatively confirmed. The spin-0 negative parity hypothesis is excluded with a confidence level exceeding 99.9%.
Decay channels (outcomes of particle decaying) are as predicted	The Standard Model predicts the decay patterns of a 125 GeV Higgs boson. Are these all being seen, and at the right rates? Particularly significant, we should observe decays into pairs of photons ($\gamma\gamma$), W and Z bosons (WW and ZZ), bottom quarks (bb), and tau leptons ($\tau\tau$), among the possible outcomes.	$\gamma\gamma$, $\tau\tau$, WW and ZZ observed; evidence for bb seen. All observed signal strengths are consistent with the Standard Model prediction.

Couples to mass (i.e., strength of interaction with Standard Model particles proportional to their mass)	Particle physicist Adam Falkowski states that the essential qualities of a Higgs boson are that it is a spin-0 (scalar) particle which *also* couples to mass (W and Z bosons); proving spin-0 alone is insufficient.	Couplings to mass strongly evidenced ("At 95% confidence level c_V is within 15% of the standard model value c_V=1").
Higher energy results remain consistent	After the LHC's 2015 restart at the higher energy of 13 TeV, searches for multiple Higgs particles (as predicted in some theories) and tests targeting other versions of particle theory continued. These higher energy results must continue to give results consistent with Higgs theories	Analysis of collisions up to July 2017 do not show deviations from the Standard Model, with experimental precisions better than results at lower energies.

Findings Since 2013

In July 2017, CERN confirmed that all measurements still agree with the predictions of the Standard Model, and called the discovered particle simply "the Higgs boson". As of April 2018, the Large Hadron Collider has continued to produce findings that confirm the 2013 understanding of the Higgs field and particle.

The LHC's experimental work since restarting in 2015 has included probing the Higgs field and boson to a greater level of detail, and confirming whether or not less common predictions were correct. In particular, exploration since 2015 has provided strong evidence of the predicted direct decay into fermions such as pairs of bottom quarks (3.6 sigma) - described as an "important milestone" in understanding its short lifetime an other rare decays - and also to confirm decay into pairs of tau leptons (5.9 sigma). This was described by CERN as being *"of paramount importance to establishing the coupling of the Higgs boson to leptons and represents an important step towards measuring its couplings to third generation fermions, the very heavy copies of the electrons and quarks, whose role in nature is a profound mystery"*. Published results as of 19 Mar 2018 at 13 TeV for ATLAS and CMS had their measurements of the Higgs mass at 124.98 ± 0.28 GeV and 125.26 ± 0.21 GeV respectively.

Technical Aspects and Mathematical Formulation

In the Standard Model, the Higgs field is a four-component scalar field that forms a complex doublet of the weak isospin SU(2) symmetry:

$$\phi = \frac{1}{\sqrt{2}}\begin{pmatrix} \phi^1 + i\phi^2 \\ \phi^0 + i\phi^3 \end{pmatrix},$$

while the field has charge +½ under the weak hypercharge U(1) symmetry.

The Higgs part of the Lagrangian is

$$\mathcal{L}_H = \left|\left(\partial_\mu - igW_\mu^a\tau^a - i\frac{1}{2}g'B_\mu\right)\phi\right|^2 + \mu^2\phi^\dagger\phi - \lambda(\phi^\dagger\phi)^2,$$

where W_μ^a and B_μ are the gauge bosons of the SU(2) and U(1) symmetries, g and g' their respective coupling constants, $\tau^a = \frac{1}{2}\sigma^a$ (where σ^a are the Pauli matrices) a complete set generators of the SU(2) symmetry, and $\lambda > 0$ and $\mu^2 > 0$, so that the ground state breaks the SU(2) symmetry. The ground state of the Higgs field (the bottom of the potential) is degenerate with different ground states related to each other by a SU(2) gauge transformation. It is always possible to pick a gauge such that in the ground state $\phi^1 = \phi^2 = \phi^3 = 0$. The expectation value of ϕ^0 in the ground state (the vacuum expectation value or VEV) is then $\langle \phi^0 \rangle = \frac{1}{\sqrt{2}} v$, where $v = \frac{1}{\sqrt{\lambda}}|\mu|$. The measured value of this parameter is ~246 GeV/c^2. It has units of mass, and is the only free parameter of the Standard Model that is not a dimensionless number. Quadratic terms in W_μ and B_μ arise, which give masses to the W and Z bosons:

$$m_W = \tfrac{1}{2} v|g|,$$

$$m_Z = \tfrac{1}{2} v\sqrt{g^2 + g'^2},$$

with their ratio determining the Weinberg angle, $\cos\theta_W = \frac{m_W}{m_Z} = \frac{|g|}{\sqrt{g^2 + g'^2}}$, and leave a massless U(1) photon, γ. The mass of the Higgs

boson itself is given by

$$m_H = \sqrt{2\mu^2} \equiv \sqrt{2\lambda v^2}.$$

The quarks and the leptons interact with the Higgs field through Yukawa interaction terms:

$$\begin{aligned}
\mathcal{L}_Y = &-\lambda_u^{ij} \frac{\phi^0 - i\phi^3}{\sqrt{2}} \overline{u}_L^i u_R^j + \lambda_u^{ij} \frac{\phi^1 - i\phi^2}{\sqrt{2}} \overline{d}_L^i u_R^j \\
&- \lambda_d^{ij} \frac{\phi^0 + i\phi^3}{\sqrt{2}} \overline{d}_L^i d_R^j - \lambda_d^{ij} \frac{\phi^1 + i\phi^2}{\sqrt{2}} \overline{u}_L^i d_R^j \\
&- \lambda_e^{ij} \frac{\phi^0 + i\phi^3}{\sqrt{2}} \overline{e}_L^i e_R^j - \lambda_e^{ij} \frac{\phi^1 + i\phi^2}{\sqrt{2}} \overline{\nu}_L^i e_R^j + h.c.,
\end{aligned}$$

where $(d,u,e,\nu)_{L,R}^i$ are left-handed and right-handed quarks and leptons of the ith generation, $\lambda_{u,d,e}^{ij}$ are matrices of Yukawa couplings where h.c. denotes the hermitian conjugate of all the preceding terms. In the symmetry breaking ground state, only the terms containing ϕ^0 remain, giving rise to mass terms for the fermions. Rotating the quark and lepton fields to the basis where the matrices of Yukawa couplings are diagonal, one gets

$$\mathcal{L}_m = -m_u^i \overline{u}_L^i u_R^i - m_d^i \overline{d}_L^i d_R^i - m_e^i \overline{e}_L^i e_R^i + h.c.,$$

where the masses of the fermions are $m_{u,d,e}^i = \frac{1}{\sqrt{2}} \lambda_{u,d,e}^i v$, and $\lambda_{u,d,e}^i$ denote the eigenvalues of the Yukawa matrices.

Alternatives to the Standard Higgs Model

The Alternative models to the Standard Higgs Model are models which are considered by many particle physicists to solve some of Higgs boson's existing problems. Two of the most currently researched models are quantum triviality, and Higgs hierarchy problem.

In particle physics, elementary particles and forces give rise to the world around us. Physicists explain the behaviors of these particles and how they interact using the Standard Model—a widely accepted framework believed to explain most of the world we see around us. Initially, when these models were being developed and tested, it seemed that the mathematics behind those models, which were satisfactory in areas already tested, would also forbid elementary particles from having any mass, which showed clearly that these initial models were incomplete. In 1964 three groups of physicists almost simultaneously released papers describing how masses could be given to these particles, using approaches known as symmetry breaking. This approach allowed the particles to obtain a mass, without breaking other parts of particle physics theory that were already believed reasonably correct. This idea became known as the Higgs mechanism, and later experiments confirmed that such a mechanism does exist—but they could not show exactly *how* it happens.

The simplest theory for how this effect takes place in nature, and the theory that became incorporated into the Standard Model, was that if one or more of a particular kind of "field" (known as a Higgs field) happened to permeate space, and if it could interact with elementary particles in a particular way, then this would give rise to a Higgs mechanism in nature. In the basic Standard Model there is one field and one related Higgs boson; in some extensions to the Standard Model there are multiple fields and multiple Higgs bosons.

In the years since the Higgs field and boson were proposed as a way to explain the origins of symmetry breaking, several alternatives have been proposed that suggest how a symmetry breaking mechanism could occur without requiring a Higgs field to exist. Models which do not include a Higgs field or a Higgs boson are known as Higgsless models. In these models, strongly interacting dynamics rather than an additional (Higgs) field produce the non-zero vacuum expectation value that breaks electroweak symmetry.

The announcement of the discovery of a new scalar particle by the Large Hadron Collider (LHC) experiments ATLAS and CMS has immediately triggered activities to determine the properties of this particle. The measurement of its couplings to other SM particles and the extraction of its spin and parity quantum numbers are steps in order to establish the scalar as the Higgs boson, i.e. the particle related to EWSB. Any deviations in these properties from the SM expectation would hint to physics beyond the SM (BSM). Although the Higgs boson looks very SM-like, there is still room for interpretations within BSM theories. In absence of any direct detection of NP particles, the Higgs sector becomes particularly interesting. The precise measurements of the Higgs properties help to reveal the underlying mechanism of EWSB and in particular may shed light on the question if the underlying dynamics is strongly or weakly interacting. An example for the latter are supersymmetric (SUSY) extensions of the SM which remain weakly interacting up to high energies. The talk, that is summarised here, focuses on non-SUSY extensions of the SM. Composite Higgs Models shall be presented as theories emerging from a strongly interacting sector, before moving on to effective theory descriptions that allow for a more model-independent investigation of the Higgs sector. The last part finally is dedicated to specific (non-SUSY) models and how they can be probed through coupling measurements.

Composite Higgs Boson

In composite Higgs models the Higgs boson arises as a pseudo Nambu-Goldstone boson from a strongly-coupled sector. As result of the Goldstone nature of the Higgs boson, in the Strongly Interacting Light Higgs (SILH) scenario there is a light narrow Higgs-like scalar, which is the bound state from some strong dynamics and which is separated by a mass gap from the other usual resonances of the strong dynamics. At low energy, the particle content is hence the same as in the SM. The Higgs couplings to the SM particles, however, are modified. In composite Higgs models the problem of the generation of fermion masses is solved by the hypothesis of partial compositeness. The SM fermions, which are elementary, couple linearly to heavy states of the strong sector with the same quantum numbers, implying in particular the top quark to be largely composite. The global symmetry of the strong sector is explicitly broken by these couplings,and the Higgs potential is generated from loops of SM particles with the dominant contribution coming from the top quark. A low-mass Higgs boson of $\sim$ 125 GeV can naturally be accommodated only if the heavy quark partners are rather light, with masses below about 1 TeV. From a phenomenological point of view, the modified Higgs couplings to the SM particles not only change the Higgs production and decay rates, but notably lead to an increase of the cross section for double Higgs production in vector boson fusion with the energy. Furthermore, composite Higgs models are challenged by electroweak precision tests (EWPTs). The tension with the S and T parameters can be weakened through the contributions from new heavy fermions.

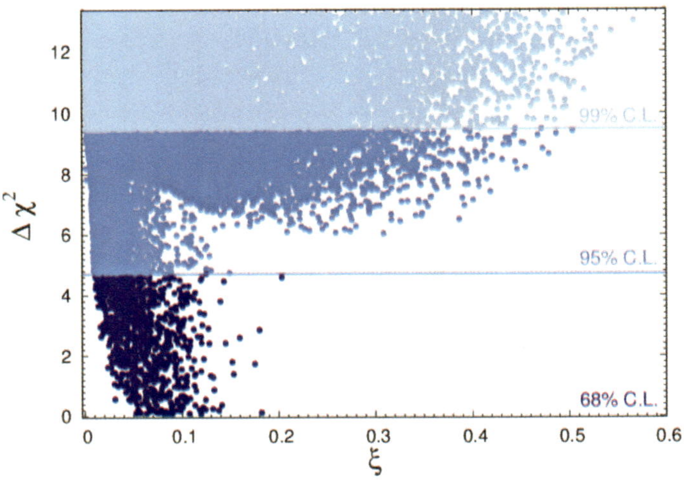

Parameters passing the X² test of electroweak precision observables, which fulfull $|V_{tb}|$ 0:92

This is shown in figure for a model with composite bottom quarks, where the fermions are embedded in the 10, the smallest possible representation of SO(5) that allows for partially composite bottom quarks while being compatible with the EWPTs by implementing custodial symmetry. Performing a χ^2 test, taking into account the EWPTs and the measurement of V_{tb}, it displays $\Delta\chi^2 = \chi^2 - \chi^2_{\min}$ as a function of ξ for the points passing the test after a scan over the model parameters. Here $\xi = v^2/f^2$, where $v \approx$ 246 GeV is the vacuum expectation value and f the typical scale of the Goldstone bosons of the strong sector.

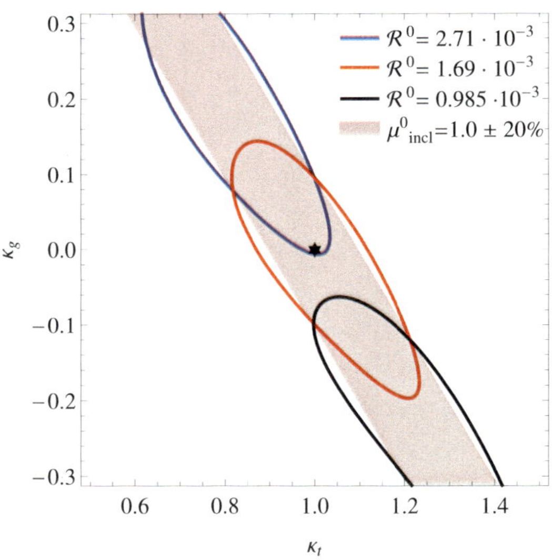

Figure: The 95% confidence level (C.L.) contours from a χ^2 test, in the plane of the top and effective gluon coupling modifiers κ_t and κ_g for different values of the inclusive signal strength μ^0_{incl} and the boosted observable R°.

As long as no heavy fermion partners have been detected directly, their inuence on loop induced processes like Higgs production through gluon fusion becomes particularly interesting. It has been shown, that the process computed by applying the Low-Energy Theorem (LET) is insensitive to the details of the couplings and masses of the strong sector. In double Higgs production, however, the LET is not reliable any more and the cross section becomes sensitive to the properties of the strong sector. Also the production of a boosted Higgs boson in association with a high-transverse momentum jet is sensitive to the details of heavy fermions and can be exploited to measure the Higgs coupling to top pairs.

Effective Lagrangian Approach

The plethora of NP extensions calls for an effective framework that captures NP effects in a model-independent way. The scale $\wedge$ at which NP becomes effective, proposed by natural mechanisms of EWSB, is not far from the TeV scale, so that a convenient framework for a model-independent analysis of deviations from the SM is given by an effective Lagrangian approach which enlarges the SM by including higher-dimensional operators built from SM fields. At the dimension-6 level, there are 59 linearly independent operators, taking into account only one fermion generation and allowing also for CP-odd operators. There has been some discussion not only about the most appropriate basis to use but also about the minimum number of operators that should be applied to best capture the NP impact on Higgs physics. The authors found that for one family there are 8 CP-even operators that, at tree-level, can only affect Higgs physics and no other SM processes. In a bottom-up approach, taking as starting point all possible new interactions among SM fields, the authors derived the set of independent new interactions at the dimension-6 level, that are presently best tested by the experiments and that give the best way to constrain NP. They found 18 of these BSM primary effects, not taking into account four-fermion deviations, minimal avour violation suppressed deviations and those arising from CP violation. There are 8 Higgs primaries, 7 EWPT primaries and 3 primaries affecting triple gauge boson couplings (TGC). All other NP effects are not independent and are correlated with the BSM primaries. In particular it is found, that large NP effects can still be revealed in the Higgs decay $H \rightarrow Z\gamma$ while the deviationsfrom the SM in the differential distributions of Higgs decays into a vector boson and a fermion pair are already constrained from TGC measurements.

Supersymmetry

Supersymmetry, in particle physics, a symmetry between fermions (subatomic particles with half-integer values of intrinsic angular momentum, or spin) and bosons (particles with integer values of spin). Supersymmetry is a complex mathematical framework based on the theory of group transformations that was developed beginning in the early 1970s to understand at a more-fundamental level the burgeoning number of subatomic particles being produced in high-energy particle accelerator experiments. It has evolved to address internal inconsistencies that arose in attempts to unify the forces in the Standard Model of particle physics. Supersymmetry is an essential feature of supergravity, the quantum field theory of the gravitational force, and of string theory, an ambitious attempt to provide a self-consistent quantum theory unifying all particles and forces in nature.

A physical entity is said to exhibit symmetry when it appears unchanged after undergoing a transformation operation. A square, for example, has a fourfold symmetry by which it appears the same when rotated about its centre through 90, 180, 270, and 360 degrees; four 90-degree rotations bring the square back to its original position. Symmetry with respect to time and space transformations is embodied within physical laws such as the conservation of energy and the conservation of momentum. With supersymmetry, fermions can be transformed into bosons without changing the structure of the underlying theory of the particles and their interactions. Thus, supersymmetry provides a relationship between the elementary particles that make up matter—quarks and leptons, which are all fermions—and the "force-carrier" particles that transmit the fundamental interactions of matter (all bosons). By showing that one type of particle is in effect a different facet of the other type, supersymmetry reduces the number of basic types of particle from two to one.

When a fermion is transformed into a boson and then back again into a fermion, it turns out that the particle has moved in space, an effect that is related to special relativity. Supersymmetry therefore relates transformations in an internal property of particles (spin) to transformations in space-time. In particular, when supersymmetry is made a "local" symmetry, so that the transformations vary over space-time, it automatically includes a particle with a spin of 2, which can be identified as the graviton, the "force carrier" associated with gravity. Theories involving supersymmetry in its local form are therefore often known as supergravity theories.

Supersymmetry also plays an important role in modern theories of particle physics because the new particles that it requires can eliminate various infinite quantities that otherwise appear in calculations of particle interactions at high energies, particularly in attempts at unified theories of the fundamental forces. These new particles are the bosons (or fermions) into which the known fermions (or bosons) are transformed by supersymmetry. Thus, supersymmetry implies a doubling of the number of the known particles. For example, fermions such as electrons and quarks should have bosonic supersymmetric partners, which have been given the names of selectrons and squarks. Similarly, known bosons such as the photon and the gluon should have fermionic supersymmetric partners, called the photino and the gluino. There has been no experimental evidence that such "superparticles" exist. If they do indeed exist, their masses could be in the range of 50 to 1,000 times that of the proton.

Applications

Extension of Possible Symmetry Groups

One reason that physicists explored supersymmetry is because it offers an extension to the more familiar symmetries of quantum field theory. These symmetries are grouped into the Poincaré group and internal symmetries and the Coleman–Mandula theorem showed that under certain assumptions, the symmetries of the S-matrix must be a direct product of the Poincaré group with a compact internal symmetry group or if there is not any mass gap, the conformal group with a compact internal symmetry group. In 1971 Golfand and Likhtman were the first to show that the Poincaré algebra can be extended through introduction of four anticommuting spinor generators (in four dimensions), which later became known as supercharges. in 1975 the Haag-Lopuszanski-Sohnius theorem analyzed all possible superalgebras in the general form, including those with an extended number of the supergenerators and central charges. This extended super-Poincaré algebra paved the way for obtaining a very large and important class of supersymmetric field theories.

The Supersymmetry Algebra

Traditional symmetries of physics are generated by objects that transform by the tensor representations of the Poincaré group and internal symmetries. Supersymmetries, however, are generated by objects that transform by the spin representations. According to the spin-statistics theorem, bosonic fields commute while fermionic fields anticommute. Combining the two kinds of fields into a single algebra requires the introduction of a Z_2-grading under which the bosons are the even elements and the fermions are the odd elements. Such an algebra is called a Lie superalgebra.

The simplest supersymmetric extension of the Poincaré algebra is the Super-Poincaré algebra. Expressed in terms of two Weyl spinors, has the following anti-commutation relation:

$$\{Q_\alpha, \bar{Q}_{\dot{\beta}}\} = 2(\sigma^\mu)_{\alpha\dot{\beta}} P_\mu$$

and all other anti-commutation relations between the Qs and commutation relations between the Qs and Ps vanish. In the above expression $P_\mu = -i\partial_\mu$ are the generators of translation and σ^μ are the Pauli matrices.

There are representations of a Lie superalgebra that are analogous to representations of a Lie algebra. Each Lie algebra has an associated Lie group and a Lie superalgebra can sometimes be extended into representations of a Lie supergroup.

The Supersymmetric Standard Model

Incorporating supersymmetry into the Standard Model requires doubling the number of particles since there is no way that any of the particles in the Standard Model can be superpartners of each other. With the addition of new particles, there are many possible new interactions. The simplest possible supersymmetric model consistent with the Standard Model is the Minimal Supersymmetric Standard Model (MSSM) which can include the necessary additional new particles that are able to be superpartners of those in the Standard Model.

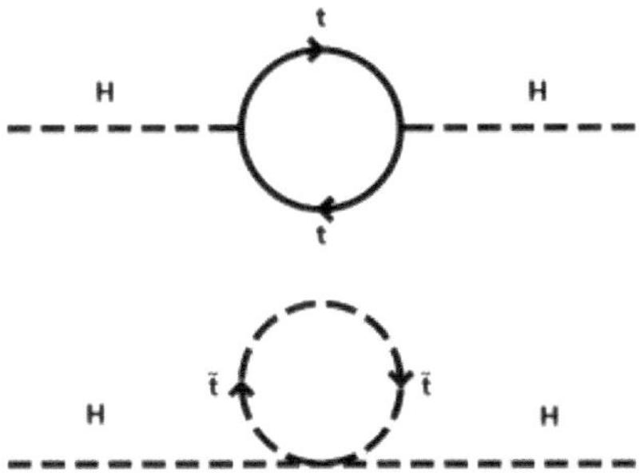

Cancellation of the Higgs boson quadratic mass renormalization between fermionic top quark loop and scalar stop squark tadpole Feynman diagrams in a supersymmetric extension of the Standard Model

One of the main motivations for SUSY comes from the quadratically divergent contributions to the Higgs mass squared. The quantum mechanical interactions of the Higgs boson causes a large renormalization of the Higgs mass and unless there is an accidental cancellation, the natural size of the Higgs mass is the greatest scale possible. This problem is known as the hierarchy problem. Supersymmetry reduces the size of the quantum corrections by having automatic cancellations between fermionic and bosonic Higgs interactions. If supersymmetry is restored at the weak scale, then the Higgs mass is related to supersymmetry breaking which can be induced from small non-perturbative effects explaining the vastly different scales in the weak interactions and gravitational interactions.

In many supersymmetric Standard Models there is a heavy stable particle (such as neutralino) which could serve as a weakly interacting massive particle (WIMP) dark matter candidate. The existence of a supersymmetric dark matter candidate is related closely to R-parity.

The standard paradigm for incorporating supersymmetry into a realistic theory is to have the underlying dynamics of the theory be supersymmetric, but the ground state of the theory does not respect the symmetry and supersymmetry is broken spontaneously. The supersymmetry break can not be done permanently by the particles of the MSSM as they currently appear. This means that there is a new sector of the theory that is responsible for the breaking. The only constraint on this new sector is that it must break supersymmetry permanently and must give superparticles TeV scale masses. There are many models that can do this and most of their details do not matter. In order to parameterize the relevant features of supersymmetry breaking, arbitrary soft SUSY breaking terms are added to the theory which temporarily break SUSY explicitly but could never arise from a complete theory of supersymmetry breaking.

Gauge-coupling Unification

One piece of evidence for supersymmetry existing is gauge coupling unification. The renormalization group evolution of the three gauge coupling constants of the Standard Model is somewhat sensitive to the present particle content of the theory. These coupling constants do not quite meet together at a common energy scale if we run the renormalization group using the Standard Model. With the addition of minimal SUSY joint convergence of the coupling constants is projected at approximately 10^{16} GeV.

Supersymmetric Quantum Mechanics

Supersymmetric quantum mechanics adds the SUSY superalgebra to quantum mechanics as opposed to quantum field theory. Supersymmetric quantum mechanics often becomes relevant when studying the dynamics of supersymmetric solitons, and due to the simplified nature of having fields which are only functions of time (rather than space-time), a great deal of progress has been made in this subject and it is now studied in its own right.

SUSY quantum mechanics involves pairs of Hamiltonians which share a particular mathematical relationship, which are called *partner Hamiltonians*. (The potential energy terms which occur in the Hamiltonians are then known as *partner potentials*.) An introductory theorem shows that for every eigenstate of one Hamiltonian, its partner Hamiltonian has a corresponding eigenstate with the same energy. This fact can be exploited to deduce many properties of the eigenstate spectrum. It is analogous to the original description of SUSY, which referred to bosons and fermions. We can imagine a "bosonic Hamiltonian", whose eigenstates are the various bosons of our theory. The SUSY partner of this Hamiltonian would be "fermionic", and its eigenstates would be the theory's fermions. Each boson would have a fermionic partner of equal energy.

Supersymmetry in Condensed Matter Physics

SUSY concepts have provided useful extensions to the WKB approximation. Additionally, SUSY has been applied to disorder averaged systems both quantum and non-quantum (through statistical mechanics), the Fokker-Planck equation being an example of a non-quantum theory. The 'supersymmetry' in all these systems arises from the fact that one is modelling one particle and as such the 'statistics' don't matter. The use of the supersymmetry method provides a mathematical rigorous alternative to the replica trick, but only in non-interacting systems, which attempts to address the so-called 'problem of the denominator' under disorder averaging.

Supersymmetry in Optics

Integrated optics was recently found to provide a fertile ground on which certain ramifications of SUSY can be explored in readily-accessible laboratory settings. Making use of the analogous mathematical structure of the quantum-mechanical Schrödinger equation and the wave equation governing the evolution of light in one-dimensional settings, one may interpret the refractive index distribution of a structure as a potential landscape in which optical wave packets propagate. In this manner, a new class of functional optical structures with possible applications in phase matching, mode conversion and space-division multiplexing becomes possible. SUSY transformations have been also proposed as a way to address inverse scattering problems in optics and as a one-dimensional transformation optics.

Supersymmetry in Dynamical Systems

All stochastic (partial) differential equations, the models for all types of continuous time dynamical systems, possess topological supersymmetry. In the operator representation of stochastic evolution, the topological supersymmetry is the exterior derivative which is commutative with the stochastic evolution operator defined as the stochastically averaged pullback induced on differential forms by SDE-defined diffeomorphisms of the phase space. The topological sector of the so-emerging supersymmetric theory of stochastic dynamics can be recognized as the Witten-type topological field theory.

The meaning of the topological supersymmetry in dynamical systems is the preservation of the phase space continuity—infinitely close points will remain close during continuous time evolution even in the presence of noise. When the topological supersymmetry is broken spontaneously, this property is violated in the limit of the infinitely long temporal evolution and the model can be said to exhibit (the stochastic generalization of) the butterfly effect. From a more general perspective, spontaneous breakdown of the topological supersymmetry is the theoretical essence of the ubiquitous dynamical phenomenon variously known as chaos, turbulence, self-organized criticality etc. The Goldstone theorem explains the associated emergence of the long-range dynamical behavior that manifests itself as 1/f noise, butterfly effect, and the scale-free statistics of sudden (instantonic) processes, e.g., earthquakes, neuroavalanches, solar flares etc., known as the Zipf's law and the Richter scale.

Supersymmetry in Mathematics

SUSY is also sometimes studied mathematically for its intrinsic properties. This is because it describes complex fields satisfying a property known as holomorphy, which allows holomorphic quantities to be exactly computed. This makes supersymmetric models useful "toy models" of more realistic theories. A prime example of this has been the demonstration of S-duality in four-dimensional gauge theories that interchanges particles and monopoles.

The proof of the Atiyah-Singer index theorem is much simplified by the use of supersymmetric quantum mechanics.

Supersymmetry in Quantum Gravity

Supersymmetry is part of superstring theory, a string theory of quantum gravity, although it could in theory be a component of other quantum gravity theories as well, such as loop quantum gravity. For superstring theory to be consistent, supersymmetry seems to be required at some level (although it may be a strongly broken symmetry). If experimental evidence confirms supersymmetry in the form of supersymmetric particles such as the neutralino that is often believed to be the lightest superpartner, some people believe this would be a major boost to superstring theory. Since supersymmetry is a required component of superstring theory, any discovered supersymmetry would be consistent with superstring theory. If the Large Hadron Collider and other major particle physics experiments fail to detect supersymmetric partners, many versions of superstring theory which had predicted certain low mass superpartners to existing particles may need to be significantly revised.

Minimal Supersymmetric Standard Model

The Minimal Supersymmetric Standard Model (MSSM) is an extension to the Standard Model that realizes supersymmetry. MSSM is the minimal supersymmetrical model as it considers only "the [minimum] number of new particle states and new interactions consistent with phenomenology". Supersymmetry pairs bosons with fermions, so every Standard Model particle has a superpartner yet undiscovered. If we find these superparticles, it equates to discovering such particles as dark matter, could provide evidence for grand unification, and provide hints as to whether string theory describes nature. The failure to find evidence for supersymmetry using the Large Hadron Collider suggests a leaning to abandon it.

Theoretical Motivations

There are three principal motivations for the MSSM over other theoretical extensions of the Standard Model, namely:

- Naturalness
- Gauge coupling unification
- Dark Matter

These motivations come out without much effort and they are the primary reasons why the MSSM is the leading candidate for a new theory to be discovered at collider experiments such as the Tevatron or the LHC.

Naturalness

The original motivation for proposing the MSSM was to stabilize the Higgs mass to radiative corrections that are quadratically divergent in the Standard Model (hierarchy problem). In supersymmetric models, scalars are related to fermions and have the same mass. Since fermion masses are logarithmically divergent, scalar masses inherit the same radiative stability. The Higgs vacuum expectation value is related to the negative scalar mass in the Lagrangian. In order for the radiative corrections to the Higgs mass to not be dramatically larger than the actual value, the mass of the superpartners of the Standard Model should not be significantly heavier than the Higgs VEV—roughly 100 GeV. In 2012, the Higgs particle was discovered at the LHC, and its mass was found to be 125-126 GeV.

Gauge-coupling Unification

If the superpartners of the Standard Model are near the TeV scale, then measured gauge couplings of the three gauge groups unify at high energies. The beta-functions for the MSSM gauge couplings are given by

Gauge Group	$\alpha^{-1}(M_{Z^0})$	b_0 MSSM
SU(3)	8.5	-3
SU(2)	29.6	$+1$
U(1)	59.2	$+6\frac{3}{5}$

where α_1^{-1} is measured in SU(5) normalization—a factor of $\frac{3}{5}$ different than the Standard Model's normalization and predicted by Georgi–Glashow SU(5).

The condition for gauge coupling unification at one loop is whether the following expression is satisfied $\frac{\alpha_3^{-1} - \alpha_2^{-1}}{\alpha_2^{-1} - \alpha_1^{-1}} = \frac{b_{03} - b_{02}}{b_{02} - b_{01}}$.

Remarkably, this is precisely satisfied to experimental errors in the values of $\alpha^{-1}(M_{Z^0})$. There are two loop corrections and both TeV-scale and GUT-scale threshold corrections that alter this condition on gauge coupling unification, and the results of more extensive calculations reveal that gauge coupling unification occurs to an accuracy of 1%, though this is about 3 standard deviations from the theoretical expectations.

This prediction is generally considered as indirect evidence for both the MSSM and SUSY GUTs. It should be noted that gauge coupling unification does not necessarily imply grand unification and there exist other mechanisms to reproduce gauge coupling unification. However, if superpartners are found in the near future, the apparent success of gauge coupling unification would suggest that a supersymmetric grand unified theory is a promising candidate for high scale physics.

Dark Matter

If R-parity is preserved, then the lightest superparticle (LSP) of the MSSM is stable and is a Weakly interacting massive particle (WIMP) — i.e. it does not have electromagnetic or strong interactions. This makes the LSP a good dark matter candidate and falls into the category of cold dark matter (CDM) particle.

Predictions of the MSSM Regarding Hadron Colliders

The Tevatron and LHC have active experimental programs searching for supersymmetric particles. Since both of these machines are hadron colliders — proton antiproton for the Tevatron and proton proton for the LHC — they search best for strongly interacting particles. Therefore, most experimental signature involve production of squarks or gluinos. Since the MSSM has R-parity, the lightest supersymmetric particle is stable and after the squarks and gluinos decay each decay chain will contain one LSP that will leave the detector unseen. This leads to the generic prediction that the MSSM will produce a 'missing energy' signal from these particles leaving the detector.

Neutralinos

There are four neutralinos that are fermions and are electrically neutral, the lightest of which is typically stable. They are typically labeled $\tilde{N}_1^0, \tilde{N}_2^0, \tilde{N}_3^0, \tilde{N}_4^0$ (although sometimes $\tilde{\chi}_1^0, \ldots, \tilde{\chi}_4^0$ is used instead). These four states are mixtures of the Bino and the neutral Wino (which are the neutral electroweak Gauginos), and the neutral Higgsinos. As the neutralinos are Majorana fermions, each of them is identical with its antiparticle. Because these particles only interact with the weak vector bosons, they are not directly produced at hadron colliders in copious numbers. They primarily appear as particles in cascade decays of heavier particles usually originating from colored supersymmetric particles such as squarks or gluinos.

In R-parity conserving models, the lightest neutralino is stable and all supersymmetric cascades decays end up decaying into this particle which leaves the detector unseen and its existence can only be inferred by looking for unbalanced momentum in a detector.

The heavier neutralinos typically decay through a Z^0 to a lighter neutralino or through a $W^{\pm}$ to chargino. Thus a typical decay is

$$\widehat{N}_2^0 \rightarrow \widehat{N}_1^0 + Z^0 \qquad \rightarrow \text{Missing energy} + \ell^+ + \ell^-$$

$$\widehat{N}_2^0 \rightarrow C\tilde{\chi}_1^{\pm} + W^{\mp} \rightarrow \widehat{N}_1^0 + W^{\pm} + W^{\mp} \rightarrow \text{Missing energy} + \ell^+ + \ell^-$$

The mass splittings between the different Neutralinos will dictate which patterns of decays are allowed.

Charginos

There are two Charginos that are fermions and are electrically charged. They are typically labeled $C\tilde{\chi}_1^{\pm}$ and $C\tilde{\chi}_2^{\pm}$ (although sometimes $\tilde{\chi}_1^{\pm}$ and $\tilde{\chi}_2^{\pm}$ is used instead). The heavier chargino can decay through Z^o to the lighter chargino. Both can decay through a $W^{\pm}$ to neutralino.

Squarks

The squarks are the scalar superpartners of the quarks and there is one version for each Standard Model quark. Due to phenomenological constraints from flavor changing neutral currents, typically the lighter two generations of squarks have to be nearly the same in mass and therefore are not given distinct names. The superpartners of the top and bottom quark can be split from the lighter squarks and are called *stop* and *sbottom*.

In the other direction, there may be a remarkable left-right mixing of the stops $\tilde{t}$ and of the sbottoms $\tilde{b}$ because of the high masses of the partner quarks top and bottom:

- $\tilde{t}_1 = e^{+i\phi}\cos(\theta)\widetilde{t_L} + \sin(\theta)\widetilde{t_R}$
- $\tilde{t}_2 = e^{-i\phi}\cos(\theta)\widetilde{t_R} - \sin(\theta)\widetilde{t_L}$

A similar story holds for bottom with its own parameters $\tilde{b}$ and ϕ. Squarks can be produced through strong interactions and therefore are easily produced at hadron colliders. They decay to quarks and neutralinos or charginos which further decay. In R-parity conserving scenarios, squarks are pair produced and therefore a typical signal is

- $\tilde{q}\tilde{\bar{q}} \rightarrow q\tilde{N}_1^0\bar{q}\tilde{N}_1^0 \rightarrow$ 2 jets + missing energy
- $\tilde{q}\tilde{\bar{q}} \rightarrow q\tilde{N}_2^0\bar{q}\tilde{N}_1^0 \rightarrow q\tilde{N}_1^0\ell\bar{\ell}\bar{q}\tilde{N}_1^0 \rightarrow$ 2 jets + 2 leptons + missing energy

Gluinos

Gluinos are Majorana fermionic partners of the gluon which means that they are their own antiparticles. They interact strongly and therefore can be produced significantly at the LHC. They can only decay to a quark and a squark and thus a typical gluino signal is

- $\tilde{g}\tilde{g} \rightarrow (q\tilde{\bar{q}})(\bar{q}\tilde{q}) \rightarrow (q\bar{q}\tilde{N}_1^0)(\bar{q}q\tilde{N}_1^0) \rightarrow$ 4 jets + Missing energy

Because gluinos are Majorana, gluinos can decay to either a quark+anti-squark or an anti-quark+squark with equal probability. Therefore, pairs of gluinos can decay to

- $\tilde{g}\tilde{g} \rightarrow (\bar{q}\tilde{q})(\bar{q}\tilde{q}) \rightarrow (q\bar{q}\tilde{C}_1^+)(q\bar{q}\tilde{C}_1^+) \rightarrow (q\bar{q}W^+)(q\bar{q}W^+) \rightarrow$ 4 jets+ $\ell^+\ell^+$ + Missing energy

This is a distinctive signature because it has same-sign di-leptons and has very little background in the Standard Model.

Sleptons

Sleptons are the scalar partners of the leptons of the Standard Model. They are not strongly interacting and therefore are not produced very often at hadron colliders unless they are very light.

Because of the high mass of the tau lepton there will be left-right mixing of the stau similar to that of stop and sbottom.

Sleptons will typically be found in decays of a charginos and neutralinos if they are light enough to be a decay product.

- $\tilde{C}^+ \to \tilde{\ell}^+ \nu$
- $\tilde{N}^0 \to \tilde{\ell}^+ \ell^-$

MSSM Fields

Fermions have bosonic superpartners (called sfermions), and bosons have fermionic superpartners (called bosinos). For most of the Standard Model particles, doubling is very straightforward. However, for the Higgs boson, it is more complicated.

A single Higgsino (the fermionic superpartner of the Higgs boson) would lead to a gauge anomaly and would cause the theory to be inconsistent. However, if two Higgsinos are added, there is no gauge anomaly. The simplest theory is one with two Higgsinos and therefore two scalar Higgs doublets. Another reason for having two scalar Higgs doublets rather than one is in order to have Yukawa couplings between the Higgs and both down-type quarks and up-type quarks; these are the terms responsible for the quarks' masses. In the Standard Model the down-type quarks couple to the Higgs field (which has Y=-1/2) and the up-type quarks to its complex conjugate (which has Y=+1/2). However, in a supersymmetric theory this is not allowed, so two types of Higgs fields are needed.

SM Particle type	Particle	Symbol	Spin	R-Parity	Superpartner	Symbol	Spin	R-parity
Fermions	Quark	q	$\frac{1}{2}$	+1	Squark	$\tilde{q}$	0	-1
	Lepton	ℓ	$\frac{1}{2}$	+1	Slepton	$\tilde{\ell}$	0	-1
Bosons	W	W	1	+1	Wino	$\tilde{W}$	$\frac{1}{2}$	-1
	B	B	1	+1	Bino	$\tilde{B}$	$\frac{1}{2}$	-1
	Gluon	g	1	+1	Gluino	$\tilde{g}$	$\frac{1}{2}$	-1
Higgs bosons	Higgs	h_u, h_d	0	+1	Higgsinos	$\tilde{h}_u, \tilde{h}_d$	$\frac{1}{2}$	-1

MSSM Superfields

In supersymmetric theories, every field and its superpartner can be written together as a superfield. The superfield formulation of supersymmetry is very convenient to write down manifestly supersymmetric theories (i.e. one does not have to tediously check that the theory is supersymmetric term by term in the Lagrangian). The MSSM contains vector superfields associated with the Standard Model gauge groups which contain the vector bosons and associated gauginos. It also contains chiral superfields for the Standard Model fermions and Higgs bosons (and their respective superpartners).

field	multiplicity	representation	Z_2-parity	Standard Model particle
Q	3	$(3,2)_{\frac{1}{6}}$	–	left-handed quark doublet
U^c	3	$(\bar{3},1)_{-\frac{2}{3}}$	–	right-handed up-type anti-quark
D^c	3	$(\bar{3},1)_{\frac{1}{3}}$	–	right-handed down-type anti-quark
L	3	$(1,2)_{-\frac{1}{2}}$	–	left-handed lepton doublet
E^c	3	$(1,1)_{1.}$	–	right-handed anti-lepton
H_u	1	$(1,2)_{\frac{1}{2}}$	+	Higgs
H_d	1	$(1,2)_{-\frac{1}{2}}$	+	Higgs

MSSM Higgs Mass

The MSSM Higgs Mass is a prediction of the Minimal Supersymmetric Standard Model. The mass of the lightest Higgs boson is set by the Higgs *quartic coupling*. Quartic couplings are not soft supersymmetry-breaking parameters since they lead to a quadratic divergence of the Higgs mass. Furthermore, there are no supersymmetric parameters to make the Higgs mass a free parameter in the MSSM (though not in non-minimal extensions). This means that Higgs mass is a prediction of the MSSM. The LEP II and the IV experiments placed a lower limit on the Higgs mass of 114.4 GeV. This lower limit is significantly above where the MSSM would typically predict it to be but does not rule out the MSSM; the discovery of the Higgs with a mass of 125 GeV is within the maximal upper bound of approximately 130 GeV that loop corrections within the MSSM would raise the Higgs mass to. Proponents of the MSSM point out that a Higgs mass within the upper bound of the MSSM calculation of the Higgs mass is a successful prediction, albeit pointing to more fine tuning than expected.

Formulas

The only susy-preserving operator that creates a quartic coupling for the Higgs in the MSSM arise for the D-terms of the SU(2) and U(1) gauge sector and the magnitude of the quartic coupling is set by the size of the gauge couplings.

This leads to the prediction that the Standard Model-like Higgs mass (the scalar that couples approximately to the vev) is limited to be less than the Z mass

$$m_{h^0}^2 \le m_{Z^0}^2 \cos^2 2\beta \,.$$

Since supersymmetry is broken, there are radiative corrections to the quartic coupling that can increase the Higgs mass. These dominantly arise from the 'top sector'

$$m_{h^0}^2 \le m_{Z^0}^2 \cos^2 2\beta + \frac{3}{\pi^2}\frac{m_t^4 \sin^4\beta}{v^2}\log\frac{m_{\tilde{t}}}{m_t}$$

where m_t is the top mass and $m_{\tilde{t}}$ is the mass of the top squark. This result can be interpreted as the RG running of the Higgs quartic coupling from the scale of supersymmetry to the top mass—however since the top squark mass should be relatively close to the top mass, this is usually a fairly modest contribution and increases the Higgs mass to roughly the LEP II bound of 114 GeV before the top squark becomes too heavy.

Finally there is a contribution from the top squark A-terms

$$\mathcal{L} = y_t m_{\tilde{t}} a\, h_u \tilde{q}_3 \tilde{u}_3^c$$

where a is a dimensionless number. This contributes an additional term to the Higgs mass at loop level, but is not logarithmically enhanced

$$m_{h^0}^2 \le m_{Z^0}^2 \cos^2 2\beta + \frac{3}{\pi^2}\frac{m_t^4 \sin^4\beta}{v^2}\left(\log\frac{m_{\tilde{t}}}{m_t} + a^2(1-a^2/12)\right)$$

by pushing $a \to \sqrt{6}$ (known as 'maximal mixing') it is possible to push the Higgs mass to 125 GeV without decoupling the top squark or adding new dynamics to the MSSM.

As the Higgs was found at around 125 GeV (along with no other superparticles) at the LHC, this strongly hints at new dynamics beyond the MSSM, such as the 'Next to Minimal Supersymmetric Standard Model' (NMSSM); and suggests some correlation to the little hierarchy problem.

The MSSM Lagrangian

The Lagrangian for the MSSM contains several pieces.

- The first is the Kähler potential for the matter and Higgs fields which produces the kinetic terms for the fields.

- The second piece is the gauge field superpotential that produces the kinetic terms for the gauge bosons and gauginos.
- The next term is the superpotential for the matter and Higgs fields. These produce the Yukawa couplings for the Standard Model fermions and also the mass term for the Higgsinos. After imposing R-parity, the renormalizable, gauge invariant operators in the superpotential are

$$W = \mu H_u H_d + y_u H_u Q U^c + y_d H_d Q D^c + y_l H_d L E^c$$

The constant term is unphysical in global supersymmetry (as opposed to supergravity).

Soft Susy Breaking

The last piece of the MSSM Lagrangian is the soft supersymmetry breaking Lagrangian. The vast majority of the parameters of the MSSM are in the susy breaking Lagrangian. The soft susy breaking are divided into roughly three pieces.

- The first are the gaugino masses

$$\mathcal{L} \supset m_{\frac{1}{2}} \tilde{\lambda}\tilde{\lambda} + h.c.$$

Where $\tilde{\lambda}$ are the gauginos and $m_{\frac{1}{2}}$ is different for the wino, bino and gluino.

- The next are the soft masses for the scalar fields

$$\mathcal{L} \supset m_0^2 \phi^\dagger \phi$$

where ϕ are any of the scalars in the MSSM and m_0 are 3×3 Hermitian matrices for the squarks and sleptons of a given set of gauge quantum numbers. The eigenvalues of these matrices are actually the masses squared, rather than the masses.

- There are the A and B terms which are given by

$$\mathcal{L} \supset B_\mu h_u h_d + A h_u \tilde{q}\widetilde{u^c} + A h_d \tilde{q}\widetilde{d^c} + A h_d \tilde{l}\widetilde{e^c} + h.c.$$

The A terms are 3×3 complex matrices much as the scalar masses are.

- Although not often mentioned with regard to soft terms, to be consistent with observation, one must also include Gravitino and Goldstino soft masses given by

$$\mathcal{L} \supset m_{3/2} \Psi^\alpha_\mu (\sigma^{\mu\nu})^\beta_\alpha \Psi_\beta + m_{3/2} G^\alpha G_\alpha + h.c.$$

The reason these soft terms are not often mentioned are that they arise through local supersymmetry and not global supersymmetry, although they are required otherwise if the Goldstino were massless it would contradict observation. The Goldstino mode is eaten by the Gravitino to become massive, through a gauge shift, which also absorbs the would-be "mass" term of the Goldstino.

Theories of Supersymmetry Breaking

A large amount of theoretical effort has been spent trying to understand the mechanism for soft supersymmetry breaking that produces the desired properties in the superpartner masses and interactions. The three most extensively studied mechanisms are:

Gravity-mediated Supersymmetry Breaking

Gravity-mediated supersymmetry breaking is a method of communicating supersymmetry breaking to the supersymmetric Standard Model through gravitational interactions. It was the first method proposed to communicate supersymmetry breaking. In gravity-mediated supersymmetry-breaking models, there is a part of the theory that only interacts with the MSSM through gravitational interaction. This hidden sector of the theory breaks supersymmetry. Through the supersymmetric version of the Higgs mechanism, the gravitino, the supersymmetric version of the graviton, acquires a mass. After the gravitino has a mass, gravitational radiative corrections to soft masses are incompletely cancelled beneath the gravitino's mass.

It is currently believed that it is not generic to have a sector completely decoupled from the MSSM and there should be higher dimension operators that couple different sectors together with the higher dimension operators suppressed by the Planck scale. These operators give as large of a contribution to the soft supersymmetry breaking masses as the gravitational loops; therefore, today people usually consider gravity mediation to be gravitational sized direct interactions between the hidden sector and the MSSM.

mSUGRA stands for minimal supergravity. The construction of a realistic model of interactions within $N = 1$ supergravity framework where supersymmetry breaking is communicated through the supergravity interactions was carried out by Ali Chamseddine, Richard Arnowitt, and Pran Nath in 1982. mSUGRA is one of the most widely investigated models of particle physics due to its predictive power requiring only 4 input parameters and a sign, to determine the low energy phenomenology from the scale of Grand Unification. The most widely used set of parameters is:

Symbol	Description
m_0	the common mass of the scalars (sleptons, squarks, Higgs bosons) at the Grand Unification scale
$m_{1/2}$	the common mass of the gauginos and higgsinos at the Grand Unification scale
A_0	the common trilinear coupling

$\tan\beta$	the ratio of the vacuum expectation values of the two Higgs doublets
$sign(\mu)$	the sign of the higgsino mass parameter

Gravity-Mediated Supersymmetry Breaking was assumed to be flavor universal because of the universality of gravity; however, in 1986 Hall, Kostelecky, and Raby showed that Planck-scale physics that are necessary to generate the Standard-Model Yukawa couplings spoil the universality of the supersymmetry breaking.

Gauge-mediated Supersymmetry Breaking (GMSB)

Gauge-mediated supersymmetry breaking is method of communicating supersymmetry breaking to the supersymmetric Standard Model through the Standard Model's gauge interactions. Typically a hidden sector breaks supersymmetry and communicates it to massive messenger fields that are charged under the Standard Model. These messenger fields induce a gaugino mass at one loop and then this is transmitted on to the scalar superpartners at two loops. Requiring stop squarks below 2 TeV, the maximum Higgs boson mass predicted is just 121.5GeV. With the Higgs being discovered at 125GeV - this model requires stops above 2 TeV.

Anomaly-mediated Supersymmetry Breaking (AMSB)

Anomaly-mediated supersymmetry breaking is a special type of gravity mediated supersymmetry breaking that results in supersymmetry breaking being communicated to the supersymmetric Standard Model through the conformal anomaly. Requiring stop squarks below 2 TeV, the maximum Higgs boson mass predicted is just 121.0GeV. With the Higgs being discovered at 125GeV - this scenario requires stops heavier than 2 TeV.

Next-to-Minimal Supersymmetric Standard Model

In particle physics, NMSSM is an acronym for Next-to-Minimal Supersymmetric Standard Model. It is a supersymmetric extension to the Standard Model that adds an additional singlet chiral superfield to the MSSM and can be used to dynamically generate the μ term, solving the μ-problem.

The Minimal Supersymmetric Standard Model does not explain why the μ parameter in the superpotential term $\mu H_u H_d$ is at the electroweak scale. The idea behind the Next to Minimal Supersymmetric Standard Model is to promote the μ term to a gauge singlet, chiral superfield S. Note that the scalar superpartner of the singlino S is denoted by $\hat{S}$ and the spin-1/2 singlino superpartner by $\tilde{S}$ in the following. The superpotential for the NMSSM is given by

$$W_{NMSSM} = W_{Yuk} + \lambda S H_u H_d + \frac{\kappa}{3} S^3$$

where W_{Yuk} gives the Yukawa couplings for the Standard Model fermions. Since the superpotential has mass dimension three, the couplings λ and κ are dimensionless, hence the μ-problem of the MSSM is solved in the NMSSM – the superpotential of the NMSSM is scale invariant. The role of the λ term is to generate an effective μ term. This is done with the scalar component of the singlet $\hat{S}$ getting a vacuum-expectation value $\langle\hat{S}\rangle$, that is, we have

$$\mu_{eff} = \lambda\langle\hat{S}\rangle$$

Without the κ term the superpotential would have a U(1)' symmetry, so-called Peccei–Quinn symmetry. This additional symmetry would alter the phenomenology completely. The role of the κ term is to break this U(1)' symmetry. The κ term is introduced trilinear such that κ is dimensionless. However, there remains a discrete $\mathbb{Z}_3$ symmetry, which is moreover broken spontaneously. In principle this leads to the domain wall problem. Introducing additional, but suppressed terms, the $\mathbb{Z}_3$ symmetry can be broken without changing phenomenology at the electroweak scale. It is assumed that the domain wall problem is circumvented in this way without any modifications except far beyond the electroweak scale.

Also alternative models have been proposed which solve the μ-problem of the MSSM. One idea is to keep the κ term in the superpotential and take the U(1)' symmetry into account. Assuming this symmetry to be local an additional Z' gauge boson is predicted in this model, called UMSSM.

Phenomenology

Due to the additional singlet S the NMSSM alters in general the phenomenology of both the Higgs sector and the neutralino sector compared to the MSSM.

Higgs Phenomenology

In the Standard Model we have one physical Higgs boson. In the MSSM we encounter five physical Higgs bosons. Due to the additional singlet $\hat{S}$ in the NMSSM we have two more Higgs bosons, that is, in total seven physical Higgs bosons. The Higgs sector is therefore much richer compared to the MSSM. In particular, the Higgs potential is in general no longer invariant under CP transformations. Typically, the Higgs bosons in the NMSSM are denoted in an order with increasing masses, that is, by $H_1, H_2, \ldots, H_7$ with H_1 the lightest Higgs boson. In the special case of a CP conserving Higgs potential we have three CP even Higgs bosons, H_1, H_2, H_3, two CP odd ones, A_1, A_2 and a pair of charged Higgs bosons, H^+, H^-. In the MSSM, the lightest Higgs boson is always Standard Model-like, and therefore its production and decays are roughly known. In the NMSSM, the lightest Higgs can be very light (even of the order of 1 GeV) and may have escaped detection so far. In addition, in the CP-conserving case, the lightest CP-even Higgs boson turns out to have an enhanced lower bound compared to the MSSM. This is one of the reasons why the NMSSM deserves much attraction in recent years.

Neutralino Phenomenology

The spin-1/2 singlino $\tilde{S}$ gives a fifth neutralino, compared to the four neutralinos of the MSSM. The singlino does not couple to gauge bosons, gauginos (the superpartners of the gauge bosons), leptons, sleptons (the superpartners of the leptons), quarks or squarks (the superpartners of the quarks). Suppose that a supersymmetric partner particle is produced at a collider, for instance at the LHC, the singlino is omitted in cascade decays and therefore escapes detection. However, in case the singlino is the lightest supersymmetric particle (LSP) all supersymmetric partner particles eventually decay into the singlino. Due to R parity conservation this LSP is stable. In this way the singlino could be detected via missing transverse energy in the detector.

Grand Unified Theory

Unified field theory, in particle physics, an attempt to describe all fundamental forces and the relationships between elementary particles in terms of a single theoretical framework. In physics, forces can be described by fields that mediate interactions between separate objects. In the mid-19th century James Clerk Maxwell formulated the first field theory in his theory of electromagnetism. Then, in the early part of the 20th century, Albert Einstein developed general relativity, a field theory of gravitation. Later, Einstein and others attempted to construct a unified field theory in which electromagnetism and gravity would emerge as different aspects of a single fundamental field. They failed, and to this day gravity remains beyond attempts at a unified field theory.

Grand unified theories describe the interactions of quarks and leptons within the same theoretical structure. This gives rise to the possibility that quarks can decay to leptons and specifically that the proton can decay. Early attempts at a GUT predicted that the proton's lifetime must be in the region of 10^{32} years. This prediction has been tested in experiments that monitor large amounts of matter containing on the order of 10^{32} protons, but there is no evidence that protons decay. If they do in fact decay, they must do so with a lifetime greater than that predicted by the simplest GUTs. There is also evidence to suggest that the strengths of the forces do not converge exactly unless new effects come into play at higher energies. One such effect could be a new symmetry called "supersymmetry."

A grand unified theory, or GUT, that would elegantly explain how the universe works by linking three of the four known forces together. Physicists first linked the electromagnetic force, which dictates the structure of atoms and the behavior of light, and the weak nuclear force, which underlies how particles decay.

But they didn't want to stop there. Scientists began working to link this electroweak theory with the strong force, which binds quarks together into things like the protons and neutrons in our atoms. (The fourth force that we know, gravity, doesn't have a complete working quantum theory, so it's relegated to the realm of Theories of Everything, or ToEs.)

Linking the different forces into a single theory isn't easy, since each behaves a different way. Electromagnetism is long-ranged, the weak force is short-ranged, and the strong force is weak in high-energy environments such as the early universe and strong where energy is low. To unify these three forces, scientists have to explain how they can be aspects of a single thing and yet manifest in radically different ways in the real world.

The electroweak theory unified the electromagnetic and weak forces by proposing they were aspects of a single interaction that is present only at very high energies, as in a particle accelerator or the very early universe. Above a certain threshold known as the electroweak scale, there is no difference between the two forces, but that unity is broken when the energy drops below a certain point.

GUTs also predicted that protons should decay into lighter particles. There was just one problem: Experiments didn't see that decay.

The Problem with Protons

GUTs predicted that all quarks could potentially change into lighter particles, including the quarks making up protons. In fact, GUTs said that protons would be unstable over a period much longer than the lifetime of the universe. To maximize the chances of seeing that rare proton decay, physicists needed to build detectors with a lot of atoms.

Unification of Matter Particles

$$\left[\begin{array}{ccccc|cccccccccc}
0 & W & X_r & X_g & X_b & 0 & 0 & 0 & 0 & 0 & 0 & 0 & 0 & 0 & 0 \\
\bar{W} & 0 & Y_r & Y_g & Y_b & 0 & 0 & 0 & 0 & 0 & 0 & 0 & 0 & 0 & 0 \\
\bar{X}_r & \bar{Y}_r & 0 & g_{r\bar{g}} & g_{r\bar{b}} & 0 & 0 & 0 & 0 & 0 & 0 & 0 & 0 & 0 & 0 \\
\bar{X}_g & \bar{Y}_g & \bar{g}_{r\bar{g}} & 0 & g_{g\bar{b}} & 0 & 0 & 0 & 0 & 0 & 0 & 0 & 0 & 0 & 0 \\
\bar{X}_b & \bar{Y}_b & \bar{g}_{r\bar{b}} & \bar{g}_{g\bar{b}} & 0 & 0 & 0 & 0 & 0 & 0 & 0 & 0 & 0 & 0 & 0 \\
\hline
0 & 0 & 0 & 0 & 0 & 0 & Y_r & Y_g & Y_b & X_r & X_g & X_b & 0 & 0 & 0 \\
0 & 0 & 0 & 0 & 0 & \bar{Y}_r & 0 & g_{r\bar{g}} & g_{r\bar{b}} & W & 0 & 0 & 0 & X_b & X_g \\
0 & 0 & 0 & 0 & 0 & \bar{Y}_g & \bar{g}_{r\bar{g}} & 0 & g_{g\bar{b}} & 0 & W & 0 & X_b & 0 & X_r \\
0 & 0 & 0 & 0 & 0 & \bar{Y}_b & \bar{g}_{r\bar{b}} & \bar{g}_{g\bar{b}} & 0 & 0 & 0 & W & X_g & X_r & 0 \\
0 & 0 & 0 & 0 & 0 & \bar{X}_r & \bar{W} & 0 & 0 & 0 & g_{r\bar{g}} & g_{r\bar{b}} & 0 & Y_b & Y_g \\
0 & 0 & 0 & 0 & 0 & \bar{X}_g & 0 & \bar{W} & 0 & \bar{g}_{r\bar{g}} & 0 & g_{g\bar{b}} & Y_b & 0 & Y_r \\
0 & 0 & 0 & 0 & 0 & \bar{X}_b & 0 & 0 & \bar{W} & \bar{g}_{r\bar{b}} & \bar{g}_{g\bar{b}} & 0 & Y_g & Y_r & 0 \\
0 & 0 & 0 & 0 & 0 & 0 & 0 & \bar{X}_b & \bar{X}_g & 0 & \bar{Y}_b & \bar{Y}_g & 0 & g_{r\bar{g}} & g_{r\bar{b}} \\
0 & 0 & 0 & 0 & 0 & 0 & \bar{X}_b & 0 & \bar{X}_r & \bar{Y}_b & 0 & \bar{Y}_r & \bar{g}_{r\bar{g}} & 0 & g_{g\bar{b}} \\
0 & 0 & 0 & 0 & 0 & 0 & \bar{X}_g & \bar{X}_r & 0 & \bar{Y}_g & \bar{Y}_r & 0 & \bar{g}_{r\bar{b}} & \bar{g}_{g\bar{b}} & 0
\end{array}\right]
\left[\begin{array}{c}
v_e \\ e \\ \bar{d}_r \\ \bar{d}_g \\ \bar{d}_b \\ \hline \bar{e} \\ d_r \\ d_g \\ d_b \\ u_r \\ u_g \\ u_b \\ \bar{u}_r \\ \bar{u}_g \\ \bar{u}_b
\end{array}\right]$$

Schematic representation of fermions and bosons in SU(5) GUT showing 5 + 10 split in the multiplets. Neutral bosons (photon, Z-boson, and neutral gluons) are not shown but occupy the diagonal entries of the matrix in complex superpositions

SU(5)

SU(5) is the simplest GUT. The smallest simple Lie group which contains the standard model, and upon which the first Grand Unified Theory was based, is

$$SU(5) \supset SU(3) \times SU(2) \times U(1).$$

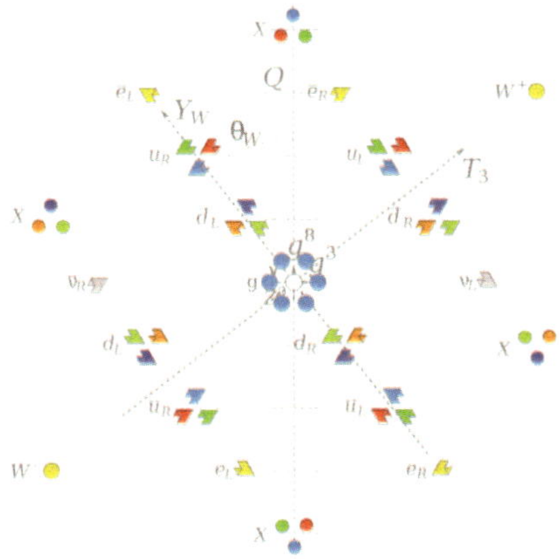

The pattern of weak isospins, weak hypercharges, and strong charges for particles in the SU(5) model, rotated by the predicted weak mixing angle, showing electric charge roughly along the vertical. In addition to Standard Model particles, the theory includes twelve colored X bosons, responsible for proton decay.

Such group symmetries allow the reinterpretation of several known particles as different states of a single particle field. However, it is not obvious that the simplest possible choices for the extended "Grand Unified" symmetry should yield the correct inventory of elementary particles. The fact that all currently known matter particles fit nicely into three copies of the smallest group representations of SU(5) and immediately carry the correct observed charges, is one of the first and most important reasons why people believe that a Grand Unified Theory might actually be realized in nature.

The two smallest irreducible representations of SU(5) are 5 and 10. In the standard assignment, the 5 contains the charge conjugates of the right-handed down-type quark color triplet and a left-handed lepton isospin doublet, while the 10 contains the six up-type quark components, the left-handed down-type quark color triplet, and the right-handed electron. This scheme has to be replicated for each of the three known generations of matter. It is notable that the theory is anomaly free with this matter content.

The hypothetical right-handed neutrinos are a singlet of SU(5), which means its mass is not forbidden by any symmetry; it doesn't need a spontaneous symmetry breaking which explains why its mass would be heavy.

SO(10)

The next simple Lie group which contains the standard model is

$$SO(10) \supset SU(5) \supset SU(3) \times SU(2) \times U(1).$$

Here, the unification of matter is even more complete, since the irreducible spinor representation 16 contains both the $\overline{5}$ and 10 of SU(5) and a right-handed neutrino, and thus the complete particle content of one generation of the extended standard model with neutrino masses. This is already the largest simple group which achieves the unification of matter in a scheme involving only the already known matter particles (apart from the Higgs sector).

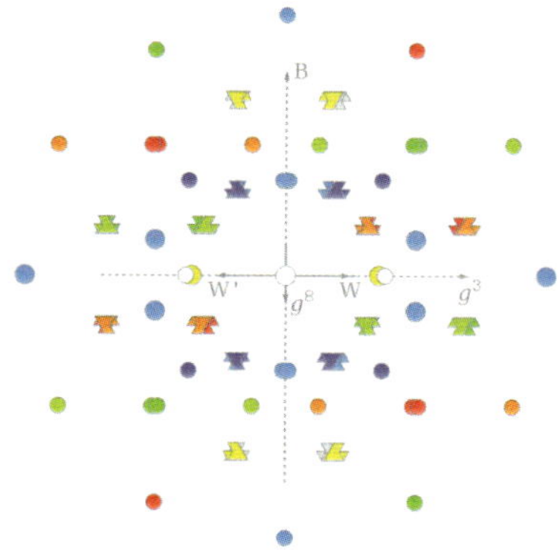

The pattern of weak isospin, W, weaker isospin, W', strong g3 and g8, and baryon minus lepton, B, charges for particles in the SO(10) Grand Unified Theory, rotated to show the embedding in E_6.

Since different standard model fermions are grouped together in larger representations, GUTs specifically predict relations among the fermion masses, such as between the electron and the down quark, the muon and the strange quark, and the tau lepton and the bottom quark for SU(5) and SO(10). Some of these mass relations hold approximately, but most don't.

The boson matrix for SO(10) is found by taking the 15 × 15 matrix from the 10 + 5 representation of SU(5) and adding an extra row and column for the right-handed neutrino. The bosons are found by adding a partner to each of the 20 charged bosons (2 right-handed W bosons, 6 massive charged gluons and 12 X/Y type bosons) and adding an extra heavy neutral Z-boson to make 5 neutral bosons in total. The boson matrix will have a boson or its new partner in each row and column. These pairs combine to create the familiar 16D Dirac spinor matrices of SO(10).

Extended Grand Unified Theories

Non-chiral extensions of the Standard Model with vectorlike split-multiplet particle spectra which naturally appear in the higher SU(N) GUTs considerably modify the desert physics and lead to the realistic (string-scale) grand unification for conventional three quark-lepton families even without using supersymmetry. On the other hand, due to a new missing VEV mechanism emerging in the supersymmetric SU(8) GUT the simultaneous solution to the gauge hierarchy (doublet-triplet splitting) problem and problem of unification of flavor can be found.

GUTs with four families / generations, SU(8): Assuming 4 generations of fermions instead of 3 makes a total of 64 types of particles. These can be put into 64 = 8 + 56 representations of SU(8). This can be divided into SU(5) × $SU(3)_F$ × U(1) which is the SU(5) theory together with some heavy bosons which act on the generation number.

GUTs with four families / generations, O(16): Again assuming 4 generations of fermions, the 128 particles and anti-particles can be put into a single spinor representation of O(16).

Symplectic Groups and Quaternion Representations

Symplectic gauge groups could also be considered. For example, Sp(8) (which is called Sp(4) in the article symplectic group) has a representation in terms of 4 × 4 quaternion unitary matrices which has a 16 dimensional real representation and so might be considered as a candidate for a gauge group. Sp(8) has 32 charged bosons and 4 neutral bosons. Its subgroups include SU(4) so can at least contain the gluons and photon of SU(3) × U(1). Although it's probably not possible to have weak bosons acting on chiral fermions in this representation. A quaternion representation of the fermions might be:

$$\begin{bmatrix} e + i\overline{e} + jv + k\overline{v} \\ u_r + i\overline{u_r} + jd_r + k\overline{d_r} \\ u_g + i\overline{u_g} + jd_g + k\overline{d_g} \\ u_b + i\overline{u_b} + jd_b + k\overline{d_b} \end{bmatrix}_L$$

A further complication with quaternion representations of fermions is that there are two types of multiplication: left multiplication and right multiplication which must be taken into account. It turns out that including left and right-handed 4 × 4 quaternion matrices is equivalent to including a single right-multiplication by a unit quaternion which adds an extra SU(2) and so has an extra neutral boson and two more charged bosons. Thus the group of left- and right-handed 4 × 4 quaternion matrcies is Sp(8) × SU(2) which does include the standard model bosons:

$$SU(4,H)_L \times H_R = Sp(8) \times SU(2) \supset SU(4) \times SU(2) \supset SU(3) \times SU(2) \times U(1)$$

If ψ is a quaternion valued spinor, A_μ^{ab} is quaternion hermitian 4 × 4 matrix coming from Sp(8) and B_μ is a pure imaginary quaternion (both of which are 4-vector bosons) then the interaction term is:

$$\overline{\psi^a}\gamma_\mu\left(A_\mu^{ab}\psi^b + \psi^a B_\mu\right)$$

Octonion Representations

It can be noted that a generation of 16 fermions can be put into the form of an octonion with each element of the octonion being an 8-vector. If the 3 generations are then put in a 3x3 hermitian matrix with certain additions for the diagonal elements then these matrices form an exceptional (Grassmann-) Jordan algebra, which has the symmetry group of one of the exceptional Lie groups (F_4, E_6, E_7 or E_8) depending on the details.

$$\psi = \begin{bmatrix} a & e & \mu \\ \bar{e} & b & \tau \\ \bar{\mu} & \bar{\tau} & c \end{bmatrix}$$

$$[\psi_A, \psi_B] \subset J_3(O)$$

Because they are fermions the anti-commutators of the Jordan algebra become commutators. It is known that E_6 has subgroup O(10) and so is big enough to include the Standard Model. An E_8 gauge group, for example, would have 8 neutral bosons, 120 charged bosons and 120 charged anti-bosons. To account for the 248 fermions in the lowest multiplet of E_8, these would either have to include anti-particles (and so have baryogenesis), have new undiscovered particles, or have gravity-like (spin connection) bosons affecting elements of the particles spin direction. Each of these possess theoretical problems.

Beyond Lie Groups

Other structures have been suggested including Lie 3-algebras and Lie superalgebras. Neither of these fit with Yang–Mills theory. In particular Lie superalgebras would introduce bosons with the wrong statistics. Supersymmetry however does fit with Yang–Mills. For example, N=4 Super Yang Mills Theory requires an SU(N) gauge group.

Unification of Forces and the Role of Supersymmetry

The unification of forces is possible due to the energy scale dependence of force coupling parameters in quantum field theory called renormalization group running, which allows parameters with vastly different values at usual energies to converge to a single value at a much higher energy scale.

The renormalization group running of the three gauge couplings in the Standard Model has been found to nearly, but not quite, meet at the same point if the hypercharge is normalized so that it is consistent with SU(5) or SO(10) GUTs, which are precisely the GUT groups which lead to a simple fermion unification. This is a significant result, as other Lie groups lead to different normalizations. However, if the supersymmetric extension MSSM is used instead of the Standard Model, the match becomes much more accurate. In this case, the coupling constants of the strong and electroweak interactions meet at the grand unification energy, also known as the GUT scale:

$$\Lambda_{GUT} \approx 10^{16}\, GeV.$$

It is commonly believed that this matching is unlikely to be a coincidence, and is often quoted as one of the main motivations to further investigate supersymmetric theories despite the fact that no supersymmetric partner particles have been experimentally observed. Also, most model builders simply assume supersymmetry because it solves the hierarchy problem—i.e., it stabilizes the electroweak Higgs mass against radiative corrections.

Neutrino Masses

Since Majorana masses of the right-handed neutrino are forbidden by SO(10) symmetry, SO(10) GUTs predict the Majorana masses of right-handed neutrinos to be close to the GUT scale where the symmetry is spontaneously broken in those models. In supersymmetric GUTs, this scale tends to be larger than would be desirable to obtain realistic masses of the light, mostly left-handed neutrinos via the seesaw mechanism.

String Theory

String theory, in particle physics, a theory that attempts to merge quantum mechanics with Albert Einstein's general theory of relativity. The name *string theory* comes from the modeling of subatomic particles as tiny one-dimensional "stringlike" entities rather than the more conventional approach in which they are modeled as zero-dimensional point particles. The theory envisions that a string undergoing a particular mode of vibration corresponds to a particle with definite properties such as mass and charge. In the 1980s, physicists realized that string theory had the potential to incorporate all four of nature's forces—gravity, electromagnetism, strong force, and weak force—and all types of matter in a single quantum mechanical framework, suggesting that it might be the long-sought unified field theory. While string theory is still a vibrant area of research that is undergoing rapid development, it remains primarily a mathematical construct because it has yet to make contact with experimental observations.

Relativity and Quantum Mechanics

In 1905 Einstein unified space and time with his special theory of relativity, showing that motion through space affects the passage of time. In 1915 Einstein further unified space, time, and gravitation with his general theory of relativity, showing that warps and curves in space and time are responsible for the force of gravity. These were monumental achievements, but Einstein dreamed of an even grander unification. He envisioned one powerful framework that would account for space, time, and all of nature's forces—something he called a unified theory. For the last three decades of his life, Einstein relentlessly pursued this vision. Although from time to time rumours spread that he had succeeded, closer scrutiny always dashed such hopes. Most of Einstein's contemporaries considered the search for a unified theory to be a hopeless, if not misguided, quest.

In contrast, the primary concern of theoretical physicists from the 1920s onward was quantum mechanics—the emerging framework for describing atomic and subatomic processes. Particles at these scales have such tiny masses that gravity is essentially irrelevant in their interactions, and so for decades quantum mechanical calculations generally ignored general relativistic effects. Instead, by the late 1960s the focus was on a different force—the strong force, which binds together the protons and neutrons within atomic nuclei. Gabriele Veneziano, a young theorist working at the European Organization for Nuclear Research (CERN), contributed a key breakthrough in 1968 with his realization that a 200-year-old formula, the Euler beta function, was capable of explaining much of the data on the strong force then being collected at various particle accelerators around the world. A few years later, three physicists—Leonard Susskind of Stanford University, Holger Nielsen of the Niels Bohr Institute, and Yoichiro Nambu of the University of Chicago—significantly amplified Veneziano's insight by showing that the mathematics underlying his proposal described the vibrational motion of minuscule filaments of energy that resemble tiny strands of string, inspiring the name *string theory*. Roughly speaking, the theory suggested that the strong force amounted to strings tethering together particles attached to the strings' endpoints.

Predictions and Theoretical Difficulties

String theory was an intuitively attractive proposal, but by the mid-1970s more-refined measurements of the strong force had deviated from its predictions, leading most researchers to conclude that string theory had no relevance to the physical universe, no matter how elegant the mathematical theory. Nevertheless, a small number of physicists continued to pursue string theory. In 1974 John Schwarz of the California Institute of Technology and Joel Scherk of the École Normale Supérieure and, independently, Tamiaki Yoneya of Hokkaido University came to a radical conclusion. They suggested that one of the supposedly failed predictions of string theory—the existence of a particular massless particle that no experiment studying the strong force had ever encountered—was actually evidence of the very unification Einstein had anticipated.

Although no one had succeeded in merging general relativity and quantum mechanics, preliminary work had established that such a union would require precisely the massless particle predicted by string theory. A few physicists argued that string theory, by having this particle built into its fundamental structure, had united the laws of the large (general relativity) and the laws of the small (quantum mechanics). Rather than merely being a description of the strong force, these physicists contended, string theory required reinterpretation as a critical step toward Einstein's unified theory.

The announcement was universally ignored. String theory had already failed in its first incarnation as a description of the strong force, and many felt it was unlikely that it would now prevail as the solution to an even more difficult problem. This view was bolstered by string theory's suffering from its own theoretical problems. For one, some of its equations showed signs of being inconsistent; for another, the mathematics of the theory demanded the universe have not just the three spatial dimensions of common experience but six others (for a total of nine spatial dimensions, or a total of ten space-time dimensions).

Dimensions and Vibrations

Because of these obstacles, the number of physicists working on the theory had dropped to two—Schwarz and Michael Green of Queen Mary College, London—by the mid-1980s. But in 1984 these two die-hard string theorists achieved a major breakthrough. Through a remarkable calculation, they proved that the equations of string theory were consistent after all. By the time word of this result had spread throughout the physics community, hundreds of researchers had dropped what they were working on and turned their full attention to string theory.

Within a few months, string theory's unified framework took shape. Much as different vibrational patterns of a violin string play different musical notes, the different vibrations of the tiny strands in string theory were imagined to yield different particles of nature. According to the theory, the strings are so small that they appear to be points—as particles had long been thought to be—but in reality they have length (about 10^{-33} cm); the mass and charge of a particle is determined by how a string vibrates. For example, string theory posits that an electron is a string undergoing one particular vibrational pattern; a quark is imagined as a string undergoing a different vibrational pattern. Crucially, among the vibrational patterns, physicists argued, would also be the particles found by experiment to communicate nature's forces. Thus, string theory was proposed as the sought-for unification of all forces and all matter.

What of the six extra spatial dimensions required by string theory? Following a suggestion made in the 1920s by Theodor Kaluza of Germany and Oskar Klein of Sweden, string theorists envisioned that dimensions come in two distinct varieties. Like the unfurled length of a long garden hose, dimensions can be big and easy to see. But like the shorter, circular girth of the garden hose, dimensions can also be far smaller and more difficult to detect. This becomes more apparent by imagining that the circular cross section of the garden hose is shrunk ever smaller, below what can be seen with the naked eye, misleading a casual observer into thinking the garden hose has only one dimension, its length. Similarly, according to string theory, the three dimensions of common experience are large and manifest, while the other six dimensions are crumpled so small that they have so far evaded detection.

During the decade from 1984 to 1994, many theoretical physicists strove to fulfill string theory's promise by developing this abstract, wholly mathematical framework into a concrete, predictive theory of nature. Because the infinitesimal size of strings has precluded their direct detection, theorists have sought to extract indirect implications of the theory that might be testable. In this regard, the extra dimensions of string theory have proved a major hurdle. Imagining these extra dimensions as small and hidden is a reasonable explanation for their apparent absence. Nevertheless, their detailed geometry is required for the theory to offer predictions. The reason is that strings are so small that they would vibrate within the tiny extra dimensions. Studies showed that, much as the shape and size of a French horn affect the vibrational patterns of airstreams coursing through the instrument, the exact shape and size of the extra dimensions would affect how strings vibrate. And since the strings' vibrations determine quantities such as particle masses and charges, predictivity requires knowledge of the geometric form of the extra dimensions. Unfortunately, the equations of string theory allow the extra dimensions to take many different geometric forms, making it difficult to extract definitive testable predictions.

M-theory and AdS/CFT Correspondence

By the mid-1990s these and other obstacles were again eroding the ranks of string theorists. But in 1995 another breakthrough reinvigorated the field. Edward Witten of the Institute for Advanced Study, building on contributions of many other physicists, proposed a new set of techniques that refined the approximate equations on which all work in string theory had so far been based. These techniques helped reveal a number of new features of string theory, including the realization that the theory has not six but seven extra spatial dimensions. The more exact equations also revealed ingredients in string theory besides strings—membranelike objects of various dimensions, collectively called branes. Finally, the new techniques established that various versions of string theory developed over the preceding decades were essentially all the same. Theorists call this unification of formerly distinct string theories by a new name, M-theory, with the meaning of M being deferred until the theory is more fully understood.

Another advance in string theory happened in 1997 when Juan Maldacena of Harvard University discovered the anti-de Sitter/conformal field theory (AdS/CFT) correspondence. Maldacena found that a string theory operating with a particular environment (involving a space-time known as an anti-de Sitter space) was equivalent to a type of quantum field theory operating in an environment with one less spatial dimension. This has proved to be one of the most profound discoveries in string theory, establishing a powerful link to the more conventional methods of quantum field theory, providing an exact mathematical formulation of string theory in certain environments, and inspiring thousands of further technical studies.

Today the understanding of many facets of string theory is still in its formative stage. Researchers recognize that, although remarkable progress has been made over the past five decades, collectively the work is burdened by its piecemeal development, with incremental discoveries having been joined like pieces of a jigsaw puzzle. That the pieces fit coherently is impressive, but the larger picture they are filling out—the fundamental principle underlying the theory—remains mysterious. Equally pressing, the theory has yet to be supported by observations and hence remains a totally theoretical construct.

Supersymmetry and Cosmological Signature

One essential quality of string theory is known as supersymmetry, a mathematical property that requires every known particle species to have a partner particle species, called a superpartner. (This property accounts for string theory often being referred to as superstring theory.) As yet, no superpartner particles have been detected experimentally, but researchers believe this may be due to their weight: they are heavier than their known counterparts and require a machine at least as powerful as the Large Hadron Collider at CERN to produce them. If the superpartner particles are found, string theory still will not be proved correct, because more-conventional point-particle theories have also successfully incorporated supersymmetry into their mathematical structure. However, the discovery of supersymmetry would confirm an essential element of string theory and give circumstantial evidence that this approach to unification is on the right track.

Even if these accelerator-based tests are inconclusive, there is another way that string theory may one day be tested. Through its impact on the earliest, most extreme moments of the universe, the physics of string theory may have left faint cosmological signatures—for example, in the form of gravitational waves or a particular pattern of temperature variations in the cosmic microwave background radiation—that may be observable by the next generation of precision satellite-borne telescopes and detectors. It would be a fitting conclusion to Einstein's quest for unification if a theory of the smallest microscopic component of matter were confirmed through observations of the largest astronomical realms of the cosmos.

Fundamentals

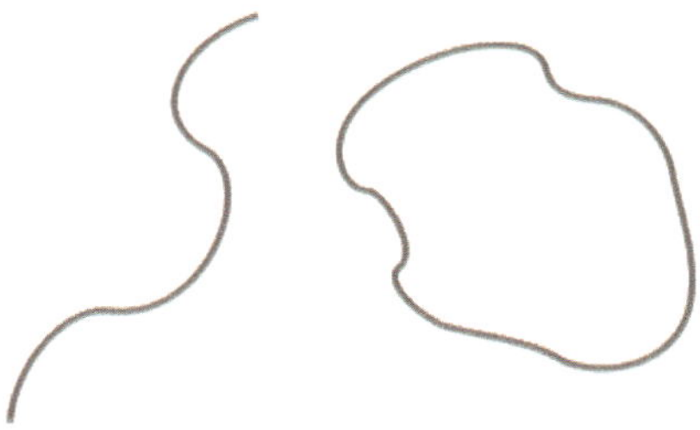

The fundamental objects of string theory are open and closed string models.

In the twentieth century, two theoretical frameworks emerged for formulating the laws of physics. The first is Albert Einstein's general theory of relativity, a theory that explains the force of gravity and the structure of space and time. The other is quantum mechanics which is a completely different formulation to describe physical phenomena using the known probability principles. By the late 1970s, these two frameworks had proven to be sufficient to explain most of the observed features of the universe, from elementary particles to atoms to the evolution of stars and the universe as a whole.

In spite of these successes, there are still many problems that remain to be solved.

One of the deepest problems in modern physics is the problem of quantum gravity. The general theory of relativity is formulated within the framework of classical physics, whereas the other fundamental forces are described within the framework of quantum mechanics. A quantum theory of gravity is needed in order to reconcile general relativity with the principles of quantum mechanics, but difficulties arise when one attempts to apply the usual prescriptions of quantum theory to the force of gravity. In addition to the problem of developing a consistent theory of quantum gravity, there are many other fundamental problems in the physics of atomic nuclei, black holes, and the early universe.

String theory is a theoretical framework that attempts to address these questions and many others. The starting point for string theory is the idea that the point-like particles of particle physics can also be modeled as one-dimensional objects called strings. String theory describes how strings propagate through space and interact with each other. In a given version of string theory, there is only one kind of string, which may look like a small loop or segment of ordinary string, and it can vibrate in different ways. On distance scales larger than the string scale, a string will look just like an ordinary particle, with its mass, charge, and other properties determined by the vibrational state of the string. In this way, all of the different elementary particles may be viewed as vibrating strings. In string theory, one of the vibrational states of the string gives rise to the graviton, a quantum mechanical particle that carries gravitational force. Thus string theory is a theory of quantum gravity.

One of the main developments of the past several decades in string theory was the discovery of certain "dualities", mathematical transformations that identify one physical theory with another. Physicists studying string theory have discovered a number of these dualities between different versions of string theory, and this has led to the conjecture that all consistent versions of string theory are subsumed in a single framework known as M-theory.

Studies of string theory have also yielded a number of results on the nature of black holes and the gravitational interaction. There are certain paradoxes that arise when one attempts to understand the quantum aspects of black holes, and work on string theory has attempted to clarify these issues. In late 1997 this line of work culminated in the discovery of the anti-de Sitter/conformal field theory correspondence or AdS/CFT. This is a theoretical result which relates string theory to other physical theories which are better understood theoretically. The AdS/CFT correspondence has implications for the study of black holes and quantum gravity, and it has been applied to other subjects, including nuclear and condensed matter physics.

Since string theory incorporates all of the fundamental interactions, including gravity, many physicists hope that it fully describes our universe, making it a theory of everything. One of the goals of current research in string theory is to find a solution of the theory that reproduces the observed spectrum of elementary particles, with a small cosmological constant, containing dark matter and a plausible mechanism for cosmic inflation. While there has been progress toward these goals, it is not known to what extent string theory describes the real world or how much freedom the theory allows in the choice of details.

One of the challenges of string theory is that the full theory does not have a satisfactory definition in all circumstances. The scattering of strings is most straightforwardly defined using the techniques of perturbation theory, but it is not known in general how to define string theory nonperturbatively. It is also not clear whether there is any principle by which string theory selects its vacuum state, the physical state that determines the properties of our universe. These problems have led some in the community to criticize these approaches to the unification of physics and question the value of continued research on these problems.

Strings

The application of quantum mechanics to physical objects such as the electromagnetic field, which are extended in space and time, is known as quantum field theory. In particle physics, quantum field theories form the basis for our understanding of elementary particles, which are modeled as excitations in the fundamental fields.

In quantum field theory, one typically computes the probabilities of various physical events using the techniques of perturbation theory. Developed by Richard Feynman and others in the first half of the twentieth century, perturbative quantum field theory uses special diagrams called Feynman diagrams to organize computations. One imagines that these diagrams depict the paths of point-like particles and their interactions.

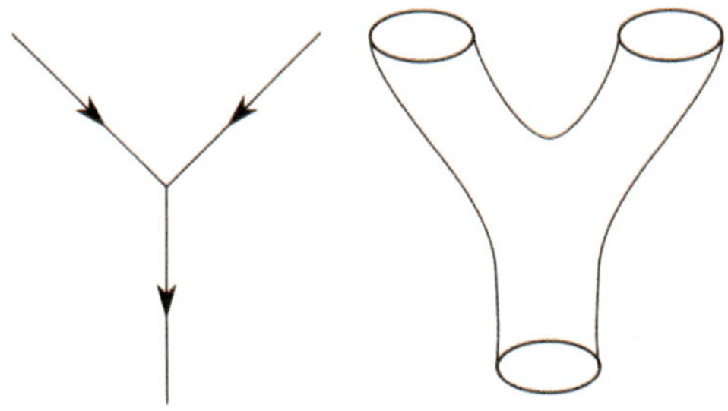

Interaction in the quantum world: worldlines of point-like particles or a worldsheet swept up by closed strings in string theory.

The starting point for string theory is the idea that the point-like particles of quantum field theory can also be modeled as one-dimensional objects called strings. The interaction of strings is most straightforwardly defined by generalizing the perturbation theory used in ordinary quantum field theory. At the level of Feynman diagrams, this means replacing the one-dimensional diagram representing the path of a point particle by a two-dimensional surface representing the motion of a string. Unlike in quantum field theory, string theory does not have a full non-perturbative definition, so many of the theoretical questions that physicists would like to answer remain out of reach.

In theories of particle physics based on string theory, the characteristic length scale of strings is assumed to be on the order of the Planck length, or 10^{-35} meters, the scale at which the effects of quantum gravity are believed to become significant. On much larger length scales, such as the scales visible in physics laboratories, such objects would be indistinguishable from zero-dimensional point particles, and the vibrational state of the string would determine the type of particle. One of the vibrational states of a string corresponds to the graviton, a quantum mechanical particle that carries the gravitational force.

The original version of string theory was bosonic string theory, but this version described only bosons, a class of particles which transmit forces between the matter particles, or fermions. Bosonic string theory was eventually superseded by theories called superstring theories. These theories describe both bosons and fermions, and they incorporate a theoretical idea called supersymmetry. This is a mathematical relation that exists in certain physical theories between the bosons and fermions. In theories with supersymmetry, each boson has a counterpart which is a fermion, and vice versa.

There are several versions of superstring theory: type I, type IIA, type IIB, and two flavors of heterotic string theory (SO(32) and $E_8 \times E_8$). The different theories allow different types of strings, and the particles that arise at low energies exhibit different symmetries. For example, the type I theory includes both open strings (which are segments with endpoints) and closed strings (which form closed loops), while types IIA, IIB and heterotic include only closed strings.

Extra Dimensions

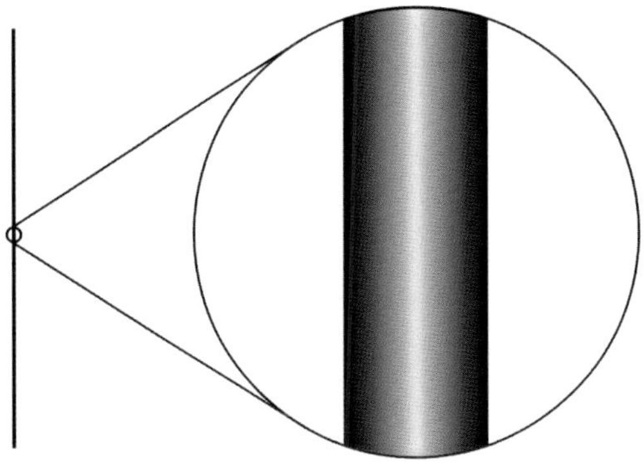

An example of compactification: At large distances, a two dimensional surface with one circular dimension looks one-dimensional.

In everyday life, there are three familiar dimensions of space: height, width and length. Einstein's general theory of relativity treats time as a dimension on par with the three spatial dimensions; in general relativity, space and time are not modeled as separate entities but are instead unified to a four-dimensional spacetime. In this framework, the phenomenon of gravity is viewed as a consequence of the geometry of spacetime.

In spite of the fact that the universe is well described by four-dimensional spacetime, there are several reasons why physicists consider theories in other dimensions. In some cases, by modeling spacetime in a different number of dimensions, a theory becomes more mathematically tractable, and one can perform calculations and gain general insights more easily. There are also situations where theories in two or three spacetime dimensions are useful for describing phenomena in condensed matter physics. Finally, there exist scenarios in which there could actually be more than four dimensions of spacetime which have nonetheless managed to escape detection.

A cross section of a quintic Calabi–Yau manifold

One notable feature of string theories is that these theories require extra dimensions of spacetime for their mathematical consistency. In bosonic string theory, spacetime is 26-dimensional, while in superstring theory it is 10-dimensional, and in M-theory it is 11-dimensional. In order to describe real physical phenomena using string theory, one must therefore imagine scenarios in which these extra dimensions would not be observed in experiments.

Compactification is one way of modifying the number of dimensions in a physical theory. In compactification, some of the extra dimensions are assumed to "close up" on themselves to form circles. In the limit where these curled up dimensions become very small, one obtains a theory in which spacetime has effectively a lower number of dimensions. A standard analogy for this is to consider a multidimensional object such as a garden hose. If the hose is viewed from a sufficient distance, it appears to have only one dimension, its length. However, as one approaches the hose, one discovers that it contains a second dimension, its circumference. Thus, an ant crawling on the surface of the hose would move in two dimensions.

Compactification can be used to construct models in which spacetime is effectively four-dimensional. However, not every way of compactifying the extra dimensions produces a model with the right properties to describe nature. In a viable model of particle physics, the compact extra dimensions must be shaped like a Calabi–Yau manifold. A

Calabi–Yau manifold is a special space which is typically taken to be six-dimensional in applications to string theory. It is named after mathematicians Eugenio Calabi and Shing-Tung Yau.

Another approach to reducing the number of dimensions is the so-called brane-world scenario. In this approach, physicists assume that the observable universe is a four-dimensional subspace of a higher dimensional space. In such models, the force-carrying bosons of particle physics arise from open strings with endpoints attached to the four-dimensional subspace, while gravity arises from closed strings propagating through the larger ambient space. This idea plays an important role in attempts to develop models of real world physics based on string theory, and it provides a natural explanation for the weakness of gravity compared to the other fundamental forces.

Dualities

One notable fact about string theory is that the different versions of the theory all turn out to be related in highly nontrivial ways. One of the relationships that can exist between different string theories is called S-duality. This is a relationship which says that a collection of strongly interacting particles in one theory can, in some cases, be viewed as a collection of weakly interacting particles in a completely different theory. Roughly speaking, a collection of particles is said to be strongly interacting if they combine and decay often and weakly interacting if they do so infrequently. Type I string theory turns out to be equivalent by S-duality to the SO(32) heterotic string theory. Similarly, type IIB string theory is related to itself in a nontrivial way by S-duality.

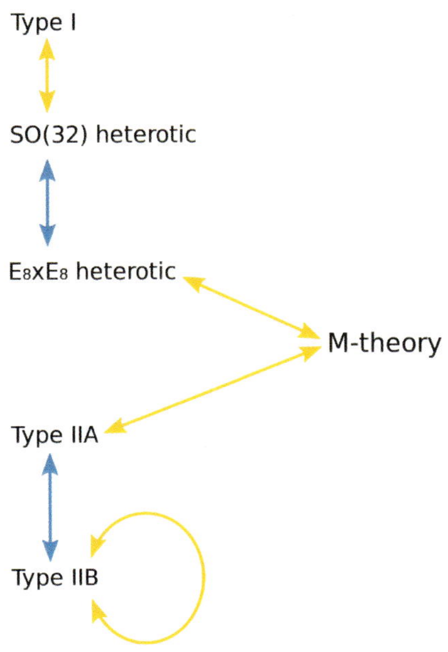

A diagram of string theory dualities. Yellow arrows indicate S-duality. Blue arrows indicate T-duality.

Another relationship between different string theories is T-duality. Here one considers strings propagating around a circular extra dimension. T-duality states that a string propagating around a circle of radius R is equivalent to a string propagating around a circle of radius 1/R in the sense that all observable quantities in one description are identified with quantities in the dual description. For example, a string has momentum as it propagates around a circle, and it can also wind around the circle one or more times. The number of times the string winds around a circle is called the winding number. If a string has momentum p and winding number n in one description, it will have momentum n and winding number p in the dual description. For example, type IIA string theory is equivalent to type IIB string theory via T-duality, and the two versions of heterotic string theory are also related by T-duality.

In general, the term *duality* refers to a situation where two seemingly different physical systems turn out to be equivalent in a nontrivial way. Two theories related by a duality need not be string theories. For example, Montonen–Olive duality is example of an S-duality relationship between quantum field theories. The AdS/CFT correspondence is example of a duality which relates string theory to a quantum field theory. If two theories are related by a duality, it means that one theory can be transformed in some way so that it ends up looking just like the other theory. The two theories are then said to be *dual* to one another under the transformation. Put differently, the two theories are mathematically different descriptions of the same phenomena.

Branes

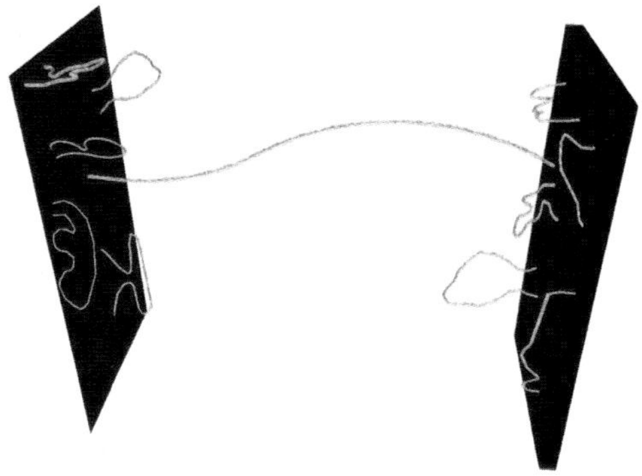

Open strings attached to a pair of D-branes.

In string theory and other related theories, a brane is a physical object that generalizes the notion of a point particle to higher dimensions. For instance, a point particle can be viewed as a brane of dimension zero, while a string can be viewed as a brane of dimension one. It is also possible to consider higher-dimensional branes. In dimension *p*, these are called *p*-branes. The word brane comes from the word "membrane" which refers to a two-dimensional brane.

Branes are dynamical objects which can propagate through spacetime according to the rules of quantum mechanics. They have mass and can have other attributes such as charge. A p-brane sweeps out a $(p+1)$-dimensional volume in spacetime called its *worldvolume*. Physicists often study fields analogous to the electromagnetic field which live on the worldvolume of a brane.

In string theory, D-branes are an important class of branes that arise when one considers open strings. As an open string propagates through spacetime, its endpoints are required to lie on a D-brane. The letter "D" in D-brane refers to a certain mathematical condition on the system known as the Dirichlet boundary condition. The study of D-branes in string theory has led to important results such as the AdS/CFT correspondence, which has shed light on many problems in quantum field theory.

Branes are frequently studied from a purely mathematical point of view, and they are described as objects of certain categories, such as the derived category of coherent sheaves on a complex algebraic variety, or the Fukaya category of a symplectic manifold. The connection between the physical notion of a brane and the mathematical notion of a category has led to important mathematical insights in the fields of algebraic and symplectic geometry and representation theory.

Black Holes

In general relativity, a black hole is defined as a region of spacetime in which the gravitational field is so strong that no particle or radiation can escape. In the currently accepted models of stellar evolution, black holes are thought to arise when massive stars undergo gravitational collapse, and many galaxies are thought to contain supermassive black holes at their centers. Black holes are also important for theoretical reasons, as they present profound challenges for theorists attempting to understand the quantum aspects of gravity. String theory has proved to be an important tool for investigating the theoretical properties of black holes because it provides a framework in which theorists can study their thermodynamics.

Bekenstein–Hawking Formula

In the branch of physics called statistical mechanics, entropy is a measure of the randomness or disorder of a physical system. This concept was studied in the 1870s by the Austrian physicist Ludwig Boltzmann, who showed that the thermodynamic properties of a gas could be derived from the combined properties of its many constituent molecules. Boltzmann argued that by averaging the behaviors of all the different molecules in a gas, one can understand macroscopic properties such as volume, temperature, and pressure. In addition, this perspective led him to give a precise definition of entropy as the natural logarithm of the number of different states of the molecules (also called *microstates*) that give rise to the same macroscopic features.

In the twentieth century, physicists began to apply the same concepts to black holes. In most systems such as gases, the entropy scales with the volume. In the 1970s, the physicist Jacob Bekenstein suggested that the entropy of a black hole is instead proportional to the *surface area* of its event horizon, the boundary beyond which matter and radiation is lost to its gravitational attraction. When combined with ideas of the physicist Stephen Hawking, Bekenstein's work yielded a precise formula for the entropy of a black hole. The Bekenstein–Hawking formula expresses the entropy S as

$$S = \frac{c^3 kA}{4\hbar G}$$

where c is the speed of light, k is Boltzmann's constant, ħ is the reduced Planck constant, G is Newton's constant, and A is the surface area of the event horizon.

Like any physical system, a black hole has an entropy defined in terms of the number of different microstates that lead to the same macroscopic features. The Bekenstein–Hawking entropy formula gives the expected value of the entropy of a black hole, but by the 1990s, physicists still lacked a derivation of this formula by counting microstates in a theory of quantum gravity. Finding such a derivation of this formula was considered an important test of the viability of any theory of quantum gravity such as string theory.

Derivation within String Theory

In a paper from 1996, Andrew Strominger and Cumrun Vafa showed how to derive the Beckenstein–Hawking formula for certain black holes in string theory. Their calculation was based on the observation that D-branes—which look like fluctuating membranes when they are weakly interacting—become dense, massive objects with event horizons when the interactions are strong. In other words, a system of strongly interacting D-branes in string theory is indistinguishable from a black hole. Strominger and Vafa analyzed such D-brane systems and calculated the number of different ways of placing D-branes in spacetime so that their combined mass and charge is equal to a given mass and charge for the resulting black hole. Their calculation reproduced the Bekenstein–Hawking formula exactly, including the factor of 1/4. Subsequent work by Strominger, Vafa, and others refined the original calculations and gave the precise values of the "quantum corrections" needed to describe very small black holes.

The black holes that Strominger and Vafa considered in their original work were quite different from real astrophysical black holes. One difference was that Strominger and Vafa considered only extremal black holes in order to make the calculation tractable. These are defined as black holes with the lowest possible mass compatible with a given charge. Strominger and Vafa also restricted attention to black holes in five-dimensional spacetime with unphysical supersymmetry.

Although it was originally developed in this very particular and physically unrealistic context in string theory, the entropy calculation of Strominger and Vafa has led to a qualitative understanding of how black hole entropy can be accounted for in any theory of quantum gravity. Indeed, in 1998, Strominger argued that the original result could be generalized to an arbitrary consistent theory of quantum gravity without relying on strings or supersymmetry. In collaboration with several other authors in 2010, he showed that some results on black hole entropy could be extended to non-extremal astrophysical black holes.

AdS/CFT Correspondence

One approach to formulating string theory and studying its properties is provided by the anti-de Sitter/conformal field theory (AdS/CFT) correspondence. This is a theoretical result which implies that string theory is in some cases equivalent to a quantum field theory. In addition to providing insights into the mathematical structure of string theory, the AdS/CFT correspondence has shed light on many aspects of quantum field theory in regimes where traditional calculational techniques are ineffective. The AdS/CFT correspondence was first proposed by Juan Maldacena in late 1997. Important aspects of the correspondence were elaborated in articles by Steven Gubser, Igor Klebanov, and Alexander Markovich Polyakov, and by Edward Witten. By 2010, Maldacena's article had over 7000 citations, becoming the most highly cited article in the field of high energy physics.

Overview of the Correspondence

In the AdS/CFT correspondence, the geometry of spacetime is described in terms of a certain vacuum solution of Einstein's equation called anti-de Sitter space. In very elementary terms, anti-de Sitter space is a mathematical model of spacetime in which the notion of distance between points (the metric) is different from the notion of distance in ordinary Euclidean geometry. It is closely related to hyperbolic space, which can be viewed as a disk as illustrated on the left. This image shows a tessellation of a disk by triangles and squares. One can define the distance between points of this disk in such a way that all the triangles and squares are the same size and the circular outer boundary is infinitely far from any point in the interior.

A tessellation of the hyperbolic plane by triangles and squares

One can imagine a stack of hyperbolic disks where each disk represents the state of the universe at a given time. The resulting geometric object is three-dimensional anti-de Sitter space. It looks like a solid cylinder in which any cross section is a copy of the hyperbolic disk. Time runs along the vertical direction in this picture. The surface of this cylinder plays an important role in the AdS/CFT correspondence. As with the hyperbolic plane, anti-de Sitter space is curved in such a way that any point in the interior is actually infinitely far from this boundary surface.

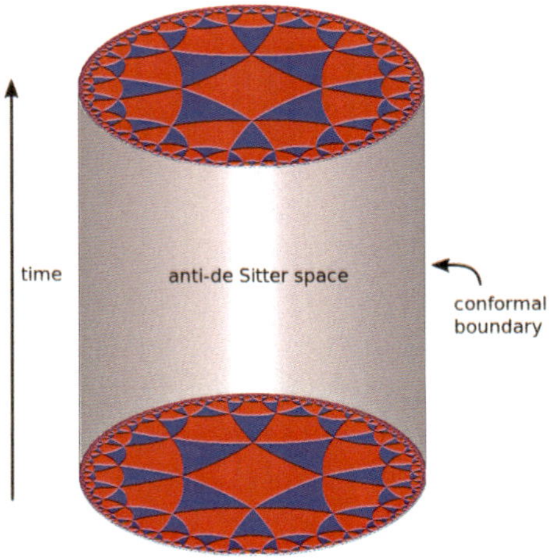

Three-dimensional anti-de Sitter space is like a stack of hyperbolic disks, each one representing the state of the universe at a given time. The resulting spacetime looks like a solid cylinder.

This construction describes a hypothetical universe with only two space dimensions and one time dimension, but it can be generalized to any number of dimensions. Indeed, hyperbolic space can have more than two dimensions and one can "stack up" copies of hyperbolic space to get higher-dimensional models of anti-de Sitter space.

An important feature of anti-de Sitter space is its boundary (which looks like a cylinder in the case of three-dimensional anti-de Sitter space). One property of this boundary is that, within a small region on the surface around any given point, it looks just like Minkowski space, the model of spacetime used in nongravitational physics. One can therefore consider an auxiliary theory in which "spacetime" is given by the boundary of anti-de Sitter space. This observation is the starting point for AdS/CFT correspondence, which states that the boundary of anti-de Sitter space can be regarded as the "spacetime" for a quantum field theory. The claim is that this quantum field theory is equivalent to a gravitational theory, such as string theory, in the bulk anti-de Sitter space in the sense that there is a "dictionary" for translating entities and calculations in one theory into their counterparts in the other theory. For example, a single particle in the gravitational theory might correspond to some collection of particles in the boundary theory. In addition, the predictions in the two theories are quantitatively identical so that if two particles have a 40 percent chance of colliding in the gravitational theory, then the corresponding collections in the boundary theory would also have a 40 percent chance of colliding.

Applications to Quantum Gravity

The discovery of the AdS/CFT correspondence was a major advance in physicists' understanding of string theory and quantum gravity. One reason for this is that the correspondence provides a formulation of string theory in terms of quantum field theory, which is well understood by comparison. Another reason is that it provides a general framework in which physicists can study and attempt to resolve the paradoxes of black holes.

In 1975, Stephen Hawking published a calculation which suggested that black holes are not completely black but emit a dim radiation due to quantum effects near the event horizon. At first, Hawking's result posed a problem for theorists because it suggested that black holes destroy information. More precisely, Hawking's calculation seemed to conflict with one of the basic postulates of quantum mechanics, which states that physical systems evolve in time according to the Schrödinger equation. This property is usually referred to as unitarity of time evolution. The apparent contradiction between Hawking's calculation and the unitarity postulate of quantum mechanics came to be known as the black hole information paradox.

The AdS/CFT correspondence resolves the black hole information paradox, at least to some extent, because it shows how a black hole can evolve in a manner consistent with quantum mechanics in some contexts. Indeed, one can consider black holes in the context of the AdS/CFT correspondence, and any such black hole corresponds to a configuration of particles on the boundary of anti-de Sitter space. These particles obey the usual rules of quantum mechanics and in particular evolve in a unitary fashion, so the black hole must also evolve in a unitary fashion, respecting the principles of quantum mechanics. In 2005, Hawking announced that the paradox had been settled in favor of information conservation by the AdS/CFT correspondence, and he suggested a concrete mechanism by which black holes might preserve information.

Applications to Nuclear Physics

A magnet levitating above a high-temperature superconductor. Today some physicists are working to understand high-temperature superconductivity using the AdS/CFT correspondence.

In addition to its applications to theoretical problems in quantum gravity, the AdS/CFT correspondence has been applied to a variety of problems in quantum field theory. One physical system that has been studied using the AdS/CFT correspondence is the quark–gluon plasma, an exotic state of matter produced in particle accelerators. This state of matter arises for brief instants when heavy ions such as gold or lead nuclei are collided at high energies. Such collisions cause the quarks that make up atomic nuclei to deconfine at temperatures of approximately two trillion kelvins, conditions similar to those present at around 10^{-11} seconds after the Big Bang.

The physics of the quark–gluon plasma is governed by a theory called quantum chromodynamics, but this theory is mathematically intractable in problems involving the quark–gluon plasma. In an article appearing in 2005, Đàm Thanh Sơn and his collaborators showed that the AdS/CFT correspondence could be used to understand some aspects of the quark–gluon plasma by describing it in the language of string theory. By applying the AdS/CFT correspondence, Sơn and his collaborators were able to describe the quark gluon plasma in terms of black holes in five-dimensional spacetime. The calculation showed that the ratio of two quantities associated with the quark–gluon plasma, the shear viscosity and volume density of entropy, should be approximately equal to a certain universal constant. In 2008, the predicted value of this ratio for the quark–gluon plasma was confirmed at the Relativistic Heavy Ion Collider at Brookhaven National Laboratory.

Applications to Condensed Matter Physics

The AdS/CFT correspondence has also been used to study aspects of condensed matter physics. Over the decades, experimental condensed matter physicists have discovered a number of exotic states of matter, including superconductors and superfluids. These states are described using the formalism of quantum field theory, but some phenomena are difficult to explain using standard field theoretic techniques. Some condensed matter theorists including Subir Sachdev hope that the AdS/CFT correspondence will make it possible to describe these systems in the language of string theory and learn more about their behavior.

So far some success has been achieved in using string theory methods to describe the transition of a superfluid to an insulator. A superfluid is a system of electrically neutral atoms that flows without any friction. Such systems are often produced in the laboratory using liquid helium, but recently experimentalists have developed new ways of producing artificial superfluids by pouring trillions of cold atoms into a lattice of criss-crossing lasers. These atoms initially behave as a superfluid, but as experimentalists increase the intensity of the lasers, they become less mobile and then suddenly transition to an insulating state. During the transition, the atoms behave in an unusual way. For example, the atoms slow to a halt at a rate that depends on the temperature and on Planck's constant, the fundamental parameter of quantum mechanics, which does not enter into the description of the other phases. This behavior has recently been understood by considering a dual description where properties of the fluid are described in terms of a higher dimensional black hole.

The Mathematics of String Theory

The Standard Model

In the standard model of particle physics, particles are considered to be points moving through space, tracing out a line called the World Line. To take into account the different interactions observed in Nature one has to provide particles with more degrees of freedom than only their position and velocity, such as mass, electric charge, color (which is the "charge" associated with the strong interaction) or spin.

The standard model was designed within a framework known as Quantum Field Theory (QFT), which gives us the tools to build theories consistent both with quantum mechanics and the special theory of relativity. With these tools, theories were built which describe with great success three of the four known interactions in Nature: Electromagnetism, and the Strong and Weak nuclear forces. Furthermore, a very successful unification between Electromagnetism and the Weak force was achieved (Electroweak Theory), and promising ideas put forward to try to include the Strong force. But unfortunately the fourth interaction, gravity, beautifully described by Einstein's General Relativity (GR), does not seem to fit into this scheme. Whenever one tries to apply the rules of QFT to GR one gets results which make no sense. For instance, the force between two gravitons (the particles that mediate gravitational interactions), becomes infinite and we do not know how to get rid of these infinities to get physically sensible results.

In String Theory, the myriad of particle types is replaced by a single fundamental building block, a 'string'. These strings can be closed, like loops, or open, like a hair. As the string moves through time it traces out a tube or a sheet, according to whether it is closed or open. Furthermore, the string is free to vibrate, and different vibrational modes of the string represent the different particle types, since different modes are seen as different masses or spins.

One mode of vibration, or 'note', makes the string appear as an electron, another as a photon. There is even a mode describing the graviton, the particle carrying the force of gravity, which is an important reason why String Theory has received so much attention. The point is that we can make sense of the interaction of two gravitons in String theory in a way we could not in QFT. There are no infinities! And gravity is not something we put in by hand. It *has* to be there in a theory of strings. So, the first great achievement of String Theory was to give a consistent theory of quantum gravity, which resembles GR at macroscopic distances. Moreover String Theory also possesses the necessary degrees of freedom to describe the other interactions! At this point a great hope was created that String Theory would be able to unify all the known forces and particles together into a single 'Theory of Everything'.

From Strings to Superstrings

The particles known in nature are classified according to their spin into bosons (integer spin) or fermions (odd half integer spin). The former are the ones that carry forces, for example, the photon, which carries electromagnetic force, the gluon, which carries the strong nuclear force, and the graviton, which carries gravitational force. The latter make up the matter we are made of, like the electron or the quark. The original String Theory only described particles that were bosons, hence *Bosonic String Theory*. It did not describe Fermions. So quarks and electrons, for instance, were not included in Bosonic String Theory.

By introducing *Supersymmetry* to Bosonic String Theory, we can obtain a new theory that describes both the forces and the matter which make up the Universe. This is the theory of *superstrings*. There are three different superstring theories which make sense, i.e. display no mathematical inconsistencies. In two of them the fundamental object is a closed string, while in the third, open strings are the building blocks. Furthermore, mixing the best features of the bosonic string and the superstring, we can create two other consistent theories of strings, Heterotic String Theories.

Extra Dimensions

One of the most remarkable predictions of String Theory is that space-time has ten dimensions! At first sight, this may be seen as a reason to dismiss the theory altogether, as we obviously have only three dimensions of space and one of time. However, if we assume that six of these dimensions are curled up very tightly, then we may never be aware of their existence. Furthermore, having these so-called compact dimensions is very beneficial if String Theory is to describe a Theory of Everything. The idea is that degrees of freedom like the electric charge of an electron will then arise simply as motion in the extra compact directions! The principle that compact dimensions may lead to unifying theories is not new, but dates from the 1920's, since the theory of Kaluza and Klein. In a sense, String Theory is the ultimate Kaluza-Klein theory.

For simplicity, it is usually assumed that the extra dimensions are wrapped up on six circles. For realistic results they are treated as being wrapped up on mathematical elaborations known as Calabi-Yau Manifolds and Orbifolds.

M-theory

Apart from the fact that instead of one there are five different, healthy theories of strings (three superstrings and two heterotic strings) there was another difficulty in studying these theories: we did not have tools to explore the theory over all possible values of the parameters in the theory. Each theory was like a large planet of which we only knew a small island somewhere on the planet. But over the last four years, techniques were developed to explore the theories more thoroughly, in other words, to travel around the seas in each of those planets and find new islands. And only then it was realized that those five string theories are actually islands on the same planet, not different ones! Thus there is an underlying theory of which all string theories are only different aspects. This was called *M-theory*. The M might stand for Mother of all theories or Mystery, because the planet we call M-theory is still largely unexplored.

The M-theory planet

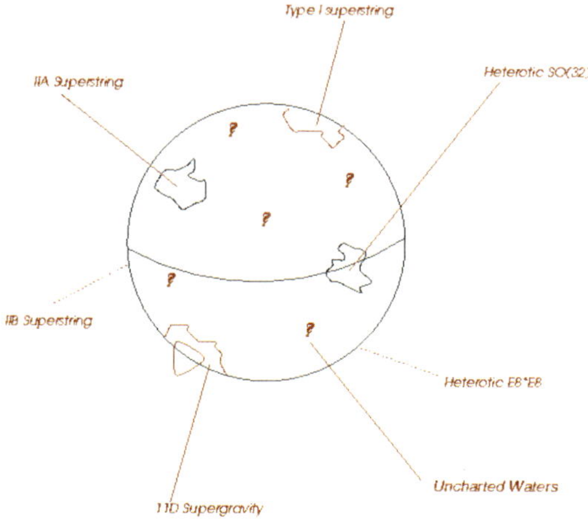

There is still a third possibility for the M in M-theory. One of the islands that was found on the M-theory planet corresponds to a theory that lives not in 10 but in 11 dimensions. This seems to be telling us that M-theory should be viewed as an 11 dimensional theory that looks 10 dimensional at some points in its space of parameters. Such a theory could have as a fundamental object a Membrane, as opposed to a string. Like a drinking straw seen at a distance, the membranes would look like strings when we curl the 11th dimension into a small circle.

Black Holes in M-theory

Black Holes have been studied for many years as configurations of spacetime in General Relativity, corresponding to very strong gravitational fields. But since we cannot build a consistent quantum theory from GR, several puzzles were raised concerning the microscopic physics of black holes. One of the most intriguing was related to the entropy of Black Holes. In thermodynamics, entropy is the quantity that measures the number of states of a system that look the same. A very untidy room has a large entropy, since one can move something on the floor from one side of the room to the other and no one will notice because of the mess - they are equivalent states. In a very tidy room, if you change anything it will be noticeable, since everything has its own place. So we associate entropy to disorder. Black Holes have a huge disorder. However, no one knew what the states associated to the entropy of the black hole were. The last four years brought great excitement in this area. Similar techniques to the ones used to find the islands of M-theory, allowed us to explain exactly what states correspond to the disorder of some black holes, and to explain using fundamental theory the thermodynamic properties that had been deduced previously using less direct arguments.

References

- R. Oerter (2006). The Theory of Almost Everything: The Standard Model, the Unsung Triumph of Modern Physics (Kindle ed.). Penguin Group. p. 2. ISBN 0-13-236678-9
- S.L. Glashow (1961). "Partial-symmetries of weak interactions". Nuclear Physics. 22 (4): 579–588. Bibcode:1961NucPh..22..579G. doi:10.1016/0029-5582(61)90469-2
- D. Haidt (4 October 2004). "The discovery of the weak neutral currents". CERN Courier. Retrieved 8 May 2008
- G.S. Guralnik (2009). "The History of the Guralnik, Hagen and Kibble development of the Theory of Spontaneous Symmetry Breaking and Gauge Particles". International Journal of Modern Physics A. 24 (14): 2601–2627. arXiv:0907.3466. Bibcode:2009IJMPA..24.2601G. doi:10.1142/S0217751X09045431
- José Luis Lucio; Arnulfo Zepeda (1987). Proceedings of the II Mexican School of Particles and Fields, Cuernavaca-Morelos, 1986. World Scientific. p. 29. ISBN 9971504340
- "Interim Summary Report on the Analysis of the 19 September 2008 Incident at the LHC" (PDF). CERN. 15 October 2008. EDMS 973073. Retrieved 28 September 2009
- CMS collaboration (2014). "Precise determination of the mass of the Higgs boson and tests of compatibility of its couplings with the standard model predictions using proton collisions at 7 and 8 TeV". The European Physical Journal C. 75 (5). arXiv:1412.8662. Bibcode:2015EPJC...75..212K. doi:10.1140/epjc/s10052-015-3351-7
- Calmet, X. (2011), "An Alternative view on the electroweak interactions", Int.J.Mod. Phys., A26 (17): 2855–2864, arXiv:1008.3780, Bibcode:2011IJMPA..26.2855C, doi:10.1142/S0217751X11053699
- Martin, Stephen P. (1997). "A Supersymmetry Primer". Advanced Series on Directions in High Energy Physics. Advanced Series on Directions in High Energy Physics. 18: 1–98. arXiv:hep-ph/9709356. doi:10.1142/9789812839657_0001. ISBN 978-981-02-3553-6
- H. Miyazawa (1966). "Baryon Number Changing Currents". Prog. Theor. Phys. 36 (6): 1266–1276. Bibcode:1966PThPh..36.1266M. doi:10.1143/PTP.36.1266
- Miri, M.-A.; Heinrich, Matthias; Christodoulides, D. N. (2014). "SUSY-inspired one-dimensional transformation optics". Optica. OSA. 1 (2): 89. arXiv:1408.0832. doi:10.1364/OPTICA.1.000089. Retrieved August 6, 2014

Permissions

All chapters in this book are published with permission under the Creative Commons Attribution Share Alike License or equivalent. Every chapter published in this book has been scrutinized by our experts. Their significance has been extensively debated. The topics covered herein carry significant information for a comprehensive understanding. They may even be implemented as practical applications or may be referred to as a beginning point for further studies.

We would like to thank the editorial team for lending their expertise to make the book truly unique. They have played a crucial role in the development of this book. Without their invaluable contributions this book wouldn't have been possible. They have made vital efforts to compile up to date information on the varied aspects of this subject to make this book a valuable addition to the collection of many professionals and students.

This book was conceptualized with the vision of imparting up-to-date and integrated information in this field. To ensure the same, a matchless editorial board was set up. Every individual on the board went through rigorous rounds of assessment to prove their worth. After which they invested a large part of their time researching and compiling the most relevant data for our readers.

The editorial board has been involved in producing this book since its inception. They have spent rigorous hours researching and exploring the diverse topics which have resulted in the successful publishing of this book. They have passed on their knowledge of decades through this book. To expedite this challenging task, the publisher supported the team at every step. A small team of assistant editors was also appointed to further simplify the editing procedure and attain best results for the readers.

Apart from the editorial board, the designing team has also invested a significant amount of their time in understanding the subject and creating the most relevant covers. They scrutinized every image to scout for the most suitable representation of the subject and create an appropriate cover for the book.

The publishing team has been an ardent support to the editorial, designing and production team. Their endless efforts to recruit the best for this project, has resulted in the accomplishment of this book. They are a veteran in the field of academics and their pool of knowledge is as vast as their experience in printing. Their expertise and guidance has proved useful at every step. Their uncompromising quality standards have made this book an exceptional effort. Their encouragement from time to time has been an inspiration for everyone.

The publisher and the editorial board hope that this book will prove to be a valuable piece of knowledge for students, practitioners and scholars across the globe.

Index

A

Atomic Nuclei, 17, 31, 48, 51, 55, 73, 75, 78, 86-87, 90, 111, 125, 140, 144, 158, 251, 255, 266

Atoms and Molecules, 2, 26, 77

B

Baryogenesis, 53, 147, 249

Baryons, 4, 18, 47, 62, 70, 83, 117, 121-122, 125, 129, 138-139, 143-148, 150-153, 171-175, 194, 207

Bekenstein-hawking Formula, 261-262

Black Holes, 147, 255, 261-263, 265-266, 269-270

Bose-einstein Condensate, 29

Bosonic String Theory, 257-258, 268

Bottom Quark, 125, 134, 137-139, 195, 247

C

Charginos, 237

Charm Quark, 124, 138, 143-144, 160, 195

Codata, 37, 59, 64-66, 120

Color Charge, 62, 71, 83, 117, 120-123, 127-128, 132, 145, 158, 168-173, 202

Condensed Matter Physics, 231, 255, 258, 266

Cosmic Rays, 2, 7, 32, 47, 92, 102, 123, 158, 186, 190

D

D-branes, 260-262

Dirac Particles, 62, 83, 116

E

Electrical Conductivity, 28-29

Electromagnetism, 17, 114, 121, 123, 140, 143, 193-195, 204, 209, 244-245, 250, 267

Electron Antineutrino, 22, 79, 94, 117, 120, 125-126, 154, 159, 179, 188, 191

Electron Diffraction, 35, 39, 41-43

Electron-neutrino, 185-186

Elementary Fermions, 123, 193

Elementary Particle, 10, 16, 59-60, 77, 109-110, 139-140, 143, 158, 160, 177, 187, 190

F

Fast Neutrons, 72-73, 90-92, 94, 101

G

Gluons, 2, 5-6, 11-12, 16-17, 48-49, 62-63, 69-71, 77, 83, 117, 121-122, 131-132, 145, 158, 168-173, 194, 196-197, 207, 213, 219, 245, 247-248

Grand Unified Theory, 192-193, 208, 234, 244-247

Gravitational Interactions, 194, 230, 241, 267

Ground State Baryons, 172, 174-175

Gyromagnetic Ratio, 60-61, 111-113

H

Hadron, 5, 17, 77, 121-122, 125-126, 129, 131-134, 137, 144-145, 169, 179, 209, 213, 217-218, 222, 225, 233, 235, 237, 254

Hadronization, 12, 14, 131

Higgs Boson, 5, 17, 129, 133, 135, 137, 156, 193, 199, 203, 205-208, 210-222, 224-225, 227, 230, 237-238, 242-243, 270

Higgs Sector, 215, 225, 243, 247

L

Large Electron-positron Collider, 183

Lattice Field Theory, 1, 9-10

Lepton, 18, 21, 32, 55, 63, 120, 123, 136, 177-187, 189, 191-192, 195, 215, 218-219, 224, 237-238, 246-247

M

M-theory, 253, 255, 258, 268-270

Massless Particles, 16, 18, 48, 179, 203, 210

Megaelectron Volts, 20, 160

Molecules, 2, 14, 26-27, 51-52, 77, 89, 261

Mssm, 11, 55-56, 137, 215, 230, 233-235, 237-244, 249

Muon Decay, 178

Muse, 65-68

N

Neutrino Mass, 5, 80, 199-200

Nmssm, 239, 242-243

Nuclear Binding Energy, 69-70, 155-156

Nuclear Physics, 85, 104, 140, 156, 159, 176, 265, 270

O

Orbital Angular Momentum, 27, 149-151, 161-162

P

Particle Decay, 3, 119

Particle Physics, 1-5, 9-11, 14-19, 21, 35, 62, 69, 79, 84, 101, 103, 116, 126, 133, 137, 159-160, 176, 184, 189, 193, 203-206, 224, 233, 244, 250, 267

Pauli Exclusion Principle, 23, 27-28, 62, 81, 83-84, 107, 117, 121-122, 146, 159, 175, 179, 193, 201

Proton Decay, 16, 48, 52-59, 63, 81, 245-246

Proton Emission, 58-59

Proton Magnetic Moment, 59

Proton Radius Puzzle, 64, 66

Q

Quantum Chromodynamics, 10, 48-49, 128-129, 140, 145, 169, 172-173, 180, 196, 207, 209, 266

Quantum Field Theory, 4-5, 9-10, 13, 170, 180-181, 196, 199, 216, 228-229, 231, 249, 253, 256, 260-261, 263-267

Quantum Gravity, 199, 233, 255, 257, 262-263, 265-267

Quantum Mechanics, 2, 4, 17, 22-23, 27, 39, 77, 114, 181, 192, 200, 213, 231-232, 250-251, 254-256, 260, 265-267

Quark-antiquark, 121, 131, 134, 170, 194, 213

R

Radioactivity, 6, 18, 58, 78, 89, 120, 195

S

Scalar Fields, 207, 240

Sea Quarks, 131

Semiconductor Materials, 20, 100

Sleptons, 237, 240-241, 244

Standard Model, 4-5, 10, 13, 17, 21, 48, 55-56, 58, 64, 79, 83-85, 116-117, 129, 133, 147, 160, 173, 178-184, 199, 200, 224, 231, 237-243, 267, 270

Stop Squark, 137, 230

String Theory, 49, 193, 200, 228, 233, 250-268

Subatomic Particles, 2-4, 16-19, 21, 69-70, 77, 94, 143-146, 160, 177, 185, 200, 227-228, 250

Supersymmetric Quantum Mechanics, 231-232

T

Theory of Relativity, 22, 250, 254-255, 257, 267

Thermal Energy, 29

Top Quark, 124, 129, 131-138, 151, 160, 196, 208, 214, 218-219, 225, 230

V

Vacuum Energy, 208-209

Virtual Particles, 22-24, 32, 62, 83, 116-117, 216

W

Wimp, 230, 235

X

X-ray Diffraction, 40-41, 82

Y

Yukawa Couplings, 136, 224, 237, 240, 242-243